CONSERVATION BIOLOGY

From The Fund In Honor Of
Thomas S. and Mary T. Hall

CONSERVATION BIOLOGY

An Evolutionary–Ecological Perspective

EDITED BY MICHAEL E. SOULÉ
UNIVERSITY OF CALIFORNIA, SAN DIEGO

AND BRUCE A. WILCOX
STANFORD UNIVERSITY

SINAUER ASSOCIATES, INC. • PUBLISHERS
SUNDERLAND, MASSACHUSETTS

Library of Congress Cataloging in Publication Data

Main entry under title:

Conservation biology.

Includes bibliographies and index.
1. Nature conservation. I. Soulé, Michael E.
II. Wilcox, Bruce A., 1948–
QH75.C66 1978 333.9′5 79-26463
ISBN 0-87893-800-1

15 14 13 12 11 10

CONSERVATION BIOLOGY: An Evolutionary-Ecological Perspective

Sunderland, Massachusetts 01375

Manufactured in the U.S.A.

ISBN: 0-87893-800-1

To Jan and Melissa

CONTENTS

Part IV Exploitation and Preservation 271

FOREWORD

While this book was being written, the 95th Congress passed legislation amending the Endangered Species Act, providing, for the first time in history, a legal mechanism for the eradication of a species. At the same time, the President's Council on Environmental Quality was engaged in a study that predicted the loss of two-thirds of all tropical forests by the turn of the century. Hundreds of thousands of species will perish, and this reduction of 10 to 20 percent of the earth's biota will occur in about half a human life span.

Clearly our impact on the biological systems of the planet is of a very different scale and rate than heretofore in history. While new species may ultimately evolve to fill the vacuum, this will take millenia. We are permanently altering the course of evolution in ways we cannot perceive and of which we are not generally aware. We are even less aware of the evolutionary impact this will have upon ourselves.

This reduction in the biological diversity of the planet is the most basic issue of our time. Careful reading of newspapers reveals bits and pieces of the story, but the major news—that the planet of tomorrow will be less fit to live on than the planet of today—never gets the banner headline it deserves. Rather, the process of biotic impoverishment is allowed to proceed increment by increment.

In spite of this, alert and concerned citizens and scientists can profoundly affect the future. The demands on science are large. New questions arise when many populations and entire biomes are being fragmented and reduced on such a scale and at such rates. These questions are a great challenge to the ingenuity of biologists and are the central focus of this volume. Unless we solve them, we will end up with less than we intend, struggling in our ignorance to protect genetically eroding populations and decaying ecosystems.

However finely honed, our scientific abilities are exercised in a large societal context. Science is trapped within society's warp and woof, much as Man is inextricably trapped in a biological context. While science can never totally free itself, it certainly can make freer progress for the benefit of society, especially if it bends some of its energies to highlighting and revealing Man's place in nature, and promoting biological literacy. When Scottish fishermen feel they must compete with seals for fish or when Chinese compete with sparrows for grain, there is obviously a need to explain that things have arrived at a marginal state.

More than ever, our effect on the biological systems of the planet will rebound to affect us. A slash and burn approach to the biosphere is no longer viable. Indeed the planet already has a reduced capacity to support Man. We need a populace and politicians aware that all decisions have a biological component, and that biology is inextricably interwoven with sociology and economics. As the planet becomes simpler biologically, it becomes more expensive economically: fish are smaller and dearer; lumber is narrower, shorter and more expensive; dwindling natural resources fuel inflation. The planet also is more vulnerable to disaster, and the quality of life inevitably declines.

Conservation is sometimes perceived as stopping everything cold, as holding whooping cranes in higher esteem than people. It is up to science to spread the understanding that the choice is not between wild places or people. Rather, it is between a rich or an impoverished existence for Man.

THOMAS E. LOVEJOY
World Wildlife Fund
Washington, D.C.
October 20, 1979

PREFACE

This book is an invitation to students and scientists; an invitation to participate in the vocation or avocation of conservation biology. If you have read this far you must have some concern for the preservation of biological diversity and its evolutionary potential. This book will allow you to deepen and broaden that interest–it is a way to become better informed and more effective.

Each of the four parts of this book addresses a particular problem or theme. The first part *(Ecological Principles of Conservation)* concerns the dynamic spatial and temporal complexity of ecosystems, particularly in the tropics. The tropical emphasis highlights our belief that tropical habitats are more sensitive, less resilient and in greater danger of complete destruction than temperate or boreal habitats. In Chapter 2 Gilbert lays the groundwork for the book by developing an intricate multidimensional picture of the exquisite specialization and complexity of plant, insect and animal interaction in tropical American forests. He emphasizes the role of "keystone" species, and the ripple effects that follow their extinction.

Eisenberg (Chapter 3) follows with a pantropical perspective on the effects of trophic level and body size on the density of mammalian species, providing a useful body of data for reserve designers and managers.

In Chapter 4 Diamond finds that patchiness in the geographic distributions of tropical birds is not only without precedents in the temperate zone, but persists with a tenacity that temperate zone biologists may find hard to accept. Relevance to theory aside, Diamond argues that the tropical habitats cannot be assumed to contain the species that are supposed to "belong" in them, and concludes that a prerequisite for a tropical nature reserve is a species inventory.

Next, Foster (Chapter 5) discusses how different kinds of disturbances (from tree falls to volcanic eruptions and ice ages) create successional habitats of certain sizes, frequencies and durations in tropical forests. This dynamic view of patchiness is emerging as one of the most important considerations in conservation planning and management. Foster calls for interventionist management practices such as controlled habitat perturbation (also discussed in Chapters 7 and 17).

The theme of Part 2 is *The Consequences of Insularization.* Throughout the world natural habitats are being cut up, cut down and cut off. This fragmentation isolates pieces of habitat and surrounds them with ever-rising seas of inhospitable terrain, mostly given over to agriculture and erosion. Wilcox (Chapter 6) reviews the theory of insular ecology and shows how it applies to isolated nature reserves. His predictions on the rates of extinction in such reserves show that even if no further encroachment occurs (once insularization is complete), even the largest reserves could potentially lose most of their large mammal species. He also discusses how taxon-specific characteristics suggest different conservation strategies and different fates among the vertebrates.

Terborgh and Winter (Chapter 7) also discuss extinction. They advance a simple and appealing hypothesis: the best predictor of extinction among birds is density. That is, "rarity is the best index of vulnerability." Echoing Gilbert, they also predict a cascade of extinction following the loss of key species, particularly top predators. Finally, they warn against a management policy of benign neglect, prescribing careful monitoring and "preventative management."

A persistent gap in the theory of conservation biology is criteria for minimum population sizes—that is, What are the threshold population sizes for the loss of short-term fitness and evolutionary plasticity? Lip service is routinely paid to this issue, but except for some suggestions based on neutralist population genetic theory, there have been no specific proposals for minimum population sizes. Three chapters (8, 9, 12) address this issue. Franklin (Chapter 8) arrives at two criteria—one for short-term fitness, applicable in captive or semi-natural programs for a small number of generations, and another criterion for long-term conservation; the minimum population sizes for the two criteria are 50 and 500, respectively. Soulé (Chapter 9) independently uses the same approach and arrives at the same answer for the short-term criterion. He also advances the proposal that, for practical purposes, the process of speciation for larger organisms in the tropics is ending.

Goodman (Chapter 10) applies demographic theory and methodology to three conservation problems: (1) a declining wild population: (2) a wild population in need of culling; and (3) a captive population that produces a surplus or progeny. His conclusions are general and should have wide application both in reserves and zoos.

Part 3 *(Captive Propagation and Conservation)* is a forum for discussing the advantages, disadvantages, technology and theory of the captive breeding of endangered species. Conway (Chapter 11) presents an authoritative overview of the subject. He stresses both the necessity for such programs, especially in cases where the species cannot survive in nature, and the genetic and economic constraints. With regard to the latter, he points out that there is only room in United States zoos to captively breed about 100 species of mammals, assuming that the zoos engage in captive breeding to the near exclusion of everything else.

In Chapter 12 Senner presents a genetic model of the survival of captive populations and tests its sensitivity to variation in its parameters. He concludes that the most important factor is effective population size, the expectation of survival being directly proportional to the maintenance number. The size of the founder group, so long as it is more than four or five, is much less important.

In Chapter 13 Benirschke, Lasley and Ryder provide an update on the advances in reproductive physiology, genetics and medical technology of zoo animals. One of the most significant advances is in the area of sexing animals, mainly birds. They argue for greater financial investment in these disciplines in order to improve the survival of newborns and to enhance the breeding potential of many species.

This theme is expanded on by Kleiman (Chapter 14). She demonstrates how a knowledge of the behavior, particularly the mating, social and rearing behavior, of animals is necessary for the propagation of many species. In addition, she points out how behavioral studies in zoos can provide novel insights and hypotheses in the more academic disciplines.

Campbell (Chapter 15) discusses one of the thorniest questions in captive propagation—the issue of reintroduction. He reviews the successes and failures of such attempts, and points out that the difficulties are in proportion to the number of generations in captivity.

In the final part, *Exploitation and Preservation,* the authors consider issues where Man is an integral variable, either as an exploiter of natural resources or as steward of nature reserves. The subject of Coe's essay (Chapter 16) is the past and present relationship between Man and animal in Africa, the last redoubt of the magnificent Pleistocene mammal fauna. Coe reviews the possibilities of game ranching, game culling and domestication as tactics for saving wildlife on that protein-starved continent.

Whitmore (Chapter 17) gives a status report on the tropical forests of the world. He discusses their past, present and future extent, the latter based on current and extrapolated deforestation rates. He concludes that in the next 20 years Man will wipe out most of the tropical forests, probably setting in motion an unprecedented chain reaction of extinctions and other environmental effects. Whitmore alludes to the role of the North American and European consumer, and how their use of imported beef and wood products contributes to this environmental vandalism.

In Chapter 18 Pyle reviews the oft-repeated arguments in favor of nature reserves, but advances the novel idea that the decisions of policy makers are made less on rational or economic grounds than in response to the relative volume and visibility of special interest groups. He also summarizes the approaches to selection and management of nature reserves taken by such organizations as the Nature Conservancy in the United States and Nature Conservancy Council in Great Britain.

In the final chapter, Ehrlich concludes that the momentum of the present growth and exploitative trends will not be changed without a dramatic and worldwide socio-political and economic revolution. He says that conservationists should not be satisfied with tactical victories when the war is being lost by a process of environmental attrition. He concludes with a call for a much more aggressive conservation strategy.

We could not possibly name all of those who have contributed to this book. Among those who provided encouragement and advice are Charles L. Bieler, Sheldon Campbell, J. Dallas Clark, Jared Diamond, Sir Otto Frankel, John Hanks, Dan Janzen, Thomas E. Lovejoy, Paul Opler, Peter Raven, Daniel Simberloff, Walter Westman and Robert Whitcomb. We are grateful to our colleagues, students and staff who assisted us at the Conservation Biology Conference held at the University of California, San Diego and the San Diego Wild Animal Park in September, 1978. In addition, the advice and contributions of those who attended the conference contributed significantly to many of the chapters in this book.

This book would not have been possible without the support of the following institutions and organizations:

Hartz Foundation
William and Flora Hewlitt Foundation
Monsanto Fund
New York Zoological Society
Regents of the University of California
Wildlife Federation Trust
World Wildlife Fund
San Diego Zoological Society.

M.E.S.
B.A.W.

CONTRIBUTORS

KURT BENIRSCHKE, San Diego Zoological Society, San Diego

SHELDON CAMPBELL, San Diego Zoological Society, San Diego

MALCOLM COE, Department of Zoology, Oxford University, Oxford

WILLIAM G. CONWAY, New York Zoological Society, Bronx, New York

JARED M. DIAMOND, Department of Physiology, UCLA Medical Center, Los Angeles

PAUL R. EHRLICH, Department of Biological Sciences, Stanford University, Stanford

JOHN F. EISENBERG, National Zoological Park, Smithsonian Institution, Washington, D.C.

ROBIN B. FOSTER, Department of Biology, University of Chicago, Chicago

IAN ROBERT FRANKLIN, Division of Animal Production, Genetics Research Laboratories, North Ryde, N.S.W., Australia

LAWRENCE E. GILBERT, Department of Zoology, University of Texas, Austin

DANIEL GOODMAN, Scripps Institute of Oceanography, University of California, San Diego

DEVRA G. KLEIMAN, National Zoological Park, Smithsonian Institution, Washington, D.C.

BILL LASLEY, San Diego Zoological Society, San Diego

ROBERT MICHAEL PYLE, The Nature Conservancy, Portland, Oregon

OLIVER RYDER, San Diego Zoological Society, San Diego

JOHN W. SENNER, Oregon Regional Primate Center, Beaverton, Oregon

MICHAEL E. SOULÉ, Department of Biology, University of California, San Diego

JOHN TERBORGH, Department of Biology, Princeton University, Princeton

T. C. WHITMORE, Department of Forestry, Oxford University, Oxford

BRUCE A. WILCOX, Department of Biological Sciences, Stanford University, Stanford

BLAIR WINTER, Department of Biology, Princeton University, Princeton

CHAPTER 1

CONSERVATION BIOLOGY: ITS SCOPE AND ITS CHALLENGE

Michael E. Soulé and Bruce A. Wilcox

CONSERVATION BIOLOGY

Conservation biology is a mission-oriented discipline comprising both pure and applied science. This book is not the first in the field; there are other excellent texts (Dasmann, 1968; Ehrenfeld, 1970). In addition there are symposia with a more traditional approach, or which give specialized treatment to particular habitats, regions or to the status and management of particular taxa (for example, Duffey and Watt, 1971; Duffey et al., 1974; Schoenfeld et al., 1978; Prance and Elias, 1977).

In this book we have attempted to expand the range of topics covered and the depth with which they are usually treated. With regard to breadth, conservation biology is as broad as biology itself. It focuses the knowledge and tools of all biological disciplines, from molecular biology to population biology, on one issue—nature conservation. Among the contributors to this book, for example, are botanists, zoologists, ecologists, geneticists, evolutionists, a statistician, a mathematician-demographer, a cytologist, a biochemist, an endocrinologist, a sociobiologist and experts in the field of natural resources.

With regard to depth, conservation biology can and should be as profound, intellectually rigorous and challenging as any field, limited only by the capacities of its practitioners. As a science, it is not strictly "pure," but neither is it purely "applied," for most of the chapters in this book contain ideas, data and conclusions that will advance basic science. This is as good a test of originality and rigor as any.

Conservation has a venerable history of scientific (Leopold, 1933; Smith, 1976) and philosophical (Passmore, 1974; Singer, 1975) discourse. Man has sought to protect wildlife at least as far back as Ashoka (around 250 BC). Journalists, poets and scholars of every stripe have written copiously about nature—its values and how to save it from the human menace.

In spite of this scholarly and literary legacy we feel that conservation biology is a new field, or at least a new rallying point for biologists wishing to pool their knowledge and techniques to solve problems. A community of interest and concern is often crystalized by a simple term. Conservation biology is such a term.

Unfortunately, the emergence of conservation biology as a respectable academic discipline has been slowed by prejudice. Until recently, few academically oriented biologists would touch the subject. While wildlife management, forestry and resource biologists (particularly in the industrialized temperate countries) struggled to buffer the most grievous or economically harmful of human impacts (deforestation, soil erosion, overhunting), the large majority of their academic colleagues thought the subject was beneath their dignity. But academic snobbery is no longer a viable strategy, if it ever was. Because many habitats, especially tropical ones, are on the verge of total destruction and many large animals are on the verge of extinction, the luxury of prejudice against applied science is unaffordable. Conservation genetics is an example of such academic disinterest. This field has been one of the weakest in conservation biology. One of the purposes of this book is to begin correcting this handicap.

Given encouragement, a forum, and, one hopes, the funding, biologists will gladly enlist to help save the world's rapidly expiring biota. As the table of contents will verify, "pure" scientists from many disciplines are eager to be conservation biologists. The issue that faces every student of biology today is not whether to be a conservationist, but how. Even if one rationalizes (as do many of our colleagues) that one's esoteric research is far removed from nature in the raw, or from the plight of tigers, gorillas, redwoods and "jungles," it seems to us that the study of life becomes a hollow, rarefied pursuit if the very animals and plants that fired our imaginations as children and triggered our curiosity as students should perish.

A WORD ON ECONOMICS

Conservation biology, strictly speaking, does not include the subject of economics.* For that reason, there are no chapters in this book on the

*Funding for established scientists who wish to do research in conservation biology is negligible. The National Science Foundation considers grant applications in this field to be "applied" (a pejorative somehow avoided by cancer researchers). Other governmental agencies use most of their funds for in-house research.

acquisition or maintenance costs of reserves (but, see Chapter 18 and Chapter 11). As shown in Table I, the industrialized countries have ten times the staff and ten times the money per unit area of national park as do the lesser developed countries. This gap in conservation support is serious enough, but the problem is compounded by the greater needs in many of the tropical countries, particularly in regions where poaching and other forms of encroachment require many skilled and dedicated wardens.

This is just another example of the well-known principle that scientific and technological expertise is worthless in the final analysis, if the money and resources required to implement the expertise is absent. Thus, the issue is not whether scientists should commit *some* of their time to public education and lobbying, but rather, how much of their time.

TABLE I. National park budgets and personnel.

Country	U.S. dollars/1000 ha	Staff/1000 ha
Industrialized		
Australia	433	0.042
Canada	48.5	0.135
Japan	975	0.358
United States	4239	0.350
Brazil	3782	3.500
Mean	1895	0.877
Nonindustrialized		
Congo	39	0.026
Ghana	610	0.612
Indonesia	31	0.025
Niger	25	0.013
Senegal	311	0.082
Thailand	48	0.000
Uganda	163	0.017
Zambia	210	0.006
Mean	180	0.098

(Data from Anon., 1977)

COMMENTS ON THE DESIGN OF NATURE RESERVES

An issue that has been the subject of controversy (see Simberloff and Abele, 1975, 1976; Diamond, 1976; Whitcomb et al., 1976; Terborgh,

1976) and that is mentioned or alluded to by nearly every author in this book is the optimal design of nature reserves—the little fragments of landscape where Man expects to preserve non-human life.* Nature reserves are the most valuable weapon in our conservation arsenal (Chapter 11), so they deserve extra attention in this introductory discussion.

When considering the preservation of a particular biota, a system of nature reserves can be described by reference to three features: number of reserves, size of reserves and density (or proximity) of reserves. With regard to number and size, some biogeographers have argued that reserves should be large and not necessarily numerous; others have argued for many, smaller reserves. Nevertheless, all agree that the best solution from the biogeographical standpoint is many, large reserves. But there are other design considerations and factors besides biogeography that should influence reserve design. Table II lists these; it also tabulates our decisions regarding reserve size, number and proximity. This is a simplistic picture, but it has a purpose. It shows that all viewpoints (factors) converge on the same conclusion—reserves should be manifold, large and (for most purposes) dispersed, as is documented in the following comments.

TABLE II. The components of nature reserve planning.

	Design Features		
Design Considerations	**Number**	**Size**	**Dispersed?**
Disease	Many	Large	Yes
Genetics	Many	Large	—
Community Ecology	Many	Large	—
Island Biogeography	—	Large	(see text)
Research Potential	Many	—	—
Politics, Economics	Many	—	Yes
Recreation	Many	Large	Yes
Summary	Many	Large	Dispersed

Disease

Can epidemics destroy one or more species in a region the size of a nature reserve (Frankel and Soulé, in press)? In recent years disease has eliminated many mammal species from large areas. These diseases in-

*We estimate from the 1975 UN List of National Parks and Equivalent Reserves that the total area or habitat to be preserved in tropical countries amounts to less than 1.5 percent of the original habitat—probably much less than this considering that many reserves are now being destroyed by "legal" and illegal logging operations, while many others are virtually unprotected from poachers, charcoal makers and squatters.

clude rinderpest, myxomatosis, anthrax, hoof and mouth disease, yellow fever and epidemic hemorrhagic disease. Botulism and avian cholera have devastated bird populations. Plant examples include chestnut blight and Dutch elm disease. The potential for disaster is aggravated by the susceptibility of domestic animals to most of the diseases that can plague wild species. Farms and villages with rabbits, sheep, cattle, swine and fowl will encircle most nature reserves in the near future, and these stocks will be a perennial source of contagion for the wild species.

Another consideration is disease resistance. Small, isolated populations will tend to lose genetic variability, including resistance genes, resulting in a gradual increase in susceptibility to diseases. It is folly, then, to keep all of the individuals of a species in a single reserve regardless of size, unless, of course, several well isolated populations exist within a reserve—an exceptional situation. We do not mean that the prime directive of conservation should be maximizing the number and isolation of reserves. But if the choice is between (1) a single reserve containing the last vestiges of one or more valuable species versus (2) several (perhaps smaller) isolated reserves, each with representative populations of the desirable forms and appropriate habitat, common sense and history give the nod to the latter choice.

Community Ecology

The world is patchy and patches come and go. Even a region as superficially homogeneous as the Amazon Basin will require many reserves to maintain examples of all of its habitats (Chapter 5; Chapter 17). Frequent climatic and geologic disturbances as well as fire can affect entire small reserves (Chapter 5). Reserves must be designed with the largest disturbance type in mind (Chapter 2). Viable populations of high trophic level species, large herbivores and many tropical plants, as well as many butterfly species require large reserves (Chapter 2; Chapter 3). Finally, only large areas will contain a balance of successional stages necessary for the survival of many plants, insects and the animals that depend on them (Chapter 2).

Island Biogeography

Extinction rates in large reserves will be lower than those in small reserves, so it would seem that large reserves are superior. They also hold more species at equilibrium. On the other hand, if extinction rates, particularly for vertebrates, were generally very low (measured in geological rather than historic time), a large number of small, complementary reserves, each harboring a small but unique portion of the biosphere, would suffice. Extinction rates appear to be high (Chapter 6) so it is best to maximize both size and number. Proximity of reserves to one another is also important if it enhances recolonization and gene flow between reserves. But this rule only applies to species that normally traverse wide stretches of inhospitable habitat (those that fly or have efficient wind dispersal mechanisms). Even among flying species, however, only a small minority venture out of their normal habitat (Ehrlich and Raven, 1969; Diamond, 1971; 1976; Terborgh, 1974) and so only a fraction of species would benefit from proximity.

Research

Replication of reserves offers advantages for pure and applied research because it permits management experiments and other kinds of deliberate perturbation such as benign neglect, controlled burning or logging, removal of predators or dominant herbivores and the addition of species. Such treatments would be considered too "dangerous" if there were only one or a few reserves.

Politics and Economics

Destructive political conflicts affecting the status of reserves can be avoided if opportunities for jealousy are minimized. One political source of jealousy is the hegemony of reserves in regions controlled by powerful interests. The danger is especially great in developing tropical countries where a game reserve can significantly increase the material well-being of the people living near it. This means that the best policy is the wide deployment of reserves. If there is to be a "conservation pork barrel," everyone should get his share.

A serious and thorough analysis of management costs probably would demonstrate the long-term economic advantages of large reserves. Small reserves are less expensive to establish, but in the long run they are undoubtedly more costly to operate. For example, it is shown in Chapter 6 that the decay of species diversity over time is much more rapid in small reserves than in large ones. This means that costly demographic and genetic rescue operations (especially for large species) will have to be initiated much sooner in small reserves, and that the intensity of such programs will have to be greater. Another factor relating to size is habitat succession and disturbance. In small reserves it will be necessary to undertake heroic and expensive operations just to maintain a viable mix of habitats. Such artificial therapy will rarely be necessary in large reserves (Chapter 5).

Recreation and Education

Reserves scattered liberally throughout a region will be more accessible to the populace, thus providing greater opportunities for education and recreation. Coe (Chapter 16) says this is an essential requirement for the survival of reserves in the poorer, tropical countries.

Genetics

The species which are the least dense (large herbivores, trees, large carnivores and scavengers) will probably suffer an attrition of fitness and evolutionary potential if effective population sizes are too low (Chapters 8, 9 and 12). Even the largest existing reserves may be too small for the long-term preservation of carnivores (Chapter 9). A multiplicity of reserves would add another dimension to the conservation of genetic diversity. For example, populations of the same species in different reserves would tend to harbor different sets of rare alleles at polymorphic loci.

Conclusion

The conclusion of this brief treatment is not surprising: reserves should be large, manifold and dispersed (except that a few highly vagile species benefit from proximity of reserves). To such questions as "how big?" or "how many?", the answers are "as big as possible" and "as many as possible." By the turn of the century, most options regarding the design, size and organization of nature reserves will be closed, especially in the tropics. It therefore behooves conservation biologists to exert, immediately and forcibly, their influence on these matters.

AN EMOTIONAL CALL TO ARMS

The green mantle of Earth is now being ravaged and pillaged in a frenzy of exploitation by a mushrooming mass of humans and bulldozers. Never in the 500 million years of terrestrial evolution has this mantle we call the biosphere been under such a savage attack. Certainly, there have been so-called "crises" of extinction in the past, but the rate of decay of biological diversity during these crises was sluggish compared to the galloping pace of habitat destruction today.

Perhaps the hardest thing to grasp is the geological and historical uniqueness of the next few decades. There simply is no precedent for what is happening to the biological fabric of this planet and there are no words to express the horror of those who love nature. In the lifetimes of many who read this book, the relentless harrying of habitats, particularly in the tropics, will reduce rain forests, reefs and savannas to vulnerable and senescent vestiges of their former grandeur and subtlety. But loss of habitat and loss of species is not the whole disaster. Perhaps even more shocking than the unprecedented wave of extinction is the cessation of significant evolution of new species of large plants and animals. Death is one thing—an end to birth is something else, and nature reserves are too small (not to mention, impermanent) to gestate new species of vertebrates (Chapter 9). There is no escaping the conclusion that in our lifetimes, this planet will see a suspension, if not an end, to many ecological and evolutionary processes which have been uninterrupted since the beginnings of paleontological time.

We hope it is only a suspension—that the horrible onslaught can be stopped before the regenerative powers of ecosystems are also killed. This book is a statement of that optimism. In one way or another, each of the authors is expressing a determination to protect biotic diversity and to do the enabling research and development. The authors also join in a plea: we can't do it alone, and we can't do it on a shoestring. More scientists must be solicited in the ranks of conservation biology. But this is only a beginning. Wars aren't won on intelligence alone. Money, troops, weapons and strategy are also required (see Chapters 16, 17, 18 and 19). This is the challenge of the millenium. For centuries to come, our descendants will damn us or eulogize us, depending on our integrity and the integrity of the green mantle they inherit.

PART ONE

ECOLOGICAL PRINCIPLES OF CONSERVATION

CHAPTER 2

FOOD WEB ORGANIZATION AND THE CONSERVATION OF NEOTROPICAL DIVERSITY

Lawrence E. Gilbert

Population biologists have been more concerned with the design of future natural areas than with their management (Sullivan and Shaffer, 1975; Simberloff and Abele, 1975; Diamond, 1976; Diamond and May, 1976*b*; Terborgh, 1976; Whitcomb et al., 1976). In this chapter I apply biological insights to problems of reserve management rather than to those of design, and stress the interface between plants and animals rather than particular vertebrate taxa. Rationale for these points of emphasis follow:

1. Biologists will probably not be given a significant role in land-use planning anywhere in the world. Ultimately most conservation programs will be the management of the scraps of nature left untouched by various governments.

2. Though biogeographers may design some natural areas, these "experiments" may or may not support their theories. While sound design reduces the need for management, in many cases management will have to deal with mistakes in design and unforeseen complications.

3. The history of land use in much of the world precludes any opportunity for large natural reserves. In countries like Britain, such constraints lead to an emphasis on problems of management rather than of design (Duffey and Watt, 1971). Thus, conservation biologists there have participated little, if at all, in the controversy over reserve design (e.g., Whitcomb et al., 1976).

4. Although conspicuous vertebrates receive most of the attention from conservationists, the bulk of terrestrial species are insects. My own research emphasis, and that of this paper, is on plants and herbivorous insects—which together probably constitute over half of all existing species (Gilbert, 1979).

If maximum diversity is what we wish to maintain in a reserve, the ecology of plants and their associated faunas should be the basis of sound management. Many birds and larger vertebrates are less sensitive to changes in plant species composition than to changes in habitat structure, productivity and land area. For example, consider two reserves in which the equilibrium diversity of insectivorous birds is deliberately maintained. The two might develop great differences in overall community diversity if insects and plants are not equally well monitored. Moreover, the assumption that reserves designed to maintain larger mammals will automatically be large enough to perpetuate smaller organisms without management is only true if natural disturbance rates or "patch dynamics" (page 27) are considered in the initial design.

In addition to these theoretical justifications, a focus on insects and plants provides biologists with powerful economic arguments for natural habitat conservation (Myers, 1976; Oldfield, 1976): (1) Research on natural populations of insects in intact ecosystems provides a theoretical basis for developing biological controls for insect pests and weeds; (2) Natural area preserves provide permanent reservoirs of potential control agents; (3) Research on patterns of natural resistance of wild plants to insects suggests directions for crop breeding programs; (4) Natural areas contain wild relatives of cultivated plants and are a permanent source of genetic resources in crop breeding (Browning, 1975).

In the last decade there has been an upsurge of interest in animal-plant interaction in the neotropics. While only a fraction of the known taxa have been examined ecologically, sufficient long-term studies of representative animal-plant systems are available to reveal some consistent patterns of food web organization relevant to managing biological diversity in neotropical forest preserves.

NEOTROPICAL FOOD WEBS AND THE MAINTENANCE OF DIVERSITY

The relatively stable and mild conditions of tropical rain forests, combined with high productivity and complex vertical structure, have often been used to explain the high species richness in such habitats (see MacArthur, 1972 for a review). While these factors may be valid explanations for observed differences between tropical and temperate regions, they are too general to be of much use to a manager faced with an eroding fauna and flora on a particular reserve. On a local level we need to understand the specific components of diversity within a habitat so that controlling elements may be discovered and manipulated if necessary.

From the perspective of insects and other animals which specialize on particular host plants, it is possible to identify four strictly biotic features, probably common to many terrestrial ecosystems, which are of key importance to understanding and managing the bulk of neotropical species diversity. The following synthesis emerges from the work of numerous independent researchers and builds on a previous paper (Gilbert, 1977). The four key organizational features of neotropical forests will be discussed in this order: (1) the chemical mosaic and coevolved food webs; (2) mobile links; (3) keystone mutualists; and (4) the ant mosaic.

The Chemical Mosaic and Coevolved Food Webs

As Janzen (1977) pointed out, the relatively uniform green color of vegetation disguises its true nature from the perspective of herbivores. Neotropical forests, for example, are exceedingly complex mosaics with respect to interspecific variety in secondary compounds (alkaloids, terpenes, etc.), chemicals thought to serve a defensive function in plants (see review by Levin, 1976). Consequently, most insect herbivores have evolved sensory and digestive specializations which restrict their diet to a relatively small, chemically similar fraction of the available plant species. For example, some beetles feed only on the seeds of a single tree (Janzen, 1977), presumably because of the extreme digestive specialization required to deal with highly toxic seeds.

An important feature of the chemical mosaic is the fact that plants of the same genus or family often share many compounds as well as those insects specialized to deal with such compounds. A large fraction of herbivorous insects, along with their own parasitoids, are thus organized into many separate food webs. These webs occur side by side, but are based on different, chemically and taxonomically delimited compartments of primary production.

Reference to such systems as "coevolved food webs" (Gilbert, 1977) reflects my assumption that patterns of the chemical mosaic (and thus of insect feeding) have been and will continue to be generated largely by a process of coevolution between plants and their parasites. It is, of course, possible that insects diversified against an already existing mosaic of plant chemistry and have no selective impact on the chemical mosaic. Root's (1973) "component community" or Cohen's (1977) "source food web" may be preferable to those wishing to use a more neutral terminology in discussing ecosystem substructure.

Regardless of how ecosystem substructure evolved, the important fact for conservation biology is that much of the total species diversity in a neotropical forest is to be found in many parallel, host-restricted food webs which are similar in trophic organization but different taxonomically. If we understand how diversity is maintained within a representative sample of the hundreds of such sub-systems, we may find general rules for managing the many species that we do not yet know of or about which we know little.

Two partially analyzed neotropical food webs will serve as examples for the major points involving reserve management. These food webs are delimited at the level of two plant families, the Passifloraceae and the Solanaceae, and are components of all neotropical forest habitats. Butterflies with aposematic (warning) coloration are the most visible manifestation of the herbivore community supported by these plant groups. Many of the general insights relevant to this paper derive from population and community studies of these insects. Only brief highlights of available information on the two food webs follow.

Passifloraceae. The passion flower family is represented by about 500 species in the American tropics, most being vines of the complex genus *Passiflora*. Broad geographical information on host relations of the major herbivores of this family, heliconiine butterflies (Figure 1A–C), and evidence for coevolution with these insects are summarized by Benson, Brown and Gilbert (1976). Other important herbivores of *Passiflora* include flea beetles, coreid bugs and dioptid moths, but these insects are ecologically and taxonomically less well known at present. The flea beetles resemble heliconiines in local diversity and host relationships (Gilbert and Smiley, 1978).

Only a small fraction of the known neotropical species of Passifloraceae occurs in any given local flora. Moreover, the local diversity of this plant family is strikingly constant across neotropical rainforest sites (Gilbert, 1975), ranging between 10 and 15 species. It appears that this consistency reflects combinations of microhabitats and pollinators which will support different *Passiflora* species locally. Gentry (1976a) noted similar patterns in neotropical Bignoniaceae. Since the numbers of heliconiines and other host-specialist insects correlate with the number of host species available (Gilbert and Smiley, 1978), we clearly need a better understanding of such geographically constant diversity within plant taxa to manage the diversity of associated food webs in reserves.

Studies of the passion flower food web at the Organization for Tropical Studies (OTS–La Selva field station in Costa Rica) by Smiley and Gilbert indicate that the Passifloraceae, like many other woody tropical plant families, is represented locally by species differing in growth form, successional stage occupied and pollination relationships (Figure 4F). In the flood plain forest at La Selva, for example, only three species become large canopy-emergent vines. One species is a small forest understory tree

FIGURE 1. Life histories of Heliconiine and ithomiine butterflies. A: *Heliconius ismenius* ovipositing on seedling of *Passiflora serratifolia* in forest understory, Sierra de los Tuxtlas, Vercruz, Mexico. B: *H. cydno* collecting pollen from *Anguria* (Cucurbitaceae) in Atlantic lowland forest, Costa Rica. C: Gregarious larvae of *H. doris* on canopy vine, *P. ambigua,* in Atlantic lowlands, Costa Rica. D: *Ithomia heraldica* male displays scent hairs while perched in a sun-fleck near the forest floor in Costa Rica. E: Males of several ithomiine species visiting *Eupatorium* (Compositae) for nectar and (presumably) sex pheromone precursors in successional vegetation, Corcovado Park, Costa Rica.

and six species can be found in various successional stages along with seedlings and suckers of canopy species. The full complement of ten Passifloraceae thus depends upon a mix of successional stages.

The insects specialized on *Passiflora* are further restricted with respect to the microhabitat occupied. Smiley (1978) found that within an area, the most closely related *Heliconius* species occupy different microhabitats and participate in different mimicry associations with more distantly related congeners. Within microhabitats, forest understory for example, Smiley found that mimetic pairs of heliconiines are of more distantly related species and utilize different host plants. Thus it appears that habitat heterogeneity has contributed to the evolutionary radiation of *Heliconius* into several color pattern complexes, each of which is closely tied to particular microhabitats.

Several other aspects of this food web are worth mentioning. First, each *Passiflora* population is typically composed of rare, scattered individuals which produce flowers and fruit infrequently and asynchronously. Consequently, these plants provide little incentive for the evolution of host-restricted pollinators. Instead they rely on pollinators (bees, moths, bats, hummingbirds) and seed dispersal agents (birds, bats) which visit rare, scattered plants of many taxonomic groups.* I call such organisms "mobile links" because through their foraging movements, they are of mutual concern to the reproduction of many different unrelated plants which, in turn, support otherwise independent food webs.

Another category of mobile link species should be mentioned briefly in reference to *Passiflora*. Most passion flower vines possess extrafloral nectaries which attract predators and parasitoids (ants, wasps, microhymenopterans) and a host of other insects. Some of these visitors are known to be significant as mortality factors to the insects which attack the plants. While highly specialized parasitoids are considered part of the coevolved food web (Gilbert, 1977), predators such as ants and wasps forage over many kinds of plants possessing leaf and petiole nectaries, thereby forming another kind of connection between host-based food webs (Figure 2A, C–E).

Another notable aspect of the passion vine system concerns the butterfly genus *Heliconius,* the most numerous and diverse genus of the heliconiines. Not only do these insects have the problem of rare and scattered larval host-plants, but most are specialists on new shoots of *Passiflora,* a resource more rare in space and time than the plants themselves. In spite of this, local populations can be remarkably constant at low density for years (Ehrlich and Gilbert, 1973). How do *Heliconius* populations persist in spite of low density and fluctuating larval food supplies? Part of the answer involves the cucurbit vines, *Gurania* and *Anguria.* Male flowers of these vines provide *Heliconius* with adult resources (pollen) known to increase both the fecundity and longevity of adults (Dunlap-Pianka et al., 1977). Adult reproductive longevity reduces the probability of local extinction by reducing the impact of fluxes in the larval food supply (Gilbert 1975; 1977). Although *Heliconius* adults can be opportunistic in plant species visited for pollen and nectar, it is the relatively constant supply of pollen from *Anguria* in most neotropical

*D. H. Janzen first applied the term "traplining" to emphasize that the foraging of such animals is similar to that of trappers running strings of traps. The terminology suggested here emphasizes their ecological role.

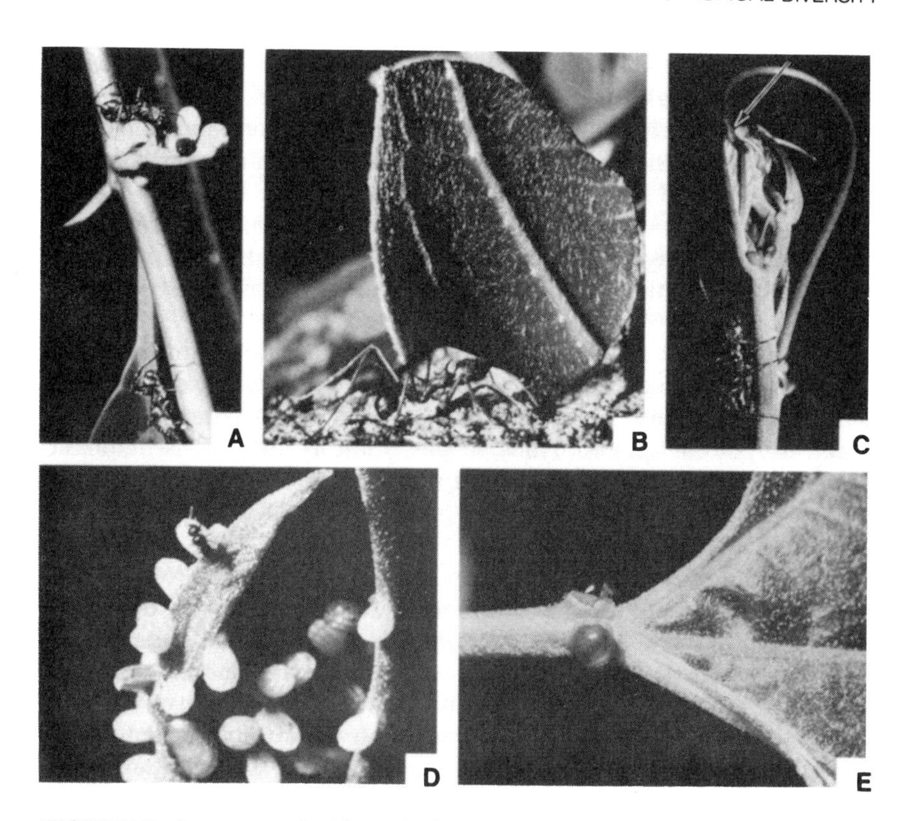

FIGURE 2. Ants, parasitoids and plant nectaries. A: *Ectatomma* tending a membracid (Homoptera) on inflorescence of *Solanum* in forest understory, Atlantic lowlands, Costa Rica. B: Leaf cutter ant *(Atta)* carrying leaf in dry forest of Tamaulipas, Mexico. C: *Ectatomma* approaches newly hatched *Heliconius* larva (arrow) on shoot of *Passiflora vitafolia* at Corcovado Park, Costa Rica. Note empty egg shell near larva and nectar glands on petioles of young leaves. D: Female chalcid parasitoid (probably Scelionidae) ovipositing on *H. sara* eggs which are always placed on *P. auriculata* in the Atlantic lowlands of Costa Rica. Eggs are less than 1 mm. tall. E: Nectar produced by petiole glands of *P. incarnata,* Braziora County, Texas.

rainforests that insures the persistance of the rarer members of the *Heliconius* community. It should be noted that *Heliconius* species which are largely separate in microhabitat preference and larval hosts can, nevertheless, be seen to overlap on the same pollen sources along forest edges and in the canopy.

Solonaceae. The tomato and eggplant family, like Passifloraceae, is represented by shrubs, vines, trees, herbs and epiphytes although shrubs and small trees are probably the dominant growth form of the family. In the neotropics, the large fauna of insects specialized on the Solanaceae includes grasshoppers, hemipterans, homopterans, chrysomelid beetles (including flea beetles) and various lepidopterans (including ithomiine butterflies). Like the heliconiines, the ithomiines form an important group of conspicuous neotropical butterflies with warning coloration (Figure 1D, E). A local community of Solanaceae might include 45 to 50 species, most of which would be of the genus *Solanum*. *Solanum* species can be specialized to any part of the forest and they comprise the major larval resources for ithomiines (although numerous other solanaceous genera are utilized).

The Solanaceae are diverse in pollination relationships, utilizing bees (most *Solanum*), bats (*Markea*), hummingbirds (*Juanulloa*), and lepidopterans (*Cestrum*). Birds and bats are known to be seed dispersal agents for certain species. Thus, as was the case in the Passifloraceae, most members of the Solanaceae rely heavily on mobile links for their continued presence in the ephemeral, patchy microhabitats to which they are restricted.

Research on ithomiine ecology by several field workers over the last decade, most notably that of Drummond (1976) and Haber (1978), provides details of larval and adult biology. Like heliconiines, ithomiine species are restricted as to suitable host plant and microhabitat. Like *Heliconius*, ithomiines depend on adult resources important to their reproductive biology. Adults regularly use fresh bird droppings and nectar plants as food, and still other plants (Boraginaceae, Compositae) which provide precursors for male courtship pheromones.*

Unlike heliconiines, ithomiines have localized courtship areas in the forest where males of several species may gather to wait for females. Adults of species with larval host plants in the forest understory can be seen visiting floral resources in early successional areas; species with larval hosts in early successional areas visit the forest understory for bird droppings and court and mate there. Thus a juxtaposition of microhabitats is important to many ithomiine species.

Even in sympatry, no species appear to be in both the Passifloraceae- and Solanaceae-based food webs (except possibly for a few mobile links). However, the diversity of each system seems to be based on similar factors. First, in both cases full diversity at the herbivore level depends upon maximum species diversity of host plants locally. In both cases, maximum plant diversity requires a balance of healthy successional stages. Second, in both cases maximum diversity of the insects depends upon important resources (adult host plants, courtship sites, and so on) other

*In the closely related danaines, pheromones necessary for mating (Pliske and Eisner, 1969) are known to be derived from the compounds (Schneider et al., 1975) which attract male ithomiines (Pliske, 1975).

than larval host plants. These secondary resources may be found in successional stages other than that occupied by the larval host. Thus, once again, a balance of successional patches often appears necessary to prevent the loss of those insect species which depend upon taxonomically and ecologically different plants in larval and adult stages. This phenomenon is not restricted to the tropics (Wicklund, 1977).

Where tropical forest has been replaced by pastures, both food webs described above are still present. However, they conspicuously lack those species dependent upon spatial heterogeneity of the habitat. Such species also appear to decline as successional stages give way to climax forest. However, since shrubs are less likely than vines to survive shading as the forest grows up, the Solanaceae food web would be expected to decline in diversity more rapidly than the Passifloraceae web. This appears to be the case at Barro Colorado Island, Panama. During the time between isolation of the island by flooding in 1914 and a census of the butterflies in the early 1930's, the ithomiines on the island dropped from about 20 species (a conservative estimate for intact forests) to 11 species (Huntington, 1932). By 1969 only six species could be found in a five month study by Emmel and Leck (1970). This drop is correlated with a decline in Solanaceae diversity as succession returned disturbed areas to forest (T. Croat, personal communication). Meanwhile, heliconiines actually increased from 12 to 13 species between 1930 and 1969. The ability of *Heliconius* to persist as small populations may also help account for the relative stability of the number of heliconiine species (Gilbert and Smiley, 1978).

Based on community samples taken by Brown (1972) heliconiines are two to five percent and ithomiines five to 11 percent of total butterfly species in neotropical rainforest. Together, they are the principal models for several major butterfly mimicry complexes, so that the loss of such species could have a drastic impact on the many edible and mimetic species in an area. Thus, a management scheme which maintains successional patchiness would help preserve a minimum of seven to 16 percent of the butterfly fauna. Furthermore, since many other plant taxa show patterns of ecological and morphological diversification similar to those of Passifloraceae and Solanaceae, it is likely that a large but unknown fraction of the plant and insect diversity would be maintained by management aimed specifically at these two food webs.

Mobile Links

Animals that are significant factors in the persistence of several plant species which, in turn, support otherwise separate food webs can be called mobile links (Figure 3). In addition to being rare and dispersed, most neotropical forest plants are obligate outcrossers (Bawa, 1974) and rely on hummingbirds (Feinsinger, 1976), bees (Delgado Salinas and Sousa Sánchez, 1977), hawk moths and bats (Heithaus et al., 1975) for pollination. Likewise, efficient seed dispersal is required for the colonization of habitat patches suitable for germination and seedling establishment.

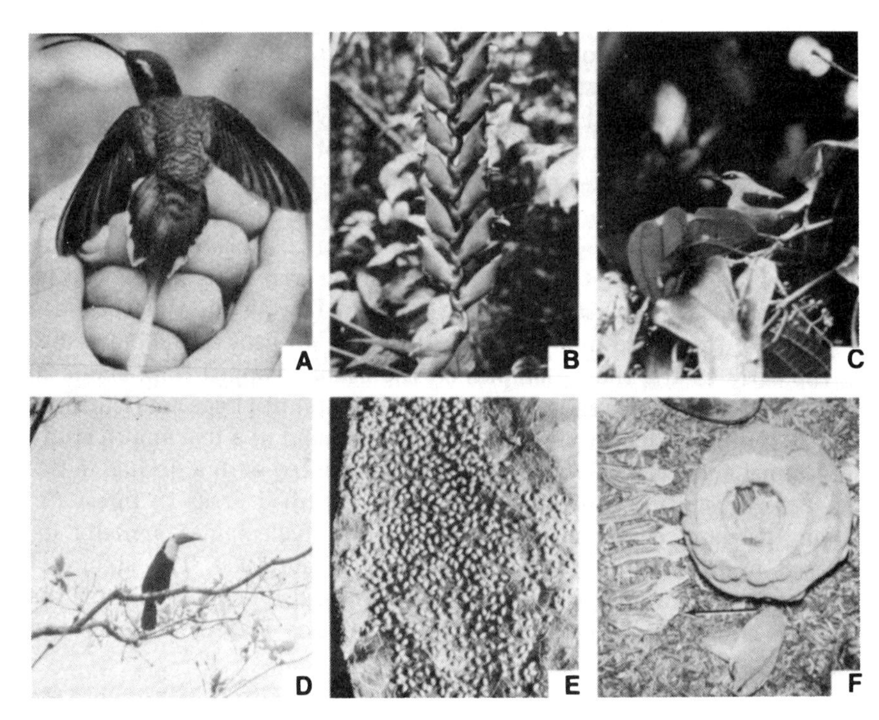

FIGURE 3. Some mobile links and plants they visit. A: Long-billed hermit, *Phaethornis superciliosus* (Osa, Costa Rica), an important pollinator in neotropical rainforests, exemplifies a mobile link. B: *Heliconia longa* (Osa, Costa Rica), frequently visited by hermit hummingbirds. Some *Heliconia* qualify as keystone mutualists. C: Blue honey creeper, *Cyanerpes cyaneus,* feeding on fruit of *Miconia* (Melastomaceae), Atlantic lowlands, Costa Rica. *Miconia* species are key resources of many small frugivorous birds. D: Toucan, *Ramphastos swainsonii,* in Atlantic lowland rainforest. This species relies on fruit of the canopy tree *Casearia* during parts of the year. At other times it and its congeners are dispersal agents for other tree species. E: Bat-dispersal cluster of *Piper* (Piperaceae) seeds. *Piper* fruits are a year-round staple for some bats. F: "Monkey pot" fruit of *Lecythis costaricensis* (Lecythidaceae), canopy tree related to brazil nut, has bat-dispersal seeds. Operculum drops from pod of mature nut. Food bodies are at end of nut (arrow) away from opening, so that bat must grab seed on its hard end, fly to a better perch and rotate it before eating food body. Seed is then dropped, completing the dispersal sequence.

FIGURE 4. Euglossine bees *(Euplusia, Euglossa, Eulema)*, a Carpenter bee and plants they visit. A: Male *Euplusia simillima* visiting *Stanhopea tigrina* for volatile chemicals in a Mexican cloud forest. B: *Euglossa* female collecting nectar from *Calathea insignis* (Marantaceae) in successional area, Atlantic lowlands, Costa Rica. C: *Euglossa* male attracted to cineole placed on tree in Costa Rica rain forest. Arrow points to orchid pollinium stuck to head. D: *Eulaema migrifacks* female which was collecting pollen (note pollen load on hind leg) on *Solanum sanctae-clarae,* a purple-flowered epiphytic shrub in Atlantic lowlands of Costa Rica. E: Flower of *Dalechampia aff. tillifolia* (Euphorbiaceae), a common successional vine in the Sierra de los Tuxtlas. Resin gland is indicated with arrow. F: Carpenter bee *(Xylocopa)* visiting *Passiflora pulchella* at Santa Rosa Park, Costa Rica.

Birds and bats are the major neotropical seed dispersers (McKey, 1975). Ants may well be important to seed movement as they appear to be in other regions (see review in Gilbert, 1979). Certain ants may also patrol and defend a variety of nectar-producing plants (Bentley, 1977). Seedling establishment as well as reproduction by adult plants may be enhanced by ant protection.

Among the most important and best researched groups of neotropical link organisms are the euglossine or orchid bees (Figure 4, A–E). It is worth elaborating the biology of these spectacular insects because they can be involved in a major way with hundreds of plant species in one area. A single euglossine species may link plant species from all stages and strata of a forest into a system of indirect mutualism.

Dodson (1966) has given a general account of euglossine-plant relationships. Euglossine females gather pollen from successional plants such as *Solanum* and *Cassia.* In some cases they are simply part of a complex of bees which visit the plants—for example, *Cassia* (Delgado Salinas and Sousa Sánchez, 1977). In other cases (for example, some *Miconia* and *Solanum* species) they may be the primary pollinators. Females are known to travel great distances in foraging and are thus important to the reproduction of low density plants (Janzen, 1971).

Resin for nest construction may be obtained opportunistically from damaged trees or, more predictably, in successional patches where certain *Dalechampia* (Euphorbiaceae) vine species provide resin as a floral reward for pollination (Armbruster and Webster, in press). Nectar is obtained from a variety of successional (Apocynaceae, Marantaceae, Rubiaceae and others), understory (Rubiaceae, Marantaceae) and canopy (Orchidaceae, Bignoniaceae, Bromeliaceae) plants by both sexes. A single euglossine species may pollinate a dozen species of nectar plants representing every level and successional stage available in a forest.

Like male ithomiine butterflies, euglossine males collect volatile chemical compounds for use in courtship behavior (Dodson, 1975). Compounds collected differ for each of the 30 to 50 species of euglossines which might occur together. Some orchids rely on a single euglossine species for pollination and provide volatile compounds such as cineole to visiting males. Mechanical methods (Dressler, 1968) and chemical divergence enforce sexual isolation of many genetically similar sympatric orchid species (Hills et. al., 1972). Other plant groups have also evolved male euglossine pollination systems. In Veracruz, Mexico, Armbruster and Webster (1979) found that while one *Dalechampia* species provides resin for and is pollinated by females, a second species produces volatile compounds and is male pollinated. Many epiphytic *Spathiphyllum* and *Anthurium* species (Araceae) have likewise specialized on male euglossines for pollination. It has been suggested that for the aroids (Araceae), as for orchids, a proliferation of species has been made possible by the availability of a diverse euglossine fauna (Williams and Dressler, 1976).

Many euglossine species rely on early successional plants for larval resources and at the same time are important, even necessary, pollinators of plants restricted to later successional stages of forest. Canopy orchids and aroids are thus indirectly dependent upon the presence of early succession patches. Such specialization evolved under conditions of continuous forest where appropriate successional stages were always available within the flight range of euglossines. In isolated habitat preserves, however, certain successional plant species might be lost during periods of low habitat disturbance and euglossine species critical to canopy plants may die off as commuting distances between diverse resources become too great.

From these considerations one might logically predict the loss of many orchids and aroids from Barro Colorado Island now that successional changes have reduced the diversity and productivity of early succession plants like some Solanaceae. Several years ago I attempted to test this prediction but learned from T. Croat, then conducting a floral survey of the island, that the orchid list had increased since the last inventory. In addition to the obvious problems (the earlier survey was probably not sufficiently intensive and orchids are long lived), it should also be pointed out that euglossines have been seen crossing Gatun Lake to the island (Dressler, 1968). Thus, my prediction may be realized in the next 30 years as adjacent mainland forest in central Panama is removed.

Rather than test such predictions with our few habitat reserves, I believe we should devise schemes to monitor and manage the diversity of such important link organisms as orchid bees. This would involve keeping track of various successional stages and some of the plants in them. In lowland areas (where land slips and other forms of natural disturbance occur at a lower rate per hectare than in steeper areas) managed disturbance will be necessary, particularly as the preserve becomes smaller and more isolated.

Representative link species should be a special focus for autecological research in conservation biology. From available information on bats, hummingbirds and bees, it appears that management rules designed for the few species we know something about will apply to many other, ecologically similar species. It is safe to say that maintaining successional variety is one general rule clearly mandated by existing information, though the details are lacking.

Keystone Mutualists

Keystone mutualists are those organisms, typically plants, which provide critical support to large complexes of mobile links. The loss of a keystone mutualist would, with some time delay, cause a loss of mobile links, followed by losses of link-dependent plants through a breakdown in reproduction and dispersal. Finally, host-specialist insect diversity should decline with the reduction in diversity of host communities. This is the sort of linked extinction predicted by Futuyma (1973).

The canopy tree *Casearia corymbosa* (Flacourtiaceae) fits the concept of keystone mutualist. Howe (1977) has shown that although only one bird, the masked tityra *(Tityra semifasciata),* is a highly effective seed dispersal agent at La Selva, 21 other fruit feeders use the tree. Howe suggests that *Casearia* is a "pivotal species" because it supports several obligate frugivores which depend on it almost entirely during the two to six week annual scarcity. Howe predicts that the loss of *Casearia* at La Selva would lead to the disappearance of *T. semifasciata* (with consequences for other trees whose seed it disperses), and *Ramphastos* toucans (with effects on *Virola* and *Protium). R. swainsonii* is shown in Figure 3D. Although *Casearia* is a canopy tree at La Selva, it supports birds which may be significant to successional as well as primary forest tree species. Perhaps the most important fruit source for smaller frugivores, *Miconia* (Melastomaceae) is shown in Figure 3C.

Two detailed hummingbird studies reveal similar keystone mutualists. Again at La Selva, Stiles (1975) found that most hummingbirds depend on *Heliconia* (Figure 3B) during the early wet season. Hermits (Figure 3A), on the other hand, rely on *Heliconia* for nectar at all times of the year. The various sympatric *Heliconia* species have displaced flowering times so that at least one or two species are always in bloom. However, as Stiles notes, flowering for some species depends on light-gap formation. In a small reserve it might, therefore, be necessary to manage light-gaps around *Heliconia* clumps to insure the survival of hermits, an important group of link species. Incidently, a study is in progress on the insect fauna of *Heliconia* (Strong, 1977).

In a mountainous locale near Monteverde, Costa Rica, Feinsinger (1976) found that *Hamelia patens* (Rubiaceae) sustains hummingbird populations during periods when other foods are scarce. *H. patens* flowers year-round but peaks from May to October when other flowers are rare.

Bats account for a large fraction of mammalian diversity in the neotropics. Like other link species, many bats require a variety of resources including insects, nectar, pollen and fruit (Figure 3E, F). Suitable roosting sites such as large hollow trees, may also be habitat restricted (Figure 5G). Heithaus et al. (1975) identified major floral sources *(Ceiba,*

FIGURE 5. Neotropical rain forest habitats. A: Successional stages along airstrip near Rincon, Osa, Costa Rica. Grasses predominate in frequently cut area. B: Forest understory, Corcovado Park, Costa Rica. C: Natural disturbance along small stream constantly generates successional habitats. Osa, Costa Rica. D: Epiphyte load of large canopy tree at La Selva, Atlantic lowlands of Costa Rica. Hundreds of plant species occur in the area shown in this photograph. E: River canyon on slopes of Vólcan Barba, Costa Rica, where land slips along steep banks maintain successional variety. Human exploitation is almost impossible here. F: Heavily grazed pasture which was recently forest in the Atlantic lowlands of Costa Rica. Few elements of natural successional vegetation can be found in such areas. G: Large hollow tree at La Selva, Costa Rica, being checked out for roosting bats.

Ochroma) and fruit sources *(Piper, Ficus, Solanum)* which would qualify as keystone mutualists for bats. It is clear from observed bat-plant interactions that few of the plant species which depend on bats for pollination or seed dispersal are sufficiently constant (in reproductive activity) or

abundant to support even a single bat species. In contrast, each keystone mutualist may support an entire bat assemblage for part of the season.

Another kind of keystone mutualist is a plant which produces large quantities of extrafloral nectar or accessible floral nectar. Smiley and Gilbert (unpublished) aimed time-lapse cameras at the leaf nectary of a large *Passiflora ambigua* vine at La Selva. In addition to *Pseudomyrmex* ants visiting the nectar glands, dozens of Hymenoptera, Hemiptera, and Diptera not otherwise involved with *P. ambigua* exploited the nectar resource. In temperate zone forests similar relationships between parasitoids of tree-feeding insects and certain nectar plants have been studied. Hassan (1967) found 148 species and 1,535 individual hymenopterous parasitoids associated with flowering herbs and shrubs within north German forests. *Daucus* (Umbelliferae) qualifies as a keystone mutualist in attracting 68 species during that study.

The study of Syme (1977) underscores the significance of Hassan's data to forest stability. Syme showed that *Asclepias* and *Daucus* nectar increases the reproductive longevity of parasitoids (which would, in turn, decrease chances for herbivore outbreak). Anderson (1976) reviewed the role of egg parasitoids in reducing rates of defoliation of forest trees. Kulman (1971) summarized experimental and observational data relating defoliation to growth and mortality of trees. Tying these observations together, it is obvious that the energetic support of the most minute Hymenoptera has demographic consequences for trees (through reduction of defoliation and other damage) and thus is of concern in forest reserve management. Key plants for the parasitoids may occur in different microhabitats from the forest "pest" insects attacked. A wise policy in temperate zone and tropical forests, therefore, would be the management of successional communities so as to maintain plants which support these tiny "control" agents (Gilbert, 1977).

Ant Mosaic

Ants defend many plants against herbivores (Bentley, 1977), but some carry about and tend plant-feeding homopterans and lepidopterans. Other ants, such as *Atta* (Figure 2B) in the neotropics, are major defoliators of forest shrubs and trees (Lugo et al., 1973). Work by Leston and others on the pests of tree crops in the old world tropics has revealed a three dimensional "ant mosaic" which results from the nonoverlap of territories of a few dominant ant species. Different assemblages of subdominant ants coexist with each dominant species so that the mosaic is a patchwork of ant communities. A consequence of the ant mosaic important to agriculture and conservation is the fact that the immunity of a plant to disease and defoliating animals is a function of its position in the ant mosaic.

Recently Leston (1978) reported the existence of a similar ant mosaic in neotropical forests. He demonstrated that whether a tree was cut by *Atta* depended upon the dominant territorial species associated with the tree. Trees occupied by *Ectatomma* (Figure 2A, C) were cut since this ant does not defend trees against *Atta*. Conversely, the presence of other species prevented leaf removal by the leaf cutters. An unexplained aspect of *Atta* behavior has been the apparent "conservation" of favored tree species near its nests (see review in Gilbert, 1979). Leston's observations may help explain such intraspecific differences in susceptibility to *Atta* attack.

It is probable that the composition of the ant fauna in a tree will also affect its suitability to foliage-feeding monkeys such as howlers (*Alouatta*). Anyone having been stung by neotropical *Paraponera* would not doubt their ability to deter other primates. For plants highly coevolved with ants (an extreme case of the ant mosaic), the defensive role of ants against vertebrates has been thoroughly studied (see review by Bentley, 1977).

Knowledge of the ant mosaic could be important to the management of small neotropical forest reserves. For example, if the number of reproductive individuals of a keystone mutualist species in an area is small, defoliation by *Atta* or howlers of all or part of the population could be sufficient to immediately eliminate important mobile links, with subsequent loss of species dependent upon the links. Just as the ant mosaic is manipulated to improve cocoa production in Africa (Majer, 1976), it could be manipulated when necessary to insure the flowering and fruiting of keystone mutualists in undersized reserves.

PLANT/ANIMAL RELATIONSHIPS AND PATCH DYNAMICS

The species diversity of animals which depend upon one or a few species of plants for critical resources is related to the population size and age structure of the appropriate plants in an area. In England, Ward and Lakhani (1977) found that 79 to 87 percent of between-site variation in species diversity of juniper feeding insects was due to variations in the numbers of host bushes available at each site. They point out that the continuous availability of reproductive aged plants is crucial for the fruit feeding fraction of the juniper fauna. Among tropical birds, fruit and flower feeders are known to be "extinction prone" (Chapter 7) presumably because the area required to assure the continuous availability of such resources is much greater than that required for the continuous availability of insect food.

Since the persistence of specialized herbivores undoubtedly depends upon the density of host plants, one must consider the factors that influence plant populations (for example, Duffey, 1977). In the case of neotropical forests where extreme microhabitat specialization has been seen in plant species, a key factor to consider is the relative abundance and continuity of various kinds of microhabitats (Figure 5). In an important paper on this subject, Pickett and Thompson (1978) suggest that successional patches of various sizes and stages within an isolated biological reserve should be considered the only source for colonizing future openings in the forest. They apply island biogeography theory to systems of microhabitat islands within a reserve, and suggest that an important factor to consider in the design of nature reserves is the size distribution, spacing and turnover rates of microhabitat patches.

The factors that influence such "patch dynamics" are of concern to both the design and management of nature reserves. These include topographic, geologic and climatic features of a region (see Chapter 5 for a detailed discussion). For example, the frequency of landslides per unit area increases from level to inclined terrain and from earthquake-free to earthquake-prone regions. In contrast, the rate of succession back to mature forest may remain relatively constant over these same gradients. Therefore, since the relative area of any particular successional stage is a dynamic balance of disturbance and succession, areas with intrinsically higher rates of disturbance (Figure 5E) will be more likely to have all stages of succession constantly available within a unit area.

Key elements of food web organization are related to habitat patchiness in Figure 6. This figure includes many of the animal/plant interactions explained in this chapter. The interactions shown in the figure are based on real cases but are oversimplified to enhance understanding of the major points in the chapter. Plant species are represented by solid hexagons (the sizes of which indicate abundance) and are organized into six chemically distinct higher taxa (shown as six "islands" in the figure). Each "island" possesses three different microhabitats: successional (light gray), forest understory (medium gray) and forest canopy (dark gray). Host specialist herbivores and their specialized parasites are shown as small solid triangles and small solid boxes, respectively. Insects are arranged above their favored hosts and microhabitat. Generalist herbivores are represented by a large triangle at the center of the figure, typified by the most habitat and host-plant generalized herbivore in the system: leaf-cutter ants *(Atta)*. Mobile link groups important in pollination, seed dispersal or plant protection are represented by the six large circles (labeled A to F). These are connected by thin lines to plants with which the mobile links interact. Interactions between mobile links and plant groups not represented in the figure are indicated by outward-radiating lines. Plants critical to the support of particular groups of mobile links, keystone mutualists, are connected to appropriate link taxa with heavy arrows. Specification of plant groups represented in Figure 6 (Roman numerals I through VI) and a brief discussion of the ecological details shown is worth emphasis.

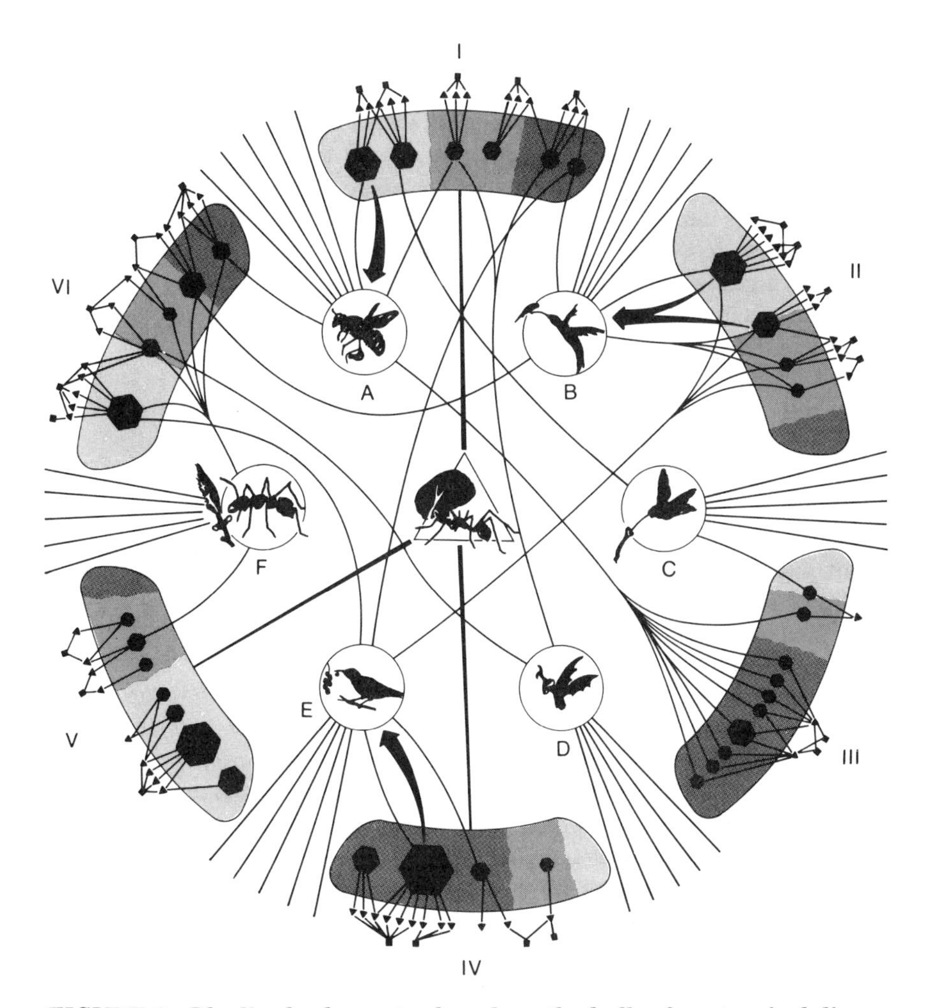

FIGURE 6. Idealized scheme to show how the bulk of neotropical diversity is organized by chemical mosaic and mutualism (see text).

I. The *Solanum* family has representatives in all microhabitats. Note that one successional species is a keystone mutualist for orchid bees (A). One canopy member, *Juanulloa,* is pollinated by hummingbirds (B), but dispersed by frugivorous birds (E). Bats (D) pollinate one canopy species *(Markea),* and disperse an understory species *(Solanum).*

II. *Heliconia* has representatives in successional habitats and in the forest understory. The two species on the left are key resources for hermit hummingbirds. None are represented in the canopy.

III. Orchids are represented primarily in the canopy and most exist at relatively low densities as shown. One species shown is pollinated by hawk moths (C), the remainder by male euglossine bees (A). Orchids have wind-dispersed seeds. Also, many are ant-defended but this is not shown.

IV. Canopy-emergent trees (family not specified) can be key resources for complexes of frugivorous birds. Juvenile stages of one species are shown in the understory. As a rule, vines and trees which reproduce in the forest canopy are nevertheless represented as juveniles in other microhabitats where leaves are available to a slightly different set of herbivores. This has been largely omitted for simplicity.

V. Grasses are poorly represented in the canopy and have relatively few mutualistic interactions. One species in the diagram has seeds dispersed by ants (F). A few (not shown) are insect pollinated.

VI. Passion flower vines are characterized by ant-visited extrafloral nectaries. *P. vitafolia,* a canopy-emergent, presents flowers in the understory which are visited by hummingbirds. Herbivores "partition" *P. vitafolia* by microhabitat as shown. *Passiflora* are largely free of *Atta* herbivory (possibly because of ant defense).

In virtually all cases the pollination, seed dispersal and defense of a plant species are carried out by different animals as indicated in plant groups I and II. For simplicity, most such multiple interactions are not indicated. Note also that most mobile links shown visit plants in several microhabitats.

As stressed in Figure 6, coevolved food webs, mobile links and keystone mutualists all depend directly or indirectly on local microhabitat heterogeneity (expressed as a balance of successional stages). This heterogeneity may be lost in two important ways. First, if a reserve is too small relative to the size of the largest disturbance patch type, elements of mature forest may all be lost at once with no ready source for recolonization (Pickett and Thompson, 1978). Second, if the reserve is too small relative to the natural disturbance rate, certain patch types might disappear temporarily through succession. While the same type patch might be returned by a later disturbance, there may be no colonization source for the plants and animals previously associated with it if the reserve is isolated.

In areas where the first problem is of concern the only practical answer seems to be to design reserves with the largest possible disturbance in mind. The second problem is more likely to occur in flat lowland forest. In this latter case, managing the disturbance rate seems a necessity in reserves too small to insure a continuum of all the important successional microhabitats. I would include the Barro Colorado Island reserves and, if it becomes further isolated, the OTS-La Selva reserve in this category. Because of important links among species which characterize different successional stages, even the temporary loss of one such stage will have a ripple effect throughout the entire system. From the perspective of plants and their closely associated animals, the prevalent idea that early successional areas need not be considered in forest reserve management is a serious misconception that is not restricted to the tropics (Wright, 1974).

CONSERVATION BIOLOGY, AGRICULTURE, AND FORESTRY

Considering both the biological complexities and the sociopolitical problems associated with the development and management of neotropical nature reserves, I believe that conservation biology should be concerned with two major questions involving the interaction of biological reserves with agriculture. First, what forms of agriculture are most compatible with the management of nature reserves? And second, what kinds of practical benefits might nature reserves have for surrounding agroecosystems?

The value of agricultural practices might be twofold. First, if primitive agricultures which maintain high crop diversity (Soemarwoto et al., 1975) or active interaction between natural habitats and crops (Wilken, 1977) are encouraged in a buffer zone around biological reserves, sources for colonization of successional patches within the reserve would be increased. Second, the primitive agricultural zone would buffer the area against harsher forms of land use and diminish political pressures for alternative uses of the reserve itself. However, it is likely that such traditional agriculture will, eventually, require subsidies from conservation organizations.

The values of biological reserves to agriculture have not been adequately addressed by conservationists. In the long run such areas are a valuable source of new cultivars, medicines and industrial products and biological control agents. In addition, new ideas to improve forestry and agriculture may originate from work in reserves. Unfortunately, at present, these values are less obvious in areas where high energy input monocultures dominate agricultural practices. It is already becoming clear that current practices cannot provide stable solutions to our food, fiber and wood needs; as energy supplies decline, this failure will be even more obvious. Conservation biology must anticipate the future needs of forestry and agriculture and develop clear connections between the preservation and management of biological diversity and these "practical" aspects of land use.

An example of the sort of situation which requires an integration of forestry, agriculture and conservation is the problem of *Passiflora mollissima* in Hawaii. *P. mollissima,* a native of the Andes, was introduced along with about 15 other *Passiflora* species. One, *P. edulis,* is a valuable crop below 1,500 ft. *P. mollissima* has become a serious pest, (the "kudzu"* of Hawaii), blanketing about 84,000 acreas of forest, killing trees and causing economic devastation to the local forest industry. The spread of *P. mollissima* threatens endemic forest species and should be of concern to national park officials and others responsible for the remnants of the Hawaiian flora. Meanwhile, the solution to the dilemma may exist in neotropical reserves where intact food webs based on *Passiflora* suggest ways to control *mollissima* without injury to the *P. edulis* industry (Gilbert, 1977; Waage, Smiley and Gilbert, in preparation). Unfortunately, the fragmented structure of our academic, research and funding institutions tends to isolate the disciplines of basic ecology, agriculture, conservation and environmental protection. At best, there is little overlap between these areas; at worst, there is active antagonism. It is incumbent upon those concerned with the long-term biological diversity of the earth to ignore the traditional boundaries of scientific subdisciplines and to seek, instead, more holistic solutions.

SUMMARY

The structure of neotropical forests as determined by animal/plant interactions is highly organized (Figure 6). The system consists of many parallel, structurally similar but taxonomically different, food webs based on particular groups of plants. Mutualism plays a crucial role in the maintenance of diversity in the system. Mobile links are animals required by many plants for reproduction and dispersal. Keystone mutualists are plants which support link organisms and indirectly support the food webs which depend upon mobile links for all or part of their species richness. Finally, because ants control rates of herbivory on many plants, the existence of a mosaic of different dominant, territorial species of ant creates a subtle but important form of patchiness.

These interactions occur against a patchwork of microhabitats defined by the size of disturbance and successional status. Many key organisms in the system are restricted to one microhabitat while others, especially mobile links, depend on the constant availability of several. The need for management of disturbance rates in neotropical reserves increases toward areas with less topographic relief and smaller size.

I suggest the following research priorities for neotropical conservation biology: (1) Further analysis of plant specific food webs; (2) Autecology of link species, keystone mutualists and dominant ant species; (3) Classification of microhabitats and study of patch dynamics; (4) Integration of conservation, agriculture and forestry.

Finally, it should be pointed out that such research should be an integral part of managing the diversity of neotropical forests.

*Kudzu is an exotic legume vine *(Pueraria)*. Introduced into the Southeastern U.S. for erosion control, it is now a serious forest pest.

SUGGESTED READINGS

Baker, H. G., 1973, Evolutionary relationships between plants and animals in American and African tropical forests, in *Tropical Forest Ecosystems in Africa and South America,* Meggers et al. (eds.), Smithsonian Inst. Press, Washington D.C., pp. 145–159. A comparison of new and old world tropical forests from the perspective of plant reproductive biology.

Brown, K. S., 1978, Heterogeneidade: fator fundamental na teoria e práctica de conservacão de ambientes tropicais, in *I Encontro Nacional Sobre a Preservacão da Fauna e Recursos Faunisticos,* (Brasilia, 1977). I.B.D.F., Brasila. The first Portuguese language discussion of some of the ideas presented in this paper.

Futuyma, D., 1973, Community structure and stability in a constant environment, *Amer. Nat.,* 107, 443–446. An important theoretical discussion of tropical diversity maintenance.

Gilbert, L. E., 1979, Development of theory in the analysis of insect-plant relationships, in *Analysis of Ecological Systems,* D. Horn, R. Mitchell and G. Stairs (eds.), Ohio State Univ. Press, Columbus. A general review of insect-plant interactions from a perspective of ecological theory.

Gilbert, L. E. and P. Raven (eds.), 1975, *Coevolution of Animals and Plants,* Univ. of Texas Press, Austin. A diverse collection of papers dealing with animal-plant interactions.

Janzen, D. H., 1968, Host plants as islands in evolutionary and contemporary time, *Amer. Nat.,* 102, 592–595. A classic paper on the evolutionary ecology of plant-based insect faunas.

Mound, D. and N. Waloff, (eds), 1978, *Diversity of Insect Faunas,* Blackwell Scientific Publications, London. The most recent symposium on insect diversity.

Pickett, S. T. A. and J. N. Thompson, 1978, Patch dynamics and the design of nature reserves, *Biol. Conserv.,* 13, 27–37. The key paper on plant community dynamics and conservation.

Price, P. W., 1977, General concepts on the evolutionary biology of parasites, *Evolution,* 31, 405–420. A good introduction to the biology of host-restricted animals.

CHAPTER 3

THE DENSITY AND BIOMASS OF TROPICAL MAMMALS

John F. Eisenberg

In this Chapter, I will consider: (1) animal numbers expressed as biomass and (2) the land area needed to sustain viable populations of species at different trophic levels. I will concentrate on the mammals, which are only a small part of most ecosystems, but are important in nutrient cycling and habitat modification (Montgomery and Sunquist, 1975). As a member of the Class Mammalia, man may well gain insight into his role in ecosystems by carefully considering the strategies of his fellow mammals.

ECOLOGICAL PRINCIPLES OF MAMMALIAN ENERGETICS

The preservation of a species necessitates a knowledge of many aspects of its life history. A species derives its nourishment and shelter from its habitat. The area necessary to preserve a self-sustaining population will vary both with the species we choose to conserve and the habitat in which the population occurs. Figure 1 illustrates the complex interrelationship between habitat, feeding adaptations, demography, social structure and population size.

Metabolic Rate, Body Size and Trophic Strategies

The rate of energy flow into a population determines its standing biomass and, therefore, its density. Density is critical because it may be correlated with extinction probability (Chapter 7). Further, because of the

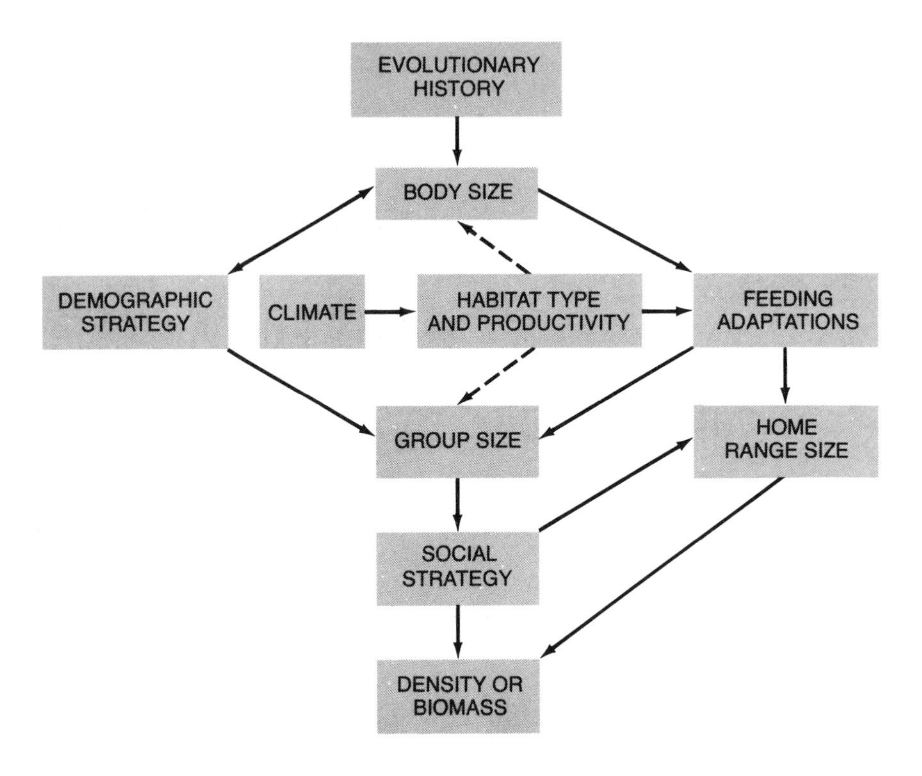

FIGURE 1. Effects of body size, environmental factors and behavioral adaptations on the density achieved by a hypothetical population of mammals.

energy tradeoff between maintenance metabolism and growth, factors that affect metabolic rate will indirectly affect biomass levels for a population.

In mammals, the relationships between body temperature and metabolic rate are too often oversimplified. Mammals are largely endothermic, so even without an environmental assist (such as sunlight, or a heated substrate) they can adjust their body temperatures. Most mammals also exhibit homeothermy (that is, they maintain a rather constant body temperature during their active periods). Of course, mammals do not show uniform adaptations with respect to the maintenance of internal body temperatures—some species indeed employ a pronounced diel fluctuation in their core temperature to conserve energy (McNab, 1978; Eisenberg, in press, *a*). Other species may show a rather constant body temperature except at certain seasons of the year. Finally, some mammalian species maintain a constant body temperature throughout their life and, thus, are true homeotherms.

Homeothermic mammals generally show a decrease in metabolic rate as absolute body size increases. Metabolic rate, when regressed logarithmically against body size, tends to decrease with a slope of −0.25 (Kleiber, 1961). Conversely, the actual amount of energy used in a given time interval tends to increase allometrically with body size at the 0.75 power. It follows that although the metabolic rate of large mammals is lower than that of small mammals, large mammals consume more food than small mammals. Given the different degrees of homeothermy and the differences in core body temperatures shown by the endothermic mammals, one would anticipate that plotting metabolic rates against body size for mammals would result in considerable scattering of data points. Indeed, this is the case (McNab, 1974). The scattering could merely represent the adaptation of specific populations and species to prevailing ambient conditions. However, McNab (1974) has proposed that some trends in the lowering of metabolic rate parallel various trophic strategies. Specializations for certain trophic strategies are often accompanied by an energetic adaptation. This involves the lowering of metabolic rate relative to the values predicted for "typical" carnivorous, granivorous or frugivorous mammals exhibiting nearly perfect homeothermy. For example, burrowing, herbivorous mammals tend to have basal metabolic rates (BMRs) slightly lower than the values predicted from the theoretical Kleiber line. Similary, ant-eating mammals have slightly lower BMRs than the theoretical norm, as do leaf-eating mammals (McNab, 1978). Nevertheless, within a given trophic specialization the same inverse relationship between size and metabolic rate is demonstrable.

Home Range, Trophic Strategy and Habitat Quality

The interactions between body size and metabolic rate are confounded by the tremendous diversity among mammals in trophic specialization. Environmental heterogeneity in habitat quality and uniformity further confound the interactions. In spite of this, some generalizations can be made. McNab (1963) showed that body size, trophic strategy, metabolic rate and the size of an individual's home range are interrelated. Excluding the bats, small mammals can cover a smaller absolute distance per unit time than larger forms. Thus, the larger terrestrial mammals have larger home ranges and a greater absolute mobility than do smaller mammals. Home range size, however, also varies according to trophic strategy. The home range size of a given herbivore is smaller than the home range of an equal weight carnivore. Given the lower BMR of folivores (McNab, 1978), the home range of a foliage-eating herbivore should be smaller than that of an equal weight grazing herbivore (Montgomery and Sunquist, 1978).

If we know a species' average metabolic rate, body size and trophic strategy could we predict its home range size and, ultimately, its space needs? The answer is no. A species' geographic range may encompass a wide variety of habitat types, each with a different theoretical carrying capacity for the species. Carrying capacity for a habitat ultimately reflects plant productivity and this can vary widely in the tropics. Thus, a species may have a smaller home range in one habitat than in another (see Kleiman and Eisenberg, 1973 for a discussion of carnivore home ranges and Leuthold, 1977 for a discussion of the home ranges of ungulates).

Social Structure and Home Range Size

As shown in Figure 1, social structure also affects density. Sedentary individuals or family groups are distributed over the landscape in one way, social groups in another. The home range size for an individual or "family" land-tenure system is far different from the home range size for a large group of individuals moving as a cohesive unit and exploiting an area in common—everything else being equal. To make comparisons between social and solitary (spaced) species, group home range size must be regressed against mean group weight (Clutton-Brock and Harvey, 1977). Moreover, in estimating home range size for group living, the adaptation of populations over a range of carrying capacities must again be considered.

A knowledge of the social behavior of a species under study is essential for an adequate interpretation of habitat utilization. Exclusive land-tenure systems may simplify calculations but too many mammals show intraspecific overlap in home range utilization. Predicting areas needed to support confined populations requires an input from many subdisciplines of zoology.

Body Size, Density and Biomass

As Figure 1 suggests, body size is a very important factor in determining density. This can be seen by examining the absolute density of a species, whether grouped or solitary, and the mean body "size" of the species in question (Figure 2). Large species tend to live at low densities. Density estimates can also be converted into biomass estimates by multiplying the average density by the average individual weight (unit weight)—an abstract figure based on the average distribution of age (size) classes within a population during a given annual cycle. Biomass gives one an appreciation of the ecological impact of a species. For instance, large mammals existing at low densities can still comprise a significant biomass with a correspondingly large impact on nutrient consumption and cycling (McKay and Eisenberg, 1974; Eisenberg and Lockhart, 1972). Biomass values thus offer a crude measure of the ecological dominance of a species.

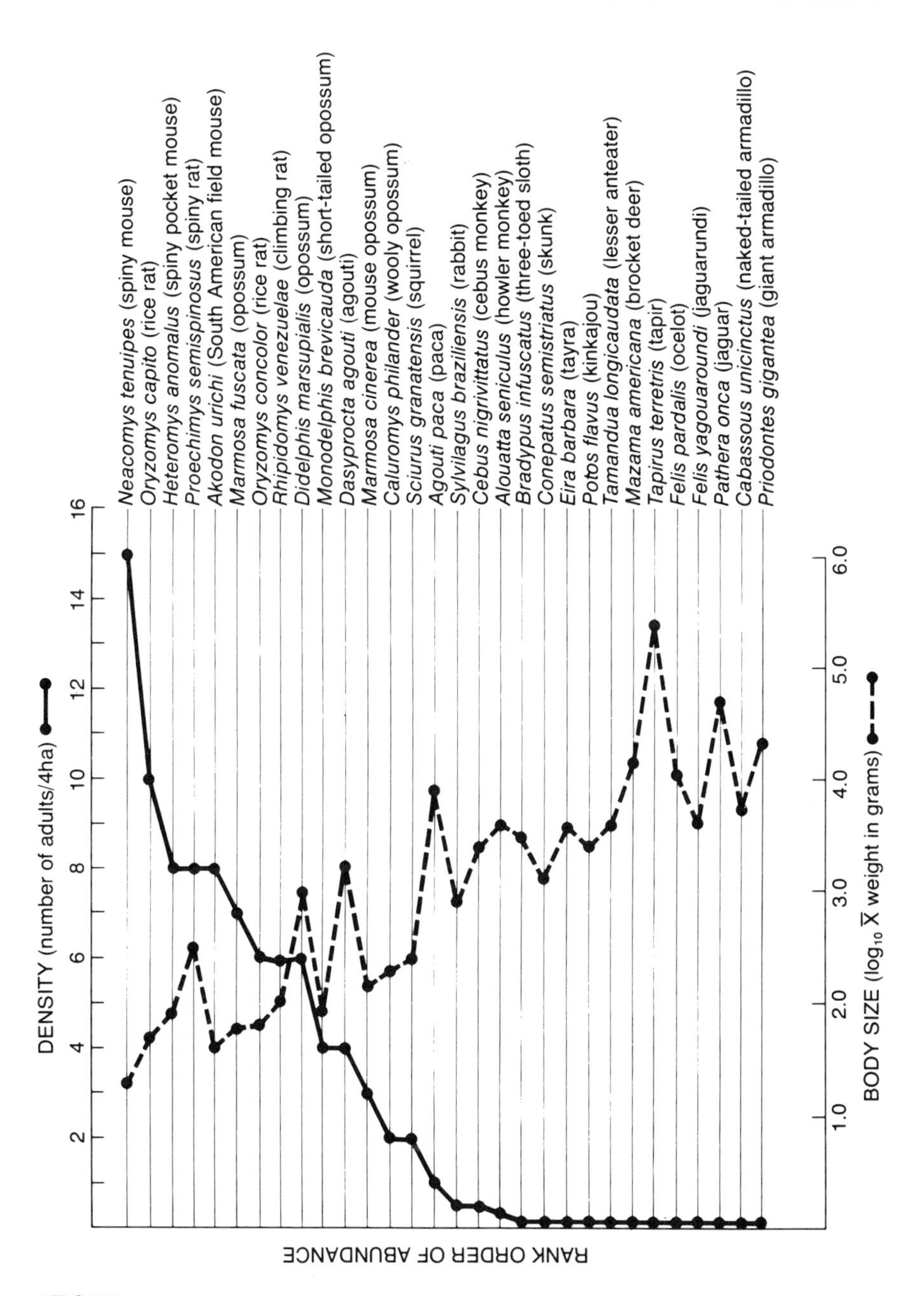

FIGURE 2. Relationship between body size and density for 30 nonvolant mammals found in Guatopo National Park, Venezuela. (From Eisenberg, O'Connell and August, 1979)

Biomass fluctuations can be considerable in species that exhibit seasonal reproduction. For mammals that reproduce seasonally, or species that approximate extreme semelparity (a single reproductive period) in their reproductive strategies, the biomass figure must be an abstract average rather than the actual biomass during any month in a given annual cycle.

Life History Patterns, Trophic Strategy and Body Size

As we have just seen, numerical density is inversely related to body size; however, biomass tends to be positively related to the absolute average size of a species (McKay, 1973; Eisenberg and Lockhart, 1972). Larger mammals tend to be long lived and usually have an iteroparous mode of reproduction (repetitive reproductive periods) with a small litter size (generally one) and a rather extended interbirth interval (Eisenberg, 1975). The reproductive activity of a given female is spread over a considerable period of time.

Longevity also tends to be tied to a trophic strategy. Large herbivores usually live longer than large carnivores. One can begin to make generalizations when contemplating a trophic pyramid: herbivores tend toward iteroparity; large herbivores tend toward extreme iteroparity; carnivores show less iteroparity and a slightly shorter potential life span. Nonvolant (nonflying), terrestrial insectivores in temperate areas are adapted for an almost semelparous reproductive pattern (single reproductive period) (Eisenberg, in press, *b)*.

At this point, one can begin to appreciate the complexity of the interrelationships between ecological energetics, trophic strategies and demography. Any analysis that considers only two or three of the above variables will be incomplete at best and totally misleading at worst. Hopefully, though, our data base is approaching the point where we can move beyond simple discussions of animal numbers, biomass and habitat requirements, and develop a holistic approach to ecology—an approach still in its infancy.

SOME GENERALIZATIONS FOR SELECTED TROPICAL HABITATS

The bats (Chiroptera) and aquatic mammals will be excluded from the remainder of this analysis. All further discussion refers to nonvolant, terrestrial mammals. It is not only conceptually useful to separate out aquatic forms, but also practical to omit the Chiroptera, since the density of the Chiroptera within the tropics is very poorly understood.

Small tropical mammals have been studied extensively in Africa, less intensively in Southeast Asia and only a small amount in the neotropics.

Fleming (1975) summarized the recent work on small (<1 kg), tropical mammal ecology. In his survey he demonstrated that small mammalian herbivores are generally not only numerically dominant but also comprise the bulk of the small mammal biomass. Small carnivores and insectivores comprise a smaller fraction of the total small mammal biomass. He further pointed out that both biomass and species diversity are at their highest in a tropical forest and are lower in a tropical savanna. Savanna populations of small mammals may occasionally show extreme fluctuations in their density.

Several authors have reported differences in the patterns of habitat utilization by large herbivorous mammals in the tropics. Bourlière and Verschuren (1960) analyzed habitat utilization trends for the larger ungulates in Albert National Park in what was then the Belgian Congo. They converted their density estimates to biomass estimates and thereby pioneered large scale population studies in tropical Africa. Similar work on ungulates was performed with great precision in Tanzania by Lamprey (1964). Recently, other African studies of large herbivores have been summarized by Hirst (1975).

Schaller (1967) attempted to calculate home ranges and biomass values for the large ungulates and carnivores found in Kanha National Park in Central India. Schaller (1972) also undertook biomass calculations for the Serengeti region in Tanzania. In general, Schaller's results demonstrate that herbivores make the greatest contribution to large mammal biomass in the tropics. The large carnivores make little contribution to the overall biomass, but have a modest effect on the herbivore biomass through their predatory activities.

Comparative Analysis of South Asian Mammal Faunas

Studies of mammalian biomass (by the author) and inferred carrying capacity of habitats have concentrated in two major zoogeographic areas: South Asia, typified by our studies in Sri Lanka, and the neotropics, with two major studies—one in Panama and the other in Venezuela. While multistratal tropical evergreen forests may sustain high biomasses of mammals, they can not support high biomasses of terrestrial herbivores. The biomass of terrestrial herbivores is therefore depressed in areas exhibiting more constant conditions of rainfall and plant productivity. In such areas the bulk of the productivity is locked in the crowns of the trees themselves, making it inaccessible to terrestrial forms (Eisenberg and Lockhart, 1972). Terrestrial species that are adapted to tropical evergreen forests usually exhibit a permanent land-tenure system, and have evolved extremely specialized strategies for existing on plant detritus and fruits that fall from the upper canopy (Smythe, 1970*b*).

The bulk of mammalian biomass in tropical rain forests is tied up in forms adapted for cropping the leaf productivity of the tree crowns. Even though arboreal leaf-eaters exist at modest densities, their total biomass is considerable and they play a dominant role in these ecosystems. Eisenberg et al. (1972) first pointed out that primate biomasses in the Old World tropics may be rather high, especially for those primates which are specialized for a folivorous diet (leaf-eating). Extreme folivores such as the langur, *Presbytis senex,* in Ceylon can reach a biomass of 1,450 kg/km^2 at maximum density. Folivorous primates occur in moderate to small troops with an average home range slightly smaller than would be predicted on the basis of studies of comparably sized troops of frugivorous primates. Folivorous adaptation on the part of primates allows them to reach high densities since they feed directly on the productivity of the plant itself and are able to harvest leaves over an entire annual cycle. Folivores do not have to range as far as species subsisting mainly on fruits (Clutton-Brock and Harvey, 1977).

Eisenberg and McKay (1974) analyzed mammalian biomass values for seven tropical habitat types. They included data only for the large mammal herbivore community, and were able to demonstrate that terrestrial herbivores showed their highest biomass levels in the grassland, grass scrub and savanna biomes. Terrestrial herbivores showed the lowest biomass levels in either extremely arid areas or in the humid, tropical evergreen-rainforest biome. McKay and Eisenberg (1974) analyzed previously published data concerning the composition of terrestrial herbivore communities for both African and South Asian habitats. They demonstrated that, regardless of the diversity of the terrestrial, large-herbivore community (which ranges from six to 18 species), only two to four species accounted for over 75 percent of the total terrestrial herbivore biomass. Only a few species for any given community emerged as "dominants" in terms of biomass.

Eisenberg and Seidensticker (1976) summarized the data for the larger terrestrial herbivores in eight South Asian national parks. The parks vary in rainfall and plant productivity. At one extreme, Udjung-Kulon in Java exhibits an abundant rainfall and a typical multistratal evergreen forest. The Gir Forest in Gujurat, India, with almost six months of drought in each annual cycle, provided a contrasting arid extreme.

Two types of density estimates were presented for the parks: (a) "crude" density, where the number of animals is expressed for the entire area surveyed; (b) "ecological" density, which is an estimate corrected for habitat differences. The latter takes into account that a given species will not occur uniformly throughout a nonhomogenous habitat. The density estimates were then converted to "crude" and "ecological" biomass values. Wherever possible, in addition to the authors' data (Eisenberg and Lockhart, 1972; Seidensticker, 1976*a*), data from the literature which also compensates for seasonal shifts in habitat utilization have been used. Such shifts are common in the large, terrestrial ungulates inhabiting areas with seasonal rainfall patterns (McKay and Eisenberg, 1974; McKay, 1973).

As one moves from dry thorn forest to moist deciduous forest one notes an increase in the biomass of mammalian herbivores. If one imagines moving along a rainfall gradient, the biomass increases progressively until the type of forest cover becomes so continuous there is very little grassland and shrub habitat. Once again, the forest will tend to support a low density of terrestrial, mammalian herbivores.

Total ungulate biomasses for selected national parks in Southeast Asia—based on crude densities—range from 383 to 2,858 kg/km^2 (Eisenberg and Seidensticker, 1976). These values are considerably lower than those determined for comparable African habitats. The source of this discrepancy is probably historical. The ungulate fauna of peninsular India and Sri Lanka is typified by a great number of browsers or partial browsers. Grazing herbivores, characteristic of the East African savannas, are not diversified in the ungulate communities of South Asia. Although the wild cattle of South Asia do graze appreciably, the natural state of monsoon India and Southeast Asia tends toward a forest climax. Under these conditions, grazing ungulates apparently have not evolved to the same extent as they have in the much older savannas of eastern Africa. Instead, browsers and especially the Cervidae show a reasonable diversity in South Asia. Home range data for the larger terrestrial herbivores of South Asia are presented in Table I.

The carnivores have received increasing attention in South Asia. Home ranges for tigers and leopards have been described by Seidensticker (1976*b*) and Schaller (1967). Density and home ranges for the sloth bear were developed by Laurie and Seidensticker (1977). The home range and habits of the leopard in Sri Lanka have been presented by Eisenberg and Lockhart (1972) and by Muckenhirn and Eisenberg (1973). These data are presented in Table I. As might be expected, carnivores of a greater size or group weight show a larger home range. When home range sizes for herbivores and carnivores, either in groups or individually, are plotted together, it can be seen that tropical herbivores have the smallest home range per unit weight and tropical carnivores the largest. Of course, arctic and boreal carnivores have even larger home ranges (Kleiman and Eisenberg, 1973).

Trends similar to those established for South Asia can be demonstrated in the American tropics, but the diversity of terrestrial mammalian herbivores in the neotropics is much lower than in Asia (Eisenberg and McKay, 1974). Although the savanna areas in the southern neotropics are quite old, the Pliocene land bridge and subsequent faunal imbalances led to widespread extinctions of endemic herbivores. Pleistocene climatic changes have also contributed to the current sterile situation in both the temperate and tropical areas of South America (Patterson and Pasqual, 1972).

In 1973 Eisenberg and Thorington published a preliminary analysis of a neotropical mammalian fauna. Combining the density data from studies in Panama involving the larger species of mammals on Barro Colorado Island (BCI) and extrapolating from studies of small mammal density from regions adjacent to the island, Eisenberg and Thorington attempted to construct the probable biomass relationships among the various trophic categories of mammals found on BCI. They relied on the small mammal work by Fleming (1971; 1973) and the work by Oppenheimer (1968), Chivers (1969), Montgomery and Sunquist (1975) and Smythe (1970*a)* who had studied primates, sloths and agoutis. We compared the best estimates for BCI with the data derived from a rescue operation in Surinam (Walsh and Gannon, 1967). We found remarkable correspondence between the proportions of various species rescued in Surinam and the presumptive proportion of various dominant species present in the fauna of BCI. It is now possible through the published work of Montgomery and Sunquist (1978), Montgomery and Lubin (1977; 1978), Smythe (1978) and Glanz (1978) together with further census work by Eisenberg and Thorington to present better estimates of the numerical abundance for the dominant terrestrial mammals comprising the community on BCI.

Figure 3 compares ecological and crude density estimates from a census of neotropical mammals conducted in the tropical evergreen forest on BCI. Figure 4 presents the corresponding crude density and biomass values. Relative abundance expressed as density is not necessarily an indicator of relative contribution to biomass. Because the tapir *(Tapirus)*, for example, exists at such a low density (less than one per km^2) it is not recorded in Figure 3, but it ranks fourth in contribution to crude biomass (Figure 4). The browsers or frugivore-browsers dominate the mammalian biomass contribution on BCI, and include the sloths *(Bradypus* and *Choloepus)* and the howler monkey *(Alouatta)*, all of which are arboreal. The paca, a mixed browsing frugivore, is a significant component of the fauna and, like the tapir, is terrestrial. The original relationships established in the paper of 1973 (Eisenberg and Thorington, 1973) seem to hold up: the herbivorous browsers comprise the dominant percentage of the

Table I. Some home range sizes for South Asian mammals.

Species	Group size and composition		Unit wt. (kg)	Home range (km^2)	kg/km^2	Locality and source
Carnivora						
Panthera tigris (tiger)	♂	1	150	52–62	2.6	a
P. tigris	♀	1	110	20–26	4.7	a
P. tigris	♀	1	110	9.3*	11.8	b
P. pardus (leopard)	♂ and ♀	2	45	10	9.0	c
P. pardus	♀	1	40	6.3*	6.3	b
P. pardus	♂ and ♀	2	45	5–12	10.3	a
Melursus ursinus (sloth bear)	♂	1	70	10	7.0	a
M. ursinus	♀ and cub	1	60	4	15.0	c
Canis aureus (golden jackal)	Family	4	7	7	4.0	c
Artiodactyla						
Sus scrofa (wild swine)	Group	9	25	12	18.7	c
Axis axis (spotted deer)	♂	1	45	2.5	18.0	c
A. axis	Group	6	45	4	67.5	c
Bubalus bubalis (water buffalo)	Group	8	272	8	272.0	c
B. bubalis	Group	6	272	5	326.4	c
B. bubalis	Group	28	250	40	190.0	c
Proboscidea						
Elephas maximus (Asiatic elephant)	♂	1	2,500	10	250.0	c
E. maximus	♂	1	2,500	17	147.1	c
E. maximus	Group	23	1,810	64 (dry season)	650.5	c
E. maximus	Group	23	1,810	25 (wet season)	1665.2	c

* ♀ with young cubs
a: Nepal; Smith and Tamang, 1977
b: Nepal; Seidensticker, 1976
c: Sri Lanka; Eisenberg and Lockhart, 1972

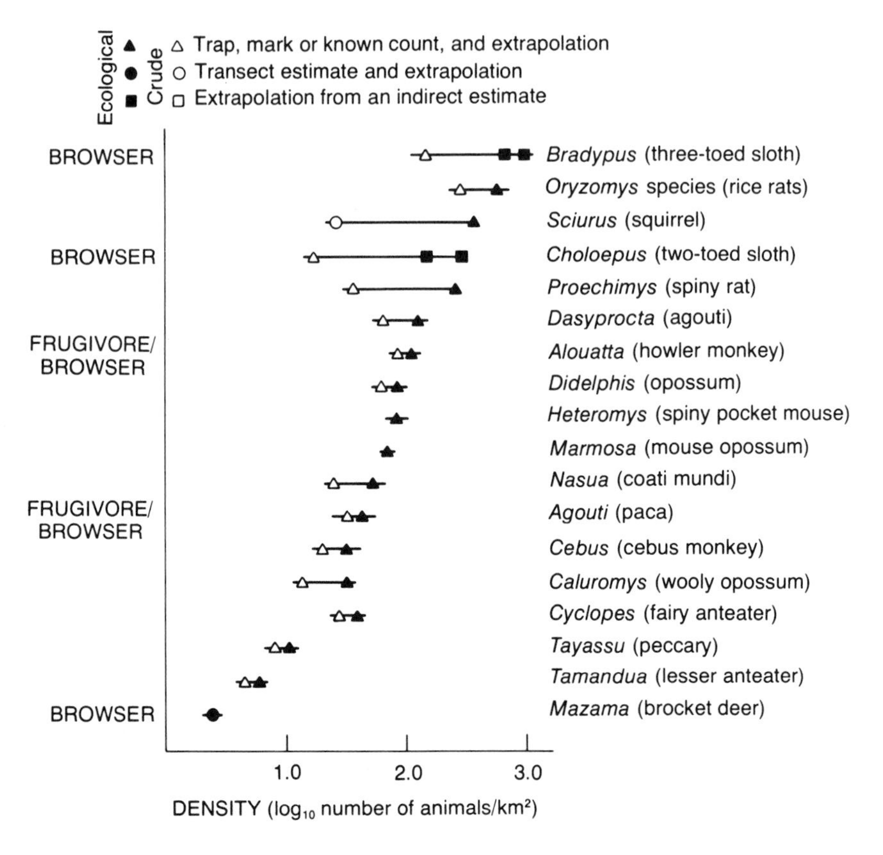

FIGURE 3. Semi-logarithmic plots for two types of density deriving from three different estimating techniques for the 18 most common nonvolant mammals on Barro Colorado Island, Panama. The species are arranged in descending order of abundance from top to bottom. For trophic levels of other species, see Figure 4. (Data modified from Eisenberg and Thorington, 1973)

mammalian biomass, followed, in descending order, by mixed browsers and frugivores, omnivores, frugivores, granivores, insectivores, myrmecophages and carnivores.

To further test the generalizations concerning ecological dominance and trophic strategy developed during investigations of Panama, intensive studies of selected mammalian species were initiated by Eisenberg and Thorington in Venezuela during 1974. They attempted to estimate the numbers of mammalian species in two major study areas: the *llanos,* which is seasonally flooded, and the submontane forest of the Venezuelan coast range. The forested habitat (Guatopo National Park) is a mixed second growth area—much of the second growth is about 30 years old. Utilizing a mixed census strategy deriving from a trap, mark and release

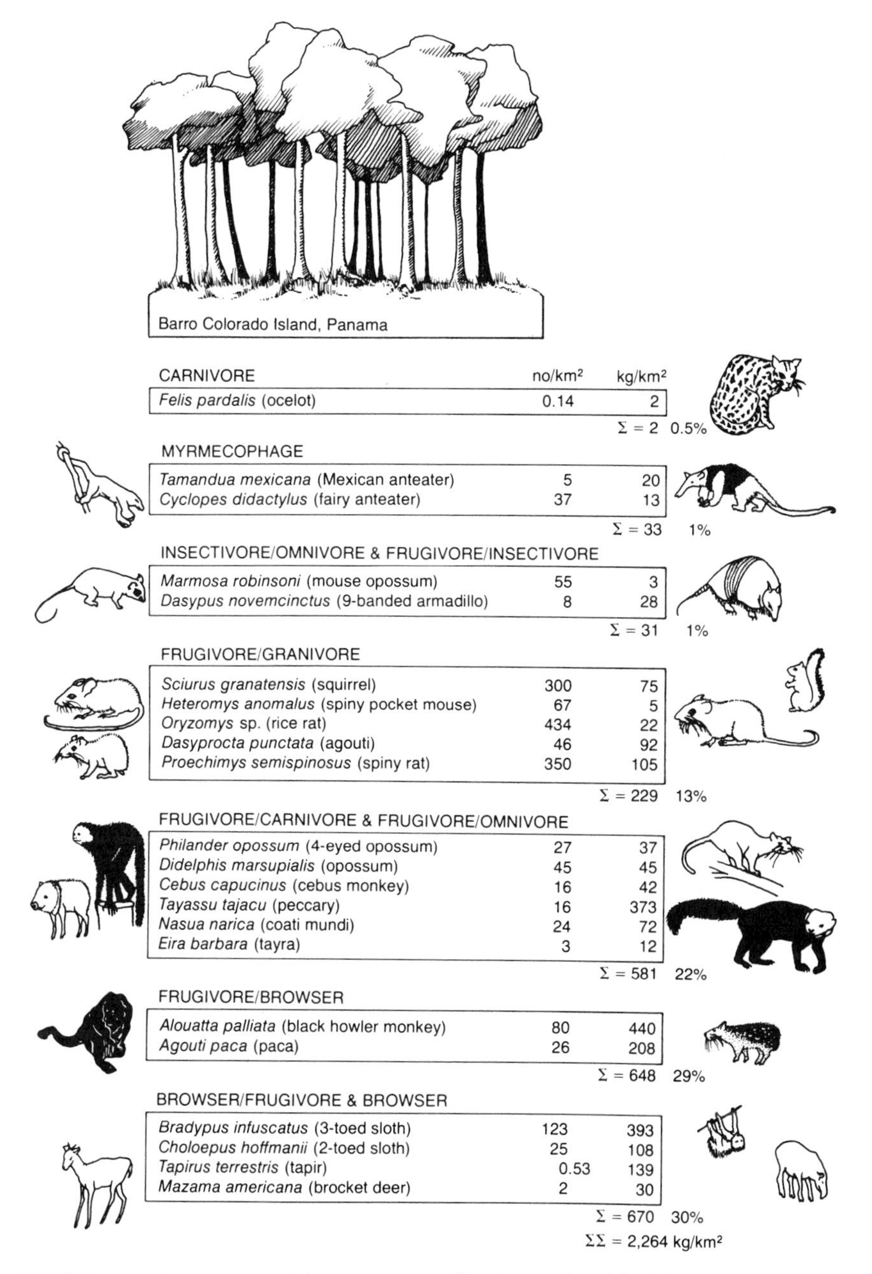

CARNIVORE	no/km²	kg/km²
Felis pardalis (ocelot)	0.14	2
		Σ = 2 0.5%
MYRMECOPHAGE		
Tamandua mexicana (Mexican anteater)	5	20
Cyclopes didactylus (fairy anteater)	37	13
		Σ = 33 1%
INSECTIVORE/OMNIVORE & FRUGIVORE/INSECTIVORE		
Marmosa robinsoni (mouse opossum)	55	3
Dasypus novemcinctus (9-banded armadillo)	8	28
		Σ = 31 1%
FRUGIVORE/GRANIVORE		
Sciurus granatensis (squirrel)	300	75
Heteromys anomalus (spiny pocket mouse)	67	5
Oryzomys sp. (rice rat)	434	22
Dasyprocta punctata (agouti)	46	92
Proechimys semispinosus (spiny rat)	350	105
		Σ = 229 13%
FRUGIVORE/CARNIVORE & FRUGIVORE/OMNIVORE		
Philander opossum (4-eyed opossum)	27	37
Didelphis marsupialis (opossum)	45	45
Cebus capucinus (cebus monkey)	16	42
Tayassu tajacu (peccary)	16	373
Nasua narica (coati mundi)	24	72
Eira barbara (tayra)	3	12
		Σ = 581 22%
FRUGIVORE/BROWSER		
Alouatta palliata (black howler monkey)	80	440
Agouti paca (paca)	26	208
		Σ = 648 29%
BROWSER/FRUGIVORE & BROWSER		
Bradypus infuscatus (3-toed sloth)	123	393
Choloepus hoffmanii (2-toed sloth)	25	108
Tapirus terrestris (tapir)	0.53	139
Mazama americana (brocket deer)	2	30
		Σ = 670 30%
		ΣΣ = 2,264 kg/km²

FIGURE 4. Percentage biomass contributions classified by trophic levels for the most abundant terrestrial mammals from Barro Colorado Island, Panama. Crude densities only are tabulated from Figure 3.

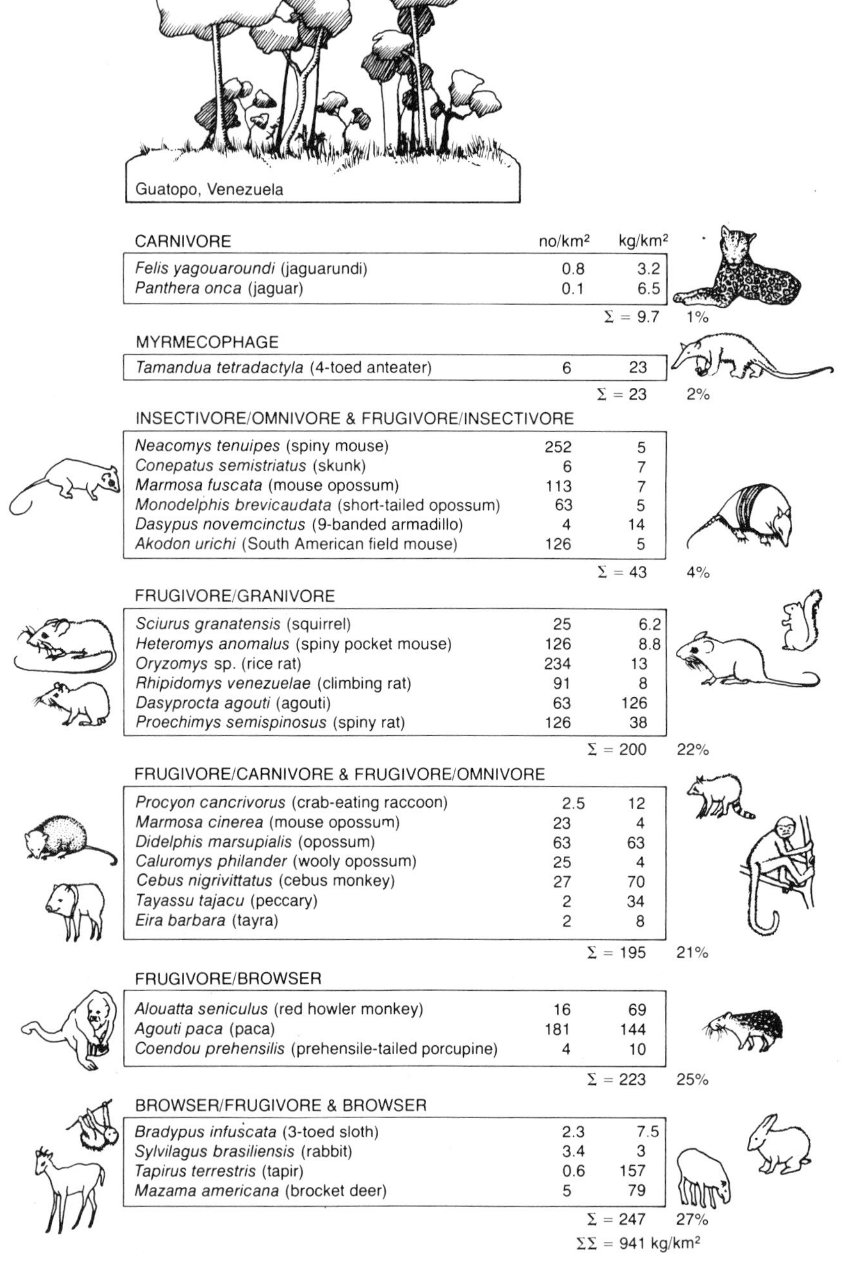

	no/km²	kg/km²	
CARNIVORE			
Felis yagouaroundi (jaguarundi)	0.8	3.2	
Panthera onca (jaguar)	0.1	6.5	
		Σ = 9.7	1%
MYRMECOPHAGE			
Tamandua tetradactyla (4-toed anteater)	6	23	
		Σ = 23	2%
INSECTIVORE/OMNIVORE & FRUGIVORE/INSECTIVORE			
Neacomys tenuipes (spiny mouse)	252	5	
Conepatus semistriatus (skunk)	6	7	
Marmosa fuscata (mouse opossum)	113	7	
Monodelphis brevicaudata (short-tailed opossum)	63	5	
Dasypus novemcinctus (9-banded armadillo)	4	14	
Akodon urichi (South American field mouse)	126	5	
		Σ = 43	4%
FRUGIVORE/GRANIVORE			
Sciurus granatensis (squirrel)	25	6.2	
Heteromys anomalus (spiny pocket mouse)	126	8.8	
Oryzomys sp. (rice rat)	234	13	
Rhipidomys venezuelae (climbing rat)	91	8	
Dasyprocta agouti (agouti)	63	126	
Proechimys semispinosus (spiny rat)	126	38	
		Σ = 200	22%
FRUGIVORE/CARNIVORE & FRUGIVORE/OMNIVORE			
Procyon cancrivorus (crab-eating raccoon)	2.5	12	
Marmosa cinerea (mouse opossum)	23	4	
Didelphis marsupialis (opossum)	63	63	
Caluromys philander (wooly opossum)	25	4	
Cebus nigrivittatus (cebus monkey)	27	70	
Tayassu tajacu (peccary)	2	34	
Eira barbara (tayra)	2	8	
		Σ = 195	21%
FRUGIVORE/BROWSER			
Alouatta seniculus (red howler monkey)	16	69	
Agouti paca (paca)	181	144	
Coendou prehensilis (prehensile-tailed porcupine)	4	10	
		Σ = 223	25%
BROWSER/FRUGIVORE & BROWSER			
Bradypus infuscata (3-toed sloth)	2.3	7.5	
Sylvilagus brasiliensis (rabbit)	3.4	3	
Tapirus terrestris (tapir)	0.6	157	
Mazama americana (brocket deer)	5	79	
		Σ = 247	27%
		ΣΣ = 941 kg/km²	

FIGURE 5. Percentage of total crude biomass for each trophic category as exemplified by the mammals of Guatopo National Park, Venezuela. Browsers and grazers contribute heavily to the total biomass value. (Data from Eisenberg, O'Connell and August, 1979)

program combined with extrapolations based on road kills and sighting frequencies, one can develop the crude biomass values shown in Figure 5. (See also Eisenberg et al., in press). Comparing the values from Guatopo with those from Panama (Figure 4) one notes a marked correspondence in the proportions of biomass occupied by each trophic level. Some interesting differences are also evident. First, the second growth forest does not support as many arboreal herbivores, in particular sloths and howler monkeys, as does the more mature forest on BCI. Furthermore, in contrast to BCI, small predators are abundant in Guatopo and are obviously having a negative impact on the biomass values of smaller diurnal rodents, such as agoutis *(Dasyprocta)* and squirrels *(Sciurus)*, and nocturnal rodents, such as the spiny rat, *Proechimys*.

In the *llanos* habitat (Hato Masaguaral), there are two separate subdivisions to be considered. Near the Guarico River is a small strip of gallery forest surrounded by a rather extensive area of semideciduous tropical forest which eventually gives way to areas of open grassland. We refer to this as the East End. The West End of the habitat is mixed grassland and palm savanna with associated *Ficus* trees as stranglers on the palms. The biomass estimates for the east end are presented in Figure 6. The raw data are derived from Eisenberg et al. (in press).

Bringing together our previous data from BCI and the current Venezuela data, we can show that arboreal mammalian biomass is high in forested areas and declines as the forest becomes discontinuous. Terrestrial mammalian biomass follows nearly the reverse pattern (Table II). If we break the mammalian species into trophic categories, carnivores show

TABLE II. Percentage of mammalian biomass divided according to substrate preferences.

Location	Arboreal	Scansorial*	Terrestrial	Remarks
Panama				
Barro Colorado[a]	70	5	25	Mature Forest (Island)
Venezuela				
Masaguaral (east)[b]	23	12	65	Riverine Llanos
Guatopo[b]	17	13	70	Second Growth Forest

* Uses both trees and ground
a: Eisenberg and Thorington, 1973
b: Eisenberg, O'Connell and August (in press)

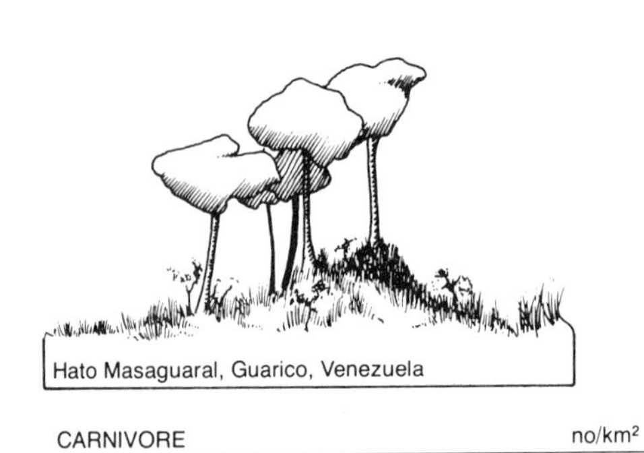

	no/km²	kg/km²	
CARNIVORE			
Felis yagouaroundi (jaguarundi)	0.25	1	
Felis pardalis (ocelot)	0.25	3	
Grison vittatus (grison)	1.2	7.6	
Puma concolor (puma)	0.09	3.6	
		$\Sigma = 15.2$	1%
MYRMECOPHAGE			
Myrmecophaga tridactyla (giant anteater)	0.12	3.2	
Tamandua tetradactyla (4-toed anteater)	2	8	
		$\Sigma = 11.2$	1%
INSECTIVORE/OMNIVORE & FRUGIVORE/INSECTIVORE			
Conepatus semistriatus (skunk)	1.2	1.4	
Marmosa robinsoni (mouse opossum)	112	5.6	
Dasypus novemcinctus (9-banded armadillo)	10	30	
		$\Sigma = 45$	4%
FRUGIVORE/GRANIVORE			
Dasyprocta agouti (agouti)	40	40	
Echimys semivillosus (spiny rat)	20	4	
Sciurus granatensis (squirrel)	27	7	
Heteromys anomalus (spiny pocket mouse)	44	3	
Oryzomys sp. (rice rat)	40	2.5	
Zygodontomys brevicauda (cane rat)	24	1.2	
		$\Sigma = 98$	9%
FRUGIVORE/CARNIVORE & FRUGIVORE/INSECTIVORE			
Procyon cancrivorus (crab-eating raccoon)	6.2	29	
Cerdocyon thous (crab-eating fox)	2.5	10	
Didelphis marsupialis (opossum)	10	10	
Cebus nigrivittatus (cebus monkey)	25	65	
Tayassu tajacu (peccary)	8.5	195	
Eira barbara (tayra)	1	2	
		$\Sigma = 311$	29%
FRUGIVORE/BROWSER			
Alouatta seniculus (red howler monkey)	20	110	
Agouti paca (paca)	12	96	
Coendou prehensilis (prehensile-tailed porcupine)	1.7	9	
		$\Sigma = 215$	20%
BROWSER/FRUGIVORE & GRAZER			
Syvilagus floridanus (rabbit)	11	9	
Hydrochoerus hydrochaeris (capybara)	10	300	
Odocoileus virginianus (white-tailed deer)	2	80	
		$\Sigma = 389$	36%
		$\Sigma\Sigma = 1{,}084$ kg/km²	

FIGURE 6. Percentage of crude biomass for each trophic category as exemplified by the mammals on the east side of Hato Masaguaral, Venezuela. Browsers and grazers contribute heavily to the total biomass value. (Data from Eisenberg, O'Connell and August, 1979)

uniformly the least percentage of total mammalian biomass; browsers and grazers constitute a large percentage (Figures 4, 5 and 6). The predictions here are in agreement with those established previously.

Mammalian biomass may reach 3,730 kg/km^2 in the more open savannas on the *llanos* of Venezuela (Hato El Frio, Apure, Venezuela) with the bulk being tied up in the semiaquatic grazer, the capybara, *Hydrochoerus* (Ojasti, 1973; Table III; Figure 5). Wild mammals are not the only significant contributors to vertebrate biomass. Birds, for which we have little data, are significant contributors and reptiles can have a relatively high contribution to vertebrate biomass in the seasonally wet *llanos*. Finally, in the *llanos* cattle ranching can reflect the potential carrying capacities of the savannas, but the data must be interpreted with caution (Table III).

TABLE III. Terrestrial, nonvolant vertebrate biomass summaries (kg/km^2) for Venezuelan habitats

	Hato Masaguaral (*llanos* habitat)				**Guatopo (forest habitat)**		
	East Side		**West Side**				
	Crude	**Ecol.**	**Crude**	**Ecol.**	**Crude**		**Ecol.**
Wild Mammals	1,084	1,747	605	965	939		1,575
Domestic Stock	7,600	10,813	7,837	8,639	—	none	—
	8,684	12,560	8,442	9,604			
Reptiles							
large	?	?	112	3,800	—	negligible	—
			Σ8,554	13,404	Σ939		1,575

Hato El Frio (*llanos* habitat)

	Crude
Wild Mammals (est.)	846
Odocoileus (deer)	320
Hydrochoerus (capybara)	2,564
Wild Mammals	3,730
Domestic Stock	18,504
Large Reptiles	171
	Σ22,405

Carrying Capacities in the Tropics

Crude biomass levels for nonvolant mammals probably decline as one comes up a latitudinal gradient. Figure 7 compares biomass values for

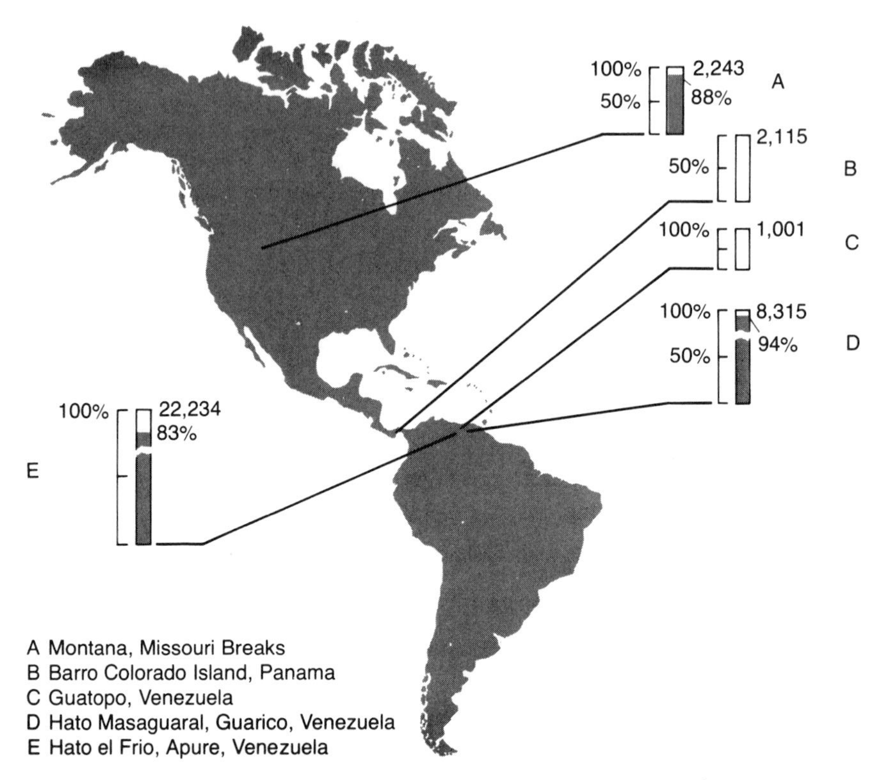

FIGURE 7. Mammalian biomass levels expressed as kg/km^2 for four neotropical sites and one temperate-zone locality in Montana. Cross hatching indicates percent contribution from domestic stock. (Data for the three neotropical habitats is taken from Table 3. The Montana data were calculated from the densities given by Mackie, 1970)

four neotropical sites with one study area in Montana. Domestic livestock values are included. Although the inclusion of domestic livestock vastly elevates mammalian biomass values in the tropics, high levels of domestic biomass may be a transient phenomenon. In both the tropics and the temperate zone grasslands, high densities of cattle are maintained only with artificial supplementary feeding in the temperate winter or during the tropical arid season. The high biomass values for livestock, then, do not reflect sustained, natural plant productivity.

There is a tendency in both the new world and old world tropics to clear land of brush and forest in hopes of stimulating grass growth to thereby extend the livestock industry. Most tropical soils are easily depleted by leaching after the forest has been cleared. Tropical soil productivity may then decline drastically with a subsequent falloff in all secondary productivity. Without the introduction of expensive synthetic fertilizers productivity can decline to an almost negligible value (Stark, 1978).

If one studies a gradient of vegetational classes from the *llanos* of central Venezuela to the multistratal evergreen forests of Panama, one sees a trend toward a reduction in the predominance of terrestrial herbivores as contributors to the total nonvolant mammalian biomass and an increase in the dominance of arboreal herbivores. The productivity of forests, however, varies greatly. Forests that are adapted to white sand soils with reduced productivity are composed of species that have a rather highly evolved set of antipredator devices which defend the plants against both vertebrate and invertebrate herbivores. The leaves store large amounts of toxins which can reduce the extent of folivore predation. This prolongs the longevity of the plant parts but may appreciably reduce the overall carrying capacity for folivorous mammals (Janzen, 1974; Montgomery and Sunquist, 1978; Chapter 5).

Any attempts to compare the tropical plant productivity of an area of relatively high fertility such as the Isthmus of Panama to the productivity of an area such as the Guiana highlands will be fraught with difficulties. A park designed to preserve faunal diversity in the Guiana highlands of the neotropics may require ten times the area of such a park in Panama (Eisenberg, in press, *c)*.

IMPLICATIONS FOR CONSERVATION

In planning reserves in the tropics, due consideration must be given to the preservation of faunal and floral diversity. The reserve should be large enough to include both elevational and microhabitat differences since tropical mammals may have strict requirements with respect to preferred habitat. Mature tropical evergreen forest does not support a diversity of terrestrial herbivores but rather a diversity of arboreal herbivores. For a defined area, mature tropical evergreen forest may, in fact, not support as high a diversity of mammals as a mosaic of mixed second growth and tropical evergreen forest (Eisenberg and Lockhart, 1972; Connell, 1978). This is explained by the fact that there are species strongly adapted to second growth conditions that will be favored in a mosaic habitat but disfavored at a full mature climax (Muul and Lim, 1978). This argument should not be used indiscriminately to encourage selective logging or to justify cutting over mature stands of tropical forest. There are species present in mature stands of tropical forest—the orangutan, *Pongo pygmaeus,* for example—which are critically adapted to that habitat and which will be eliminated if the mature forest or "climax condition" is destroyed (Rijksen, 1978). If it is impossible for governments responsible for reserve planning to maintain a sufficiently continuous tract of habitat to preserve floristic diversity—and with it animal diversity—then different types of habitat must be preserved individually. It is hoped that such tracts will either be contiguous or connected by corridors in order to maximize the area available to animal populations, thus minimizing the probability of extinction (Chapter 6).

The desired size of a reserve is, of course, a function of: (1) the particular species (or set of species) to be preserved, (2) the amount of suitable habitat in the reserve and (3) the numbers of the species in question that one wishes to maintain. The question of numbers is important because genetic variability can be quickly lost if populations are too small.

Some actual experiences in Sri Lanka provide examples. Wilpattu National Park is well managed. As a result of our studies, it was determined that the park (580 km^2) and the West Sanctuary (some 214 km^2) supported a resident population of about 70 elephants. Inclusion of the West Sanctuary in 1973 raised the area of the park to 794 km^2 and secured a resident elephant population with some 30 to 35 breeding females. This is a perilously small population from the standpoint of preserving genetic diversity, especially given the unknowns of habitat utilization trends and succession (Chapters 8 and 9).

In Wilpattu, it was estimated that a resident population of some twenty adult leopards existed within a 580 km^2 boundary (Muckenhirn and Eisenberg, 1973). A population of some ten females (without the introduction of transient specimens) is obviously incompatible with long-term survival with a high level of genetic diversity. Large carnivores are sensitive indicators of carrying capacity and define the minimum area necessary to preserve an intact ecosystem. Guatopo National Park in Venezuela includes approximately 930 km^2. Given the density estimates for terrestrial herbivores it is difficult to see how more than 20 to 25 jaguars could survive within this park in spite of the fact that it is almost twice as large as Wilpattu. The carrying capacity for large carnivores of a forested park in the neotropics is quite different from that of a forest in South Asia.

Clearly, more research in many tropical habitats must be done before sound management plans can be developed. I feel that top carnivores, such as the lion *(Panthera leo)*, the leopard *(Panthera pardus)*, the jaguar *(Panthera onca)* or the large myrmecophages, such as the aardvark *(Orycteropus)*, giant anteater *(Myrmecophaga)* and the giant armadillo *(Priodontes)*, are very sensitive indicators of the amount of disturbance inflicted on a habitat. These species all exist at low densities and will have a small effective breeding population even in large reserves. Disturbances in the food chain will further reduce densities. Low densities may result in the loss of genetic variability with a resultant loss in fecundity through inbreeding (Chapter 12). It is these large, hard to observe, numerically rare and often slow breeding species which may be studied at great profit in order to gauge the "health" and extent of an environment to be preserved.

SUGGESTED READINGS

The following selections provide further discussion on large mammal ecology:

Dasmann, R.F., 1964, *Wildlife Biology,* John Wiley and Sons, Inc., New York.

Eisenberg, J.F. and J. Seidensticker, 1976, Ungulates in southern Asia: a consideration of biomass estimates for selected habitats, *Biol. Cons.,* 10, 293–308.

Hirst, S.M., 1975, Ungulate-habitat relationships in South African woodland savanna ecosystem, *Wildl. Monogr.,* 44, 1–60. This is the classic paper describing the energetics of space utilization in mammals.

McNab, B.K., 1963, Bioenergetics and the determination of home range size, *Am. Natur.,* 97, 130–140.

The following are benchmark studies in the ecology of large mammals in the tropics:

Schaller, G., 1967, *The Deer and the Tiger,* University of Chicago Press, Chicago.

Schaller, G., 1972, *The Serengeti Lion,* University of Chicago Press, Chicago.

CHAPTER 4

PATCHY DISTRIBUTIONS OF TROPICAL BIRDS

Jared M. Diamond

Some conservation programs are aimed at saving a particular species (for example, the tiger), others at saving the many species characteristic of a particular habitat (for example, the lowland rainforest fauna and flora of Malaya). In the latter approach one sets aside a large surviving piece of the habitat in question and hopes that it contains most of the species characteristic of that habitat. Is this hope justified?

This question can be answered by comparing lists of species for different patches of the same habitat. If the patches are on different islands, we find that certain species are present on only a fraction of the islands offering apparently suitable habitat. For instance, the three large Indonesian islands of Java, Sumatra, and Borneo share similar habitats, were joined as a single land mass in the Pleistocene (see Figure 5, Chapter 6) and share many species today. Nevertheless, many large mammals and other species are patchily distributed among these three islands (Table I). For example, the tiger is on Java and Sumatra but not on Borneo; the orangutan is on Sumatra and Borneo but not on Java; and the banteng is on Java and Borneo but not on Sumatra. This distributional patchiness on islands no longer seems surprising. Most biologists accept that it must arise frequently from the random element in immigrations and extinctions. (For example, species *X* happens to have immigrated to island *A* but not to island *B*, or happens to have gone extinct on island *A* but not on island *B*.)

Within an island or continent, many habitats occur in patches (mountaintops, lakes, rivers, swamps, woodland patches in a savanna) rather than as continuous expanses. Species characteristic of such habitats would be patchily distributed even if they occupied every habitat

TABLE I. Present distributions of large mammal species on Indonesia's Greater Sundan islands.

Species	Java	Sumatra	Borneo
Tiger *(Felis tigris)*	+	+	−
Java Rhino *(Rhinoceros sondaicus)*	+	+	−
Banteng *(Bibos javanicus)*	+	−	+
Panther *(Felis pardus)*	+	−	−
Siamang Gibbon *(Symphalangus syndactylus)*	−	+	−
Malay Tapir *(Tapirus indicus)*	−	+	−
Sumatran Rhino *(Dicerorhinus sumatrensis)*	−	+	+
Orangutan *(Pongo pygmaeus)*	−	+	+
Malay Bear *(Helactos malayanus)*	−	+	+
Elephant *(Elephas maximus)*	−	+	+

+ : present; − : absent.
(From Terborgh, 1974)

patch. Furthermore, there is often an additional component of patchiness: the species occupy only certain patches of the habitat. For example, similar boreal habitats occur on the tops of 19 mountain ranges that rise out of the Great Basin deserts of western North America. The hare *Lepus townsendii,* the weasel *Mustela erminea,* the vole *Microtus longicaudus,* and the chipmunk *Eutamias umbrinus* occur on only one, four, 13, and 17 of these 19 mountain ranges, respectively, although there is no obvious reason why each species should not also occur on some of the mountains from which it is missing (J.H. Brown, 1971; 1978). Again, this type of patchiness no longer seems surprising. Disjunct pieces of habitat behave as islands, in the sense that species confined to them are subject to the vagaries of immigration and extinction (see Chapter 6 for a more detailed discussion of island theory).

Are species lists from different areas within a large, continuous expanse of habitat more or less equivalent? If so, the conservationist's task of obtaining a suitable site for a nature reserve would be simplified. One large reserve might suffice to save most of the characteristic biota of a habitat. The site of the reserve could be chosen to avoid areas earmarked for development, without sacrificing conservation value. If species are patchily distributed within a habitat, multiple reserves are needed to preserve the habitat's biota, even for a short time. Furthermore, the location of these reserves becomes more critical. They must be carefully chosen so as to contain populations of most of the habitat's species.

It turns out that distributional patchiness of this last sort is common in the tropics, much more so than in the temperate zones. The magnitude of this tropical patchiness (and of the problems that it poses for conservationists) is insufficiently appreciated. Many temperate-zone biologists are unaware of tropical patchiness; are reluctant to believe in it because they see no explanation for it; or are quick to dismiss reported cases of it as artifacts of inadequate biological surveys or subtle, unfulfilled ecological requirements. To convince skeptics of the reality of the problem, this chapter will attempt to: (1) show that patchy distributions are more frequent in species-rich areas than in species-poor areas; (2) establish that patchy distributions within continuous tropical habitat expanses are a real and common phenomenon; and (3) suggest several plausible explanations for how patchiness arises.

EVIDENCE FOR PATCHINESS FROM SHORT-TERM, EQUAL-EFFORT SURVEYS

Anecdotal evidence suggests that proportionately more species have strikingly patchy distributions in species-rich tropical areas (such as Amazonia, southeast Asia and New Guinea) than in the temperate zones or species-poor tropical areas (such as tropical islands smaller than New Guinea). To support this impression more objectively, I will show that local surveys in the tropics record a decreasing fraction of the island's lowland avifauna on increasingly species-rich islands (Figure 1). Thus, the area needed to conserve a given proportion of a regional fauna increases with the species richness of the fauna.

I surveyed the land and freshwater avifauna under standardized con-

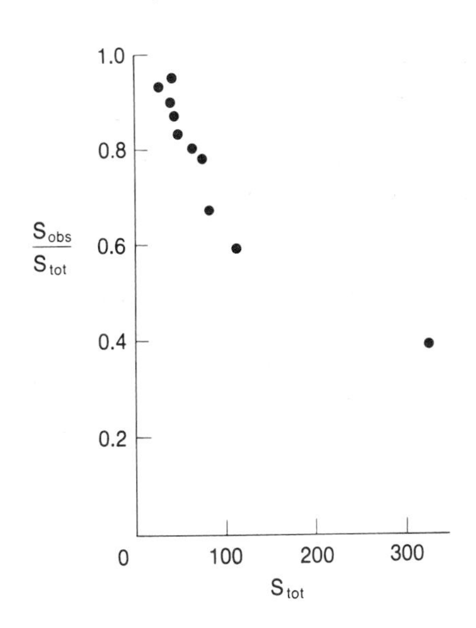

FIGURE 1. Abscissa: S_{tot}, the total number of resident lowland land and freshwater bird species on various Pacific islands. Ordinate: S_{obs}/S_{tot}, where S_{obs} is the number of those species observed in a short-term, "equal-effort" survey at one lowland locality.

ditions at lowland localities on 10 Pacific islands. The survey party consisted of myself and about 10 New Guinea or Solomon Island men. Five of the men constituted a team that has worked with me for years and traveled with me from island to island. Five other men were recruited locally. During a lifetime of observation and hunting these men have acquired a knowledge of the local fauna and flora (including birds) that compares favorably with that of the keenest western observers. We spent about a week at each village or camp, operating about 50 mist-nets daily. We spent most of each day observing birds in a radius of up to eight kilometers from the camp. The surveys were confined to elevations at or near sea-level. Our goal was to record, as completely as possible, the bird species present during the time of our visit. By the end of a week, we had virtually ceased to encounter new species. The number of species on our list (S_{obs}) was compared with the comprehensive list of resident lowland bird species on the whole island (S_{tot}). The comprehensive list included species observed by us or by other collectors at different localities on the island and by local residents at the same locality at other times. The list did not include winter visitors from higher latitudes.

In Figure 1 the ratio S_{obs}/S_{tot} is plotted against S_{tot}. The ratio is nearly 1.0 for the most species-poor islands, and decreases with increasing S_{tot} to a value of 0.39 for the richest island, New Guinea. All of these islands are within 800 kilometers of New Guinea. S_{obs} and S_{tot} refer only to species occurring at sea level, so that most of the variation in S_{tot} is correlated with variation in island area rather than with island isolation or elevation. Nine of the 10 islands surveyed exceed 40 square kilometers in area, and all have tropical rainforest as their major lowland habitat.

There are several reasons why S_{obs}/S_{tot} decreases with S_{tot}:

1. The richer the fauna, the lower will be the population density of each species on the average, and the greater will be the number of species likely to be present locally during the survey period but overlooked because of rareness. Observations by local residents over a lifetime, or by me or other visitors for longer periods or on different visits, show that this factor is only part of the explanation for the pattern in Figure 1. Most of the missing species are either never observed at the locality, or else are said to occur irregularly or at predictably different seasons.

2. Temporal patchiness of species occurrences (that is, presence of a species at a locality at certain times but not at other times) may also increase with S_{tot}. If this were true, then with increasing S_{tot} a decreasing fraction of the long-term local species list would actually be recorded in any given short survey period. While I am certain that temporal patchiness is significant for Pacific island birds, I do not know whether it increases with S_{tot}.

3. The richer the fauna, the narrower is the range of habitats occupied by the average species, and the greater will be the number of species missed because they are confined to habitats not represented within the

survey radius. Species-poor islands have few species specialized to habitats such as savanna, brooks, rivers and swamps. Species-rich islands not only have more species restricted to these nonforest habitat types, but also have more pairs or sets of species that predictably divide up forest types (such as forest on flat and hilly terrain, forest of high- and low-rainfall areas, forest with dense and less dense understory). This, in effect, is part of the explanation for Figure 1: since species of richer faunas divide habitats more finely, recognized habitats themselves (as defined by bird distributions) become more patchily distributed.

4. Geographical patchiness increases with S_{tot}. By geographical patchiness, I mean that a species which has been absent at one locality throughout living and written human history is resident in a similar habitat at another locality on the same island. The distances between occupied and empty patches on New Guinea are only a few tens of kilometers in some cases, and may actually be less, since patch boundaries have not been specifically studied. This is the type of patchiness illustrated in Figures 2 to 5.

Figure 1, then, shows that the richest faunas require the largest reserves for their preservation. Data that would yield a plot similar to Figure 1 are not available for temperate-zone islands. It is likely, however, that temperate-zone islands would fit the pattern of Figure 1, since they have lower species totals than equal-sized tropical islands and (from anecdotal evidence) provide many fewer examples of patchy distributions.

GEOGRAPHICAL PATCHINESS IN NEW GUINEA BIRDS

Many famous and baffling cases of patchy distributions have been reported in the tropics. Why should the Tongan fruit bat *(Pteropus tonganus)* be widespread on tropical Pacific islands over a 3,200 kilometer stretch from Samoa west to New Caledonia, then reappear 1,050 kilometers to the northwest on a single island (Rennell) in the Solomons and again 1,700 kilometers further to the northwest on a single island (Karkar) off of New Guinea (skipping over hundreds of intervening Solomon and Bismarck islands)? Why are so many neotropical bird species confined to single centers of distribution, or else present at scattered sites hundreds of miles apart (but not at ecologically similar intervening sites)? To consider the range of forms that such patchy distributions may assume, this section will focus on New Guinea's avifauna.

Ornithological exploration of New Guinea began in 1818. Since about 1860 most areas of the island have been surveyed repeatedly. These accumulated records show that, on a geographical scale, the distributions of many species are patchy within a continuous expanse of suitable habitat little disturbed by man, and that many of the patches have retained their gross positions for over a century (Diamond, 1972*a;* 1975*a;* 1979). Instances of geographical patchiness on the scale to be described are much rarer in temperate-zone birds, except as obvious products of recent human interference.

Distributional Gaps in the Mountains

The Central Dividing Range runs east to west along the backbone of New Guinea for 1,600 kilometers. It has elevations up to 5,000 meters and no passes under 1,500 meters. Most of New Guinea's approximately 180 montane bird species extend to both the western and eastern ends of the range. Nevertheless, each portion of the range lacks some otherwise widespread old species (Table II). For example, the sole New Guinea species of tree creeper (Climacteridae), *Climacteris leucophaea,* is common in forest between elevations of 1,500 and 3,000 meters in western and eastern New Guinea. This species, however, has a distributional gap of about 400 kilometers between longitudes 143° and 146° E in the middle of the Central Dividing Range (Figure 2), although mountains with similar forest extend uninterrupted for 1,600 kilometers. There is no other New Guinea bird species that is in the same family as, or is ecologically close to, *C. leucophaea.* Eight other montane species have distributions similar to

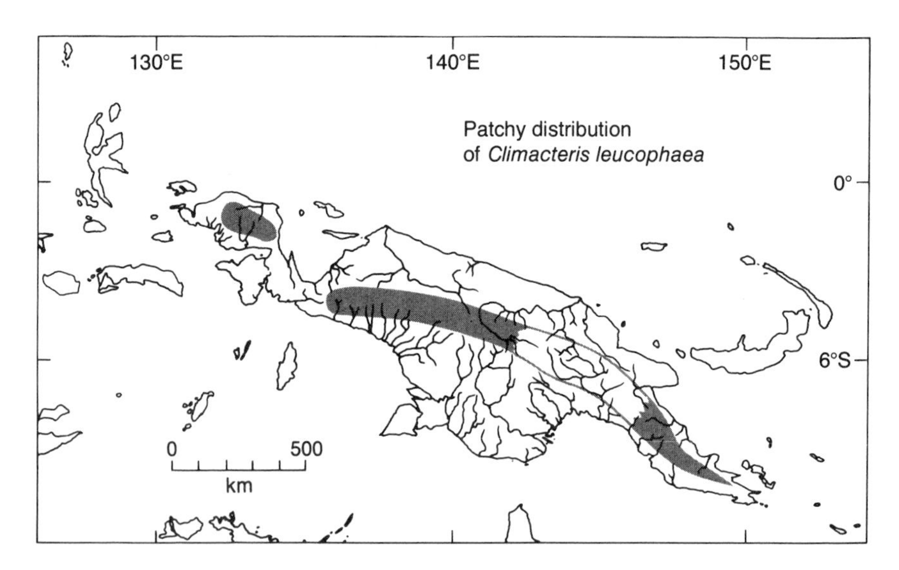

FIGURE 2. Distribution of the tree creeper *Climacteris leucophaea* in the mountains of New Guinea. The Central Dividing Range is outlined. This species is present in the gray areas and absent from the blank area. (From Diamond, 1972*a*)

TABLE II. Patchy distributions of bird species along New Guinea's Central Dividing Range.

	132–134°E	136–139°E	140–142°E	143–146°E	147–150°E
Anurophasis monorthonyx (Snow Mountain Quail)	–	+	–	–	–
Androphobus viridis (Green-backed Babbler)	–	+	–	–	–
Petroica archboldi (Snow Mountain Robin)	–	+	–	–	–
Oreornis chrysogenys (Orange-cheeked Honey-eater)	–	+	–	–	–
Lonchura teerinki (Balim Valley Mannikin)	–	+	–	–	–
Melampitta gigantea (Greater Melampitta)	+	+	–	–	–
Peneothello cryptoleucus (Grey Thicket-Flycatcher)	+	+	–	–	–
Pachycephalopsis hattamensis (Green Thicket-Flycatcher)	+	+	–	–	–
Ptiloprora erythropleura (Red-sided Streaked Honey-eater)	+	+	–	–	–
Aegotheles archboldi (Archbold's Owlet-Nightjar)	–	+	+	–	–
Pachycephala lorentzi (Lorentz's Whistler)	–	+	+	–	–
Rallicula rubra (Chestnut Rail)	+	+	+	–	–
Zosterops fuscicapilla (Yellow-bellied Mountain White-eye)	+	+	+	–	–
Porzana tabuensis (Sooty Rail)	+	+	+	+	–
Epimachus fastosus (Black Sickle-billed Bird of Paradise)	+	+	+	+	–
Paradigalla carunculata, P. brevicauda (Paradigalla)	+	+	+	+	–
Pachycephala tenebrosa (Sooty Whistler)	–	+	+	+	–
Pteridophora alberti (King of Saxony Bird of Paradise)	–	+	+	+	–
Melidectes nouhuysi, M. princeps (Bearded Honey-eater)	–	+	+	+	–
Charmosyna josefinae (Josephine's Lory)	+	+	–	+	–

(Continued)

TABLE II. (*Continued*)

	132–134°E	136–139°E	140–142°E	143–146°E	147–150°E
Archboldia papuensis (Archbold's Bowerbird)	−	+	−	+	−
Orthonyx temminckii (Logrunner)	+	+	−	−	+
Melidectes ochromelas (Cinnamon-browed Honey-eater)	+	+	−	−	+
Amalocichla sclateriana (Greater New Guinea Thrush)	−	+	−		+
Lonchura montana, L. monticola (Alpine Mannikin)	−	+	−	−	+
Climacteris leucophaea (New Guinea Tree Creeper)	+	+	+	−	+
Macgregoria pulchra (Macgregor's Bird of Paradise)	−	+	+	−	+
Ptiloprora plumbea (Leaden Honey-eater)	−	+	+	−	+
Coracina lineata (Yellow-eyed Cuckoo-Shrike)	+	−	−	−	+
Melanocharis arfakiana (Obscure Berrypecker)	+	−	−	−	+
Myzomela adolphinae (Mountain Red-headed Myzomela)	+	−	−	+	+
Zosterops novaeguineae (New Guinea Mountain White-eye)	+	−	−	+	+
Erythrura papuana (Large-billed Parrot-Finch)	+	−	−	+	+
Pachycephala modesta (Brown-backed Whistler)	−	−	+	+	+
Accipiter meyerianus (Meyer's Goshawk)	−	−	−	+	+
Accipiter buergersi (Buerger's Goshawk)	−	−	−	+	+
Paradisaea rudolphi (Blue Bird of Paradise)	−	−	−	+	+
Cnemophilus macgregorii (Crested Bird of Paradise)	−	−	−	+	+
Ptiloprora guisei (Red-backed Streaked Honey-eater)	−	−	−	+	+
Amblyornis subalaris (Eastern Gardener Bowerbird)	−	−	−	−	+

+ : present; − : absent, at the indicated longitude. The range runs east to west from 150°E to 136°E, with the mountains of the Vogelkop lying detached at 132° to 134°E.

that of *C. leucophaea,* in being absent from the middle of the Central Range. The eastern part of the Central Range lacks eight montane species present in the west and middle (for example, the King-of-Saxony bird of paradise, *Pteridophora alberti).* The western part lacks five species present to the east on the Central Range and also present on the isolated mountain range of the Vogelkop further west (for example, the cuckoo-shrike *Coracina lineata* and the honey-eater *Myzomela adolphinae),* plus six other species present in the east and middle of the Central Range (for example, the goshawk *Accipiter buergersi,* and the whistler *Pachycephala modesta).* Thus, about 22 percent of New Guinea's montane bird species have major distributional gaps of several hundred kilometers along the Central Range. (These totals consider only full species and exclude representative members of superspecies—very closely related species that replace each other in different portions of the range.)

Distributional Gaps in the Hills

Hill forest forms an uninterrupted ring around the foot of the Central Dividing Range in New Guinea. Numerous bird species confined to elevations of 300 to 1,200 meters in these hill forests are distributed over most of the hill ring, but a few exhibit major gaps. The trumpet bird of paradise, *Manucodia keraudrenii,* has a gap of several hundred kilometers in the northwest of the ring and another gap of 160 kilometers in the northeast. The sickle-billed bird of paradise, *Epimachus albertisii,* has a gap of 850 kilometers in the north and a gap of about 1,050 kilometers in the south.

Distributional Gaps in the Lowlands

The lowlands of New Guinea form a broad ring about the island's periphery which surrounds the mountain backbone. Most of New Guinea's approximately 325 lowland bird species are distributed, at least grossly, over most of the lowland ring. However, there are some major distributional gaps. For example, the starling, *Mino anais,* is widespread in the lowlands except for a stretch of 725 kilometers in northeast New Guinea (Figure 3). The otherwise widespread paradise kingfisher *Tanysiptera [galatea],* crowned pigeon *Goura [victoria],* whistler *Pitohui kirhocephalus,* fairy-wren *Todopsis cyanocephalus* and the fruit dove *Ptilinopus aurantiifrons* have similar, though not coincident, distributional gaps in this same part of the lowland ring. Other species have gaps elsewhere in the ring.

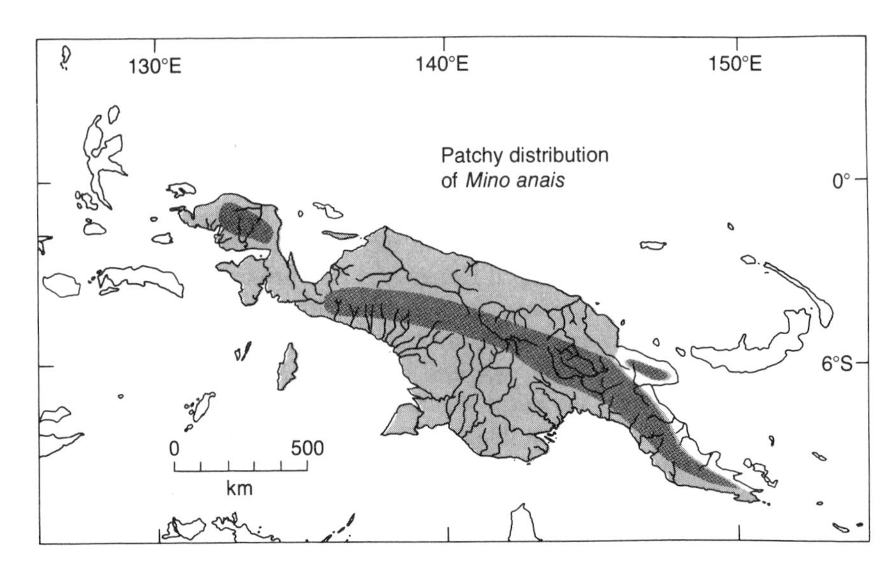

FIGURE 3. Distribution of the starling *Mino anais* in the lowlands of New Guinea. Unsuitable mountainous areas are light gray. The species is present in the dark gray area and absent from the blank area. (From Diamond, 1975*a*)

Scattered Populations

The previous cases involved species with one or two major distributional gaps in an otherwise continuous range. For other species the gaps are much larger and more numerous than the occupied areas, so that the species' range consists of several large but scattered blocks. For example, the range of the logrunner, *Cinclosoma ajax,* the sole New Guinea species of its genus, consists of four large blocks (Figure 4). The flycatcher *Poecilodryas placens* occurs in five large blocks (south slopes of southeast New Guinea, upper foothills of Fly and Purari drainages, Batanta Island, Geelvink Bay and Astrolobe Bay), where it lives in the shaded, open understory of tall forest at 300 to 1,220 meters. The bowerbird *Archboldia papuensis* is confined to two blocks on Mts. Hagen and Giluwe in the middle of the Central Range and other blocks in the west of that range.

Very Scattered Populations

The berrypecker *Melanocharis arfakiana,* which exhibits the ultimate in avian patchiness, is known from two localities at opposite ends of New Guinea, 1,600 kilometers apart (Figure 5). The honey-eater *Philemon brassi* belongs to a superspecies of which one member is on the Indonesian island of Timor, the other being widespread in Australia and adjacent parts of south New Guinea. *P. brassi,* the remaining member of

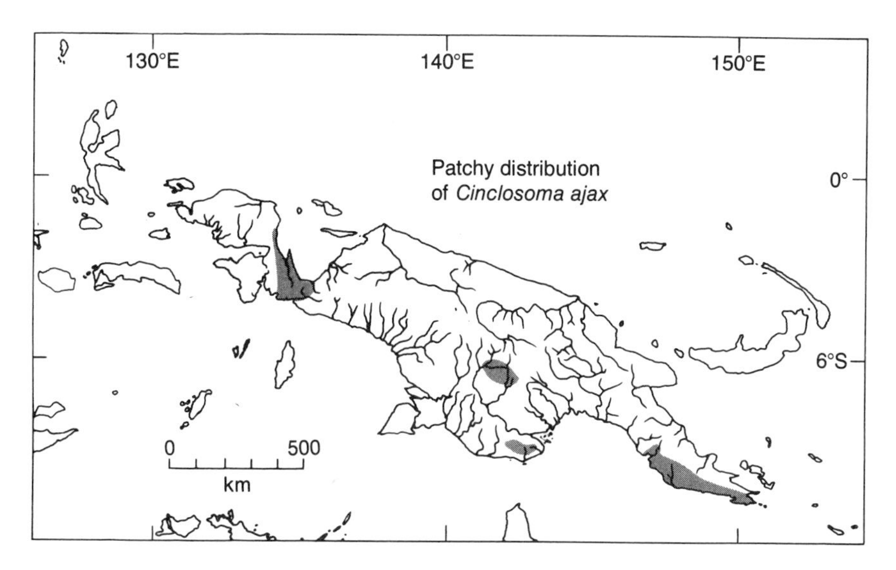

FIGURE 4. Distribution of the logrunner *Cinclosoma ajax* in the lowlands of New Guinea. The species is confined to the four blocks shown in gray. (From Diamond, 1975*a*)

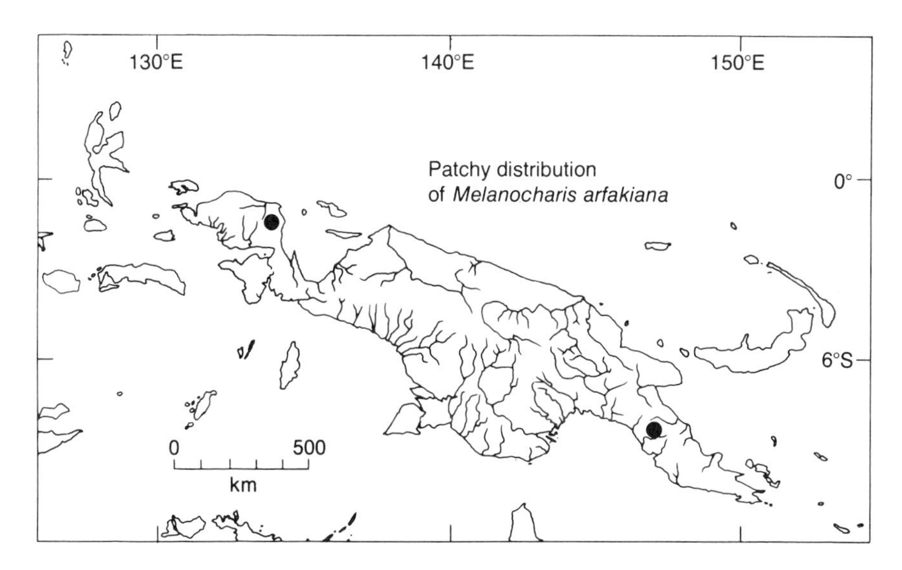

FIGURE 5. Distribution of the berrypicker *Melanocharis arfakiana* in the mountains of New Guinea. The species is found in two localities, shown in black, 1,000 miles apart. (From Diamond, 1975*a*)

the superspecies, was found to be common on one lagoon of the Idenburg River in north New Guinea, but has never been found elsewhere.

Are Patchy Species Also Rare?

The tendency toward geographical patchiness varies among species. This tendency seems to be correlated inversely with a species' population density, but the correlation is not tight. For example, among the species in hill forest areas, the cuckoo *Chrysococcyx meyerii,* the thrush *Drymodes superciliaris,* and the honey-eater *Meliphaga polygramma* are always uncommon, but they nevertheless turn up in low numbers almost everywhere. Conversely, the lowland whistler *Pitohui kirhocephalus* and the alpine finch *Lonchura montana* are generally among the most abundant species of their respective habitats—where they occur—but have large distributional gaps. On the whole, though, I think geographically patchy species often (but certainly not always) prove to be uncommon species where they occur.

THEORIES OF GEOGRAPHICAL PATCHINESS

How did the geographically patchy distributions illustrated in Figures 2 through 5 (and contributing to the relationship in Figure 1) arise? At first sight these patchy distributions are baffling. This lack of obvious explanations increases the readiness of temperate-zone biologists to dismiss reported cases of patchiness as artifacts of inadequate collecting. I suggest four explanations for these patchy distributions, explanations that will be posed as alternatives, though they may often act in combination.

Subtle Patchiness of Habitat

I believe that Figures 2 through 5 and numerous similar patterns represent species that are patchily distributed with respect to available habitat. This belief arises from the fact that habitats and conditions in the gaps seem to me to be well within the range of habitats and conditions in which the species live outside the gap.

Nevertheless, one can always object: "How can you be sure that you haven't overlooked a difference in some subtle ecological factor that is crucial to the bird but invisible to you?" One can never be sure. Many tropical species are indeed confined to a narrow range of habitats, though this may be the result of competition rather than of physiological limitations. Sometimes one does recognize a habitat variable that differentiates the occupied and empty patches. For example, the local distributions of two similar-sized fruit doves, *Ptilinopus pulchellus* and *P. coronulatus,* complement each other in the New Guinea forest to a striking and initially perplexing degree. Out of 31 localities in New Guinea surveyed by the Archbold Expeditions and the Denison-Crockett Expedition in the 1930's, 18 supported only *P. coronulatus,* 9 only *P. pulchellus,* and 4 supported both species. Eventually I noticed that (where rainfall records were available), the *pulchellus* localities generally had somewhat higher rainfall than the *coronulatus* localities.

However, other patchily distributed species occupy such a great variety of habitats and climatic zones in their blocks of occurrence that a postulate of undetected habitat requirements becomes implausible. For instance, on satellite islands where the whistler *Pachycephala melanura* is the sole species of its genus, it is abundant in all elevational zones from sea level to the summits. It occurs in a range of habitats from gardens to coconut groves to savanna to rainforest to montane cloud forest to subalpine shrubbery, and in all vertical layers from the understory to the canopy. On New Guinea it is confined to forest between 1,370 and 2,350 meters in one valley. This confinement surely has more to do with the presence of 14 congeneric whistlers on New Guinea than with adaptive limitations of this ecologically catholic bird. Similarly, the wide range of forested or alpine habitats occupied by the tree creeper *Climacteris leucophaea* or the alpine finch *Lonchura montana,* respectively, in west and east New Guinea is matched in vegetational physiognomy as well as in principal plant species by habitats lacking these bird species in central New Guinea. In general, undetected habitat correlates are less likely to emerge for bird species that are patchy at the level of large geographical blocks, like the species in Figures 2 through 5, than for species that are patchy on a much more local level, such as the *Ptilinopus* doves mentioned above.

Historical Effects

An environment that is homogeneous today may have been patchy in the past. If bird distributions respond more slowly than forest structure to environmental change, present bird distributions may continue to reflect an earlier state of patchiness. A clear example is provided by 21 New Guinea montane forest bird species represented by lowland populations in a patch near the mouth of the Fly River. If we did not realize that these Fly River forests had been joined to the cooler, subtropical forests of Australia during the Pleistocene periods of low sea level, and that these isolated populations are, therefore, relicts, their occurrence would be a mystery. The numerous neotropical species still confined to the Pleistocene site of the Napo forest refuge in South America provide a similar example (Haffer, 1969). However, practically any area in New

Guinea proves to be the site of a patch for some species, so it seems unlikely that all distributional patchiness can be explained in terms of relict populations on sites of Pleistocene refuges. We expect present-day distributional patchiness of birds resulting from Pleistocene habitat patchiness to be more marked in the tropics than in the temperate zones, as the greater vagility of temperate-zone bird species would cause such historical effects to disappear more quickly.

Immigration-Extinction Equilibria

It is a familiar finding of island biogeography that a species may, at any instant, be present on some but not all ecologically suitable islands. These patchy insular distributions can be readily interpreted as immigration-extinction equilibria. Island populations occasionally go extinct but can be reestablished by immigrations. The fraction of islands occupied at any instant increases with island area because of decreasing extinction rates (Diamond and Marshall, 1977). Similarly, even in a continuous mainland habitat a local population may occasionally disappear. The number and extent of vacant patches at any instant should increase with extinction rates and decrease with immigration rates.

Most temperate-zone bird species disperse so readily that vacated territories are likely to be recolonized by the next breeding season. Tropical bird species, however, are notorious for their low dispersal ability, so recolonization of a vacated territory on a tropical island may be slow. In addition, extinction rates increase with local species number because of competition. Hence, we expect patchiness resulting from immigration-extinction equilibria to be most marked in the species-rich tropics.

By this interpretation, patches and gaps should not be fixed but should shift with time. While many of the patches and gaps described in this chapter for New Guinea birds are known to have retained their gross position over the past century, we have no idea whether small shifts have occurred. Roughgarden (1978) has treated uneven spatial patterns in abundance by a mathematical model that could be extended, through consideration of local extinctions, to account for patchiness.

Lockouts

A species may be permanently excluded from an area with suitable habitat by established populations of competitors. Checkerboard distribution patterns on islands are familiar manifestations of competitive exclusion. For example, two ecologically similar flycatcher species, *Pachycephala pectoralis* and *P. melanura,* occur with mutally exclusive distributions on islands of the Bismarck Archipelago. Out of 50 ornithologically surveyed Bismarck islands, 11 support *P. pectoralis,* 18 P. *melanura* and 21 support neither. Each species is able to exclude the other competitively from islands on which it is established. The 21 islands lacking both *Pachycephala* species prove to support one or more other species of flycatchers. Apparent lockouts on islands off lower California have been described for lizards of the genera *Uta, Sceloporus, Urosaurus,* and *Sator* by Soulé (1966), and between the mice *Peromyscus eremicus* and *P. maniculatus* by Wilcox (personal communication). Analyses of species distributions, and records of vagrant individuals on islands lacking breeding populations of the species, show that a given species may be "locked out" of an island by certain combinations of competing species as well as by a single close competitor (Diamond, 1975*a*).

Figure 6 illustrates a similar competitive lockout pattern that occurred on the New Guinea mainland as various finch species of genus *Lonchura* colonized the mid-montane grasslands. The patchy distribution of each species suggests that the identity of the locally successful colonists was determined by chance on a first-come, first-served basis. Each established colonist was then able to exclude subsequent invaders indefinitely. Figure 7 depicts a more complicated example involving three closely related species of montane honey-eater species of genus *Melidectes*. When considered individually, each species has a peculiarly dis-

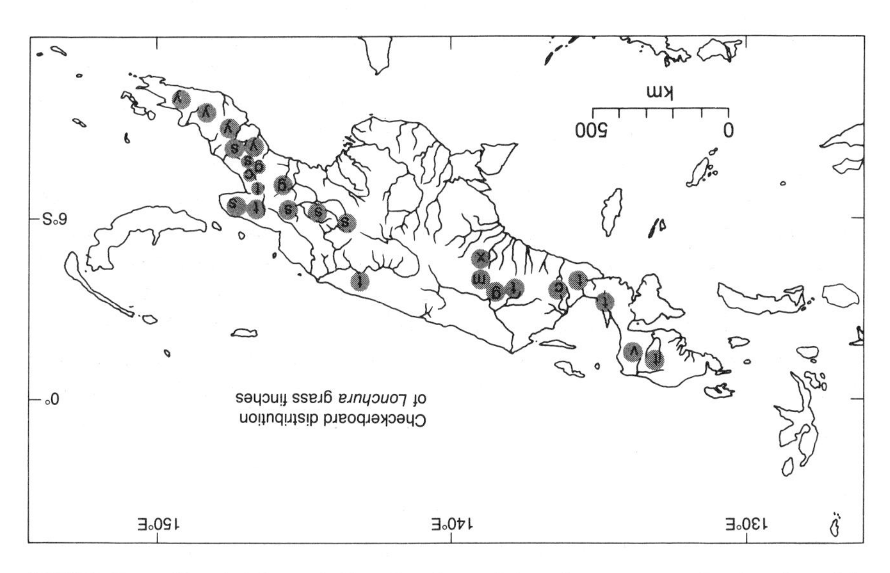

FIGURE 6. Checkerboard distribution of *Lonchura* grass finches in mid-montane grasslands of New Guinea. c: *Lonchura castaneothorax;* g: *L. grandis;* m: *L. montana;* s: *L. spectabilis;* t: *L. tristissima;* v: *L. vana;* x: *L. teerinki;* y: *L. caniceps.* (From Diamond, 1975*a*)

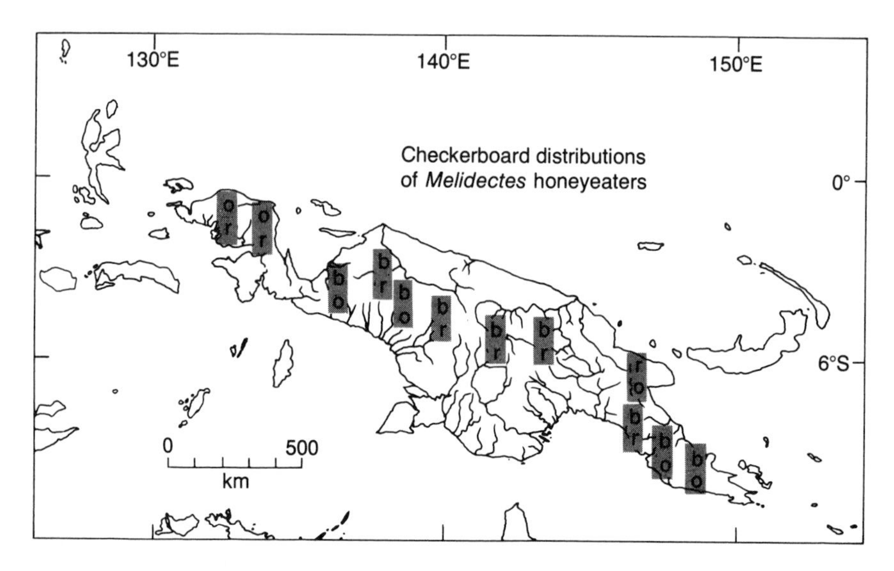

FIGURE 7. Distributions of three *Melidectes* honey-eaters in the mountains of New Guinea. (o: *M. ochromelas;* b: *M. belfordi;* r: *M. rufocrissalis* superspecies.) Most montane areas of New Guinea support two species with mutally exclusive altitudinal ranges. At each locality depicted on the map, the letters above and below indicate the species present at higher and lower altitudes, respectively. (From Diamond, 1975*a*)

junct range and is absent from several portions of the New Guinea cordillera. When the ranges of the three species are considered together, however, it is clear that (1) each mountainous area supports two species that exclude the third; (2) the identity of the locally successful combination varies in irregular checkerboard fashion; and (3) each of the three possible combinations occurs in several areas.

I suspect that similar lockouts involving combinations of several species underlie many cases of patchiness found in New Guinea—for example, in the genera *Charmosyna, Tanysiptera, Coracina, Poecilodryas, Pachycephala, Meliphaga, Ptiloprora, Pycnopygius* and *Melanocharis.* Patches arising from lockouts may occupy the same location for long times. Gilpin and Case (1976) have argued on theoretical grounds that competitive lockouts in multispecies guilds are increasingly likely with increasing species richness in a community.

SUMMARY

In the tropics—especially in species-rich tropical areas—many bird species are patchily distributed with respect to available habitat. Likely explanations are based on former (e.g. late Pleistocene) patchiness of the habitat itself, immigration-extinction equilibria, and competitive lockouts. Each of these factors is more likely to produce patchiness in the tropics than in the temperate zones, and the latter two are more significant in species-rich communities.

A practical conclusion for conservationists concerned with the future of a tropical habitat type that has a unique fauna or flora is that it may not suffice to set aside any large piece of the habitat. Instead, it may be necessary to go to the labor of obtaining a species inventory of a prospective site, and ascertaining whether the site does contain populations of most of the habitat's endemic species.

Patchiness poses, in addition, the following difficult dilemma to conservationists. Suppose that political considerations limit the total area of a habitat that can be set aside for refuges. To save species, what is the optimum allocation of this area that should go to each individual refuge? Opposite extremes of strategy would be to allocate the whole area to a single large refuge, or to many small refuges. A compelling reason why individual refuges should be as large as possible is that species have minimum area requirements, and that the species most in need of refuges tend to require large areas for survival (Terborgh, 1974*a* and *b*; 1976; Diamond, 1975*b;* 1976; Wilson and Willis, 1975; Whitcomb et al., 1976; Pickett and Thompson, 1978; Soulé et al., 1979). However, distributional patchiness may argue against allocating all the available area to a single large refuge. It is unlikely that the patches of all species in need of protection will coincide. Several refuges at different locations may be required to have patches of all critical species represented. The need for multiple refuges will arise especially in cases where distributional patchiness is due to competitive lockouts, since multiple refuges would then be required to support the different competing species or combinations of species (compare Figures 6 and 7). Given some upper limit to the total achievable area of a reserve system, a compromise must somehow be found between the need for large individual refuges and the need for multiple refuges.

In summary, habitat maps showing broad expanses of green or red colors can mislead one to suppose that species are as widely distributed as the map colors. Actually, many species are patchily distributed, for several reasons. Patchiness is exaggerated in the tropics as compared to the temperate zones, and is disproportionately more marked in areas with more diverse faunas. Appreciation of patchiness is essential in devising reserve systems.

SUGGESTED READINGS

Diamond, J.M., 1972, *Avifauna of the Eastern Highlands of New Guinea,* Nuttall Ornithological Club, Cambridge, Massachusetts. Discusses patchiness in continuous habitats.

Brown, J.H., 1971, Mammals on mountaintops: nonequilibrium insular biogeography, *Amer. Natur.* 105, 467–478. Discusses patchiness in disjunct habitats.

Haffer, J., 1969, Speciation in Amazonian forest birds, *Science* 165, 131–137. An example of patchy relict distribution.

Diamond, J.M., 1975, Assembly of species communities, in Cody, M.L. and J.M. Diamond (eds.), *Ecology and Evolution of Communities,* Harvard University Press, Cambridge, Massachusetts. Documentation of lockouts.

Gilpin, M.E. and T.J. Case, 1976, Multiple domains of attraction in competition communities, *Nature* 261, 40–42. Discusses the theory of lockouts.

Pickett, S.T.A. and J.N. Thompson, 1978, Patch dynamics and the design of nature reserves, *Biol. Conserv.* 13, 27–37. A discussion of patch dynamics and conservation.

CHAPTER 5

HETEROGENEITY AND DISTURBANCE IN TROPICAL VEGETATION

Robin B. Foster

Opportunity to select areas for biological reserves from a large region of near-natural habitats is a luxury rarely known today. Such opportunities are now primarily restricted to the poleward extremes of several "developed" countries and to some "less-developed" (or, more appropriately, "less-exploited") countries of the world, especially in South America. In such countries or regions, when the opportunity to select alternative sites for reserves still exists, it is essential to consider edaphic and climatic heterogeneity.

Heterogeneity results in resources for organisms which are patchily distributed in space and time. Patchiness of resources does not always explain patchiness in the distribution of species (Chapter 4), but when resources are patchy, species that exploit these resources must also be patchy in distribution and behavior.

The implications for designers of nature reserves are obvious. For example, species restricted to mangrove swamps will not be preserved unless mangrove swamps are preserved. But, there is also a more subtle factor involved—patches are transitory. Various kinds of disturbances—and various patterns of succession following those disturbances—constantly alter the landscape. A time-lapse movie of a large region would show continuous movement and change. This change is partly random (from storms and landslides, for example), partly cyclical (from flooding, for example), and partly directional (for example, from long-term climatic change).

An understanding of disturbance-recovery cycles is particularly important in management practices following the establishment of nature reserves. Other forms of environmental heterogeneity, such as those caused by edaphic conditions and climate, are of paramount importance only in the choice of reserves. These will be discussed first. The examples will be drawn from the tropics—a major focus area of current conservation efforts—and especially from tropical forests. The same principles should apply in most parts of the earth.

EDAPHIC HETEROGENEITY

Reserves are often chosen as representative examples of certain biological formations or ecosystems. In the tropics, lowland "rainforests" are usually considered a single type of vegetation, and rainforest reserves are often thought to be homogeneous, at least by the layman. Choices of sites are often made without considering patchy habitats, thus risking the noninclusion of a tremendous number of plant, insect and even some vertebrate species. The edaphically caused differences in forest can usually be distinguished by aerial survey, especially when they are extreme (Figure 1).

The poor, red, lateritic soils of the tropics are not as uniformly distributed as many believe. Within tropical areas there is tremendous variance in soil nutrients and many other edaphic characteristics (Burnham, 1975; Walter, 1971). The result is significant heterogeneity in the conditions for plant growth.

Direct evidence for differences in productivity is meager, but the circumstantial and anecdotal evidence is impressive (Fittkau and Klinge, 1973; Janzen, 1974). Plants growing on the demonstrably poor "white-sand" soils (spodosols) of the tropics are low in nutrient content (Stark, 1971) and are heavily endowed with digestibility-reducing compounds (McKey et al., 1978). Such plants often have thick, leathery leaves. The flora on these "poorer" sites may have less than one-fourth of their species in common with adjacent "richer" sites (Brunig, 1973; Kinzey et al., unpublished). By definition there will be the same lack of commonality among the host-specific herbivorous insects, and one would expect little overlap of the more generalist herbivorous insects as well. There are few vertebrates specializing on the white-sand habitats. Frogs are the group most frequently limited to these sites (Inger, 1975), but one large mammal, the yellow-handed titi monkey of South America *(Callicebus torquatus)*, is now thought to be primarily confined to white-sand forests (Kinzey and Gentry, 1979).

The vegetation growing on the poorest soils would appear to be limiting as a resource for animals (that is, inimical to consumer productivity). Evidence of the reduced animal production, density or biomass found in these areas is still scanty (Emmons, unpublished; McKey et al., 1978),

FIGURE 1. Aerial photograph of an isolated pocket of forest on a bleached sand soil, Sarawak, Borneo. The small tree crowns and closed canopy readily distinguish the forest on the sandy areas.

but anecdotal information abounds (Janzen, 1974; personal observations). Palm swamps and seasonally inundated forest provide other examples of vegetation patches which may be significantly different in productivity (higher or lower) than the surrounding matrix of forest. But, without regard to total productivity, the resources of such areas are unavailable for much of the year to many terrestrial animals (for example, detritivores and other nonarboreal feeders).

The point here is that reserves in low productivity habitats require much more area to support the same population sizes of animals that can be supported in smaller, high productivity habitats. If the choice of a reserve is arbitrary, the area chosen could, unwittingly, be that with the smallest resource base, lowest population densities of most animal species and greatest risk for extinctions (Chapters 6 and 7).

It is unfortunate, though likely, that the lowest productivity habitats will be deliberately chosen for conservation, especially in countries where planning for agricultural development has high priority. Low productivity habitats often have fascinating biological communities–specialized floras full of elaborate plant-animal mutualisms and monocarpic trees. But they are also very vulnerable communities in that their recovery rate is slow when severely damaged (by logging, for example). Deliberate emphasis on protection of such areas is certain to delight those concerned primarily with the preservation of plant species and of especially vulnerable ecosystems. Those more concerned with vertebrates will be less pleased since the smaller population sizes increases the risk of extinction.

If large blocks of the highest productivity habitat cannot be set aside, the best strategy is to incorporate many small patches of it within the low productivity matrix of the reserve. This is only a crude guideline, since the resources of any given species (including humans) are subject to the constraints of other kinds of heterogeneity.

WEATHER AND CLIMATE

Seasonality

Most of the tropics have alternating wet and dry seasons (Walter, 1971). The magnitude of the normal drought is of significance to the survival of plants, particularly during the plant's seedling stage (Garwood, 1979). The drought also affects the temporal availability of resources for animals. For reasons not entirely clear, areas with only a slight drought during the driest season have a peak of fruiting near the end of the following wet season (Frankie et al., 1974; Medway, 1972). Areas with a significant drought period have a peak of fruiting at the beginning of the following wet season (Frankie et al., 1974; Koelmeyer, 1959). Intermediate areas (Figure 2) often have both peaks of fruiting (Foster, 1973; Koelmeyer, 1959; Snow and Snow, 1964).

Thus, much more fruit is available at some times of year than at others; this may depend on seemingly trivial changes in the seasonality of rainfall. Inconstancy in response to climate is also the rule for the emergence of new leaves and flower resources (Frankie et al., 1974), and the availability of freshly fallen logs (Brokaw, unpublished) and leaf litter (Haines and Foster, 1977).

The availability of leaves, fruit and flowers might, at first, seem important only to plant-feeders. The rhythm of insect populations in a forest may be especially tied to plant seasonality (Wolda, 1978), particularly new leaf availability. But, by extension, insectivores and carnivores are also affected by the seasonal behavior of plants. Even the fish populations of the Amazon River are heavily dependent on fruit and other plant matter from terrestrial sources (Marlier, 1967; Gottsberger, 1978).

The way plants respond to climatic patterns is complex and will de-

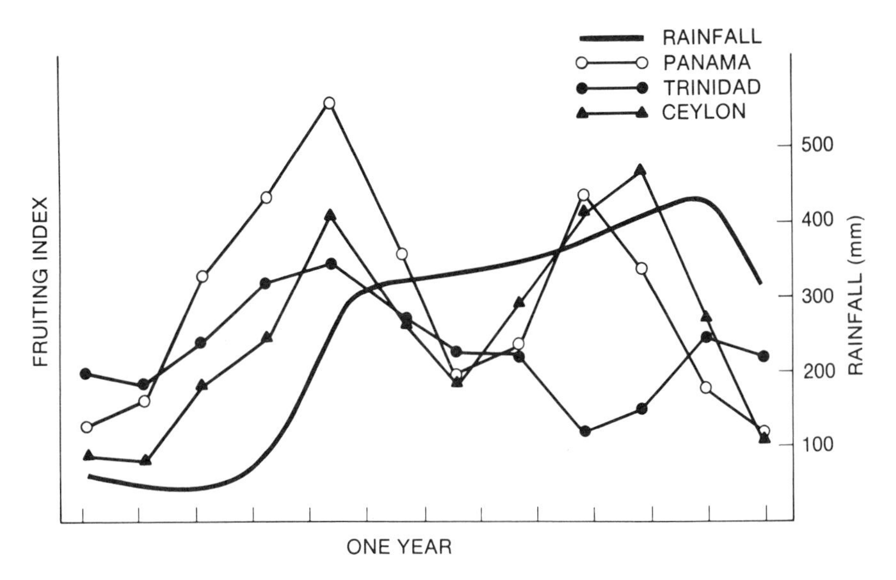

FIGURE 2. Seasonal trends in fruit availability for tropical forests with a recognizable, but not severe, drought season. The thick solid line represents a generalized rainfall pattern for all three sites. The availability of fruit resources is not at all constant during the year for fruit-eating animals. (Panama data from Foster, 1973; Trinidad data from Snow and Snow, 1964; Ceylon data from Koelmeyer, 1959)

pend on edaphic conditions. Flooding is one such important form of climate-induced seasonal heterogeneity. The periodicity of germination, flowering, fruiting and leafing tracks the rise and fall of water in regularly inundated forests of the lowland tropics (Spruce, 1908; Gottsberger, 1978.) The plant seasonality of a flooded forest is rarely in phase with adjacent uplands. Vegetation flooded by water from distant sources may be inundated at a completely different time of year than vegetation a few kilometers distant and flooded by local rivers. The plant seasonality of areas with different duration or frequency of inundation are not in synchrony. Similarly the plants in areas with different soil structure or soil nutrients do not behave the same. Topography adds another dimension to the heterogeneity. Differences in climatic pattern are gradual and large scale in broad, flat areas (as in most of the Amazon Basin), while mountainous regions frequently have abrupt, small-scale differences in seasonal pattern due to local rainshadow effects (Johnson, 1976).

For plants the "solution" to edaphic and climatic heterogeneity is the production of different species or genetic races in the areas subject to different seasonal stresses. But for many animals the solution is migration. When important resources become seasonally unavailable in one habitat, the animals move to a second habitat in which the resources are temporally out of phase with the first. For some birds and butterflies this means leaving the climatic region as a whole, which may require just a short move up mountain slopes or a flight of thousands of miles. For less vagile animals, long distance travel is out of the question. They are limited to the climatic variation in mountainous regions, the local edaphic variation in areas of uniform climate (Klein and Klein, 1976) or the variation provided by successional stages.

In a seasonally dry deciduous forest (Figure 3) both insect populations and leaf-eating primates find refuge in the evergreen riparian vegetation when the uplands are bare of leaves (Janzen and Schoener, 1968; Freese, 1976). South American frugivores such as macaws and white-lip peccaries, often considered characteristic of undisturbed, species-rich tropical forest, apparently depend heavily in some seasons on the fruit resources of the low diversity palm swamps (though information of this kind is conspicuously anecdotal). If a reserve restricts normal migrations, the result is an increase in competitive pressures or highly reduced and highly fluctuating population size. Neither of these alternatives enhances the probability of survival (see Chapters 6 and 7).

To enable animals to follow resources from place to place during regular yearly cycles, a well-chosen reserve should include habitats in which the temporal behavior of vegetation is not synchronous. Mountainous areas fit this criterion as do areas studded with lakes, swamps and rivers. Fortunately, such areas often have little economic value and are aesthetically pleasing to humans. There is also a tradition of including such areas in parks with the idea that habitat diversity increases species diversity. What has not been expressly considered is that even if only one habitat is thought to be the home of endangered species, adjacent, different habitats may provide the resources at a short but critical time of year to maintain the species' populations.

For organisms that can migrate by flying, contiguous vegetation heterogeneity is not required. Certainly on a whole-earth scale, many temperate-zone birds and butterflies are temporarily accomodated in disjunct tropical areas (Karr, 1976), and presumably this could operate on a smaller scale within the tropics as well. A discontinuous system of reserves chosen for diversity in seasonal pattern might provide for many of these species even within a single country. However, our current knowledge of the migrations of species within tropical areas is so meager that the prospects for the success of this scheme are unknown.

Year to Year Heterogeneity

Year to year climatic and resource conditions are far from constant in the tropics. At irregular intervals the flooding of large rivers reaches ex-

FIGURE 3. Deciduous forest in Costa Rica during the dry season. There are few leaf or insect resources during this time and many animals retreat to evergreen stream-side forest.

treme levels and land that is usually above high-water level is inundated (Meggars, 1971). In areas with low-nutrient soils, such as in Surinam and Malaya, massive fruiting of many of the important tree species occurs synchronously only at multi-year intervals (Schultz, 1960; Burgess, 1972). Severe droughts occur in rain forest areas (Dale, 1959), and temperatures can drop to 4°C—in lowland areas within 12° latitude of the equator—as a result of polar air mass outbreaks (Eidt, 1968). Such unusual extremes usually represent an accentuation of normal seasonal patterns. But the important difference for organisms is that the occurrence of rare events is not predictable—at least not with the same accuracy as annual cycles.

The unpredictability of climate through long periods of time is often considered a cause of the occasional extinction of small populations. For Barro Colorado Island, Panama, there are fairly good records of year to year climatic variability. Though no one has documented any extinctions resulting from any one extreme year, animal populations can suffer major fluctuations in number and behavior in response to such years. For example, in 1970 heavy dry-season rains in the Panama Canal area triggered a crash in fruit production (Foster, 1973; Billings, 1974). Mortality among the mammals was greatly increased and many switched to nonpreferred food resources. Toucans and parrots virtually disappeared. Large numbers were seen migrating east toward Colombia (N. G. Smith, personal communication). To maintain characteristic "resident" birds, such as toucans on Barro Colorado Island, may require another reserve a few hundred kilometers away, (for example, in the Darien of Panama). As with accommodation to seasonal fluctuations, a reserve or system of reserves must have adequate vegetation heterogeneity to accomodate such rare events.

A good case could be made for the preservation of vegetation in continuously wet areas. Among examples of local migration, I have noticed a pattern of movement from areas of resource stress to areas that are not only out of phase, but which normally have weaker seasonal fluctuations in moisture. It has already been proposed that continuously wet areas should receive priority in conservation because of the higher frequency of endemics (Prance, 1976; Wetterberg, 1976) and higher density of species (Gentry, 1976*b*).

SUCCESSION

Any disturbance of an area that creates new (uncolonized) habitat or significantly alters the structure of an existing biological community will result in a pattern of changes in species composition (for example, Horn, 1974). The explanation of these successional patterns is not of specific concern here. But the short-term physical and biological conditions that accompany succession in the habitat greatly expand the diversity of resources. These resources are often more important than those of the more temporally stable "climax" condition, although the kind and amount of resources available following a disturbance depend on the kind and severity of the disturbance.

Successional Areas as Resources

Following a disturbance, the first plants to arrive experience reduced competition for sunlight and soil nutrients. This is the only opportunity for an estimated 75 percent of canopy tree individuals to reach maturity (Hartshorn, 1978). In turn, plants characteristic of successional areas have several qualities which enhance their importance as resources for animals. One of these is density.

The successional plant species usually occur in dense patches, as large as the area of disturbance itself. This contrasts with mature forest in which the species, though usually patchily distributed, are at much lower densities (Hubbell, 1979). For two reasons, quantity and palatability, patches of successional plants are especially attractive to herbivores. A high density resource should minimize foraging time and increase feeding efficiency. With respect to palatability, the products of photosynthesis for a successional species are diverted to "quantity" production rather than "quality" production (Odum, 1971). For trees, this high investment in growth rate and reproduction, (lots of leaves, fruit, and "cheap" wood) occurs in a short time period. The high rate of biomass production is also associated with high palatability of leaves (Cates and Orians, 1975). The leaves are edible not only when they first emerge but throughout the life of the leaf, unlike mature forest species on which very little feeding occurs after the initial flush (Coley, unpublished).

The importance of successional resources to animals is apparent from animal migratory patterns. For example, in Amazonian Peru many of the monkeys migrate into the river-bend successional areas at times when fruit availability is lowest in the forest (J. Terborgh, personal communication). One reason for this is the relative lack of seasonality in such habitats. Successional species actively grow most of the year and their fruiting seasons are usually extended relative to mature forest species (Gomez-Pompa and Vazquez-Yanes, 1976).

Successional resources are a strict necessity for some invertebrates. A single species with distinct juvenile and adult phases may require both a successional host plant and a mature forest host plant to complete its life-cycle (Chapter 2).

Figure 4 illustrates the frequency and geographic scale of the principal kinds of disturbances. On the smallest scale, individual tree death and fall in a forest is a continuing source of microsuccession. Light penetration to the forest floor and removal of root competition from the dying trees permit a few seeds and seedlings of successional species and juveniles of mature forest species to grow up into the gap. Thus an area of "undisturbed" continuous forest often has successional resources built into it. However, different kinds of forest have different rates and conditions of gap formation. Poorly drained or wind exposed forests have a turnover rate of 50 to 100 years. Sheltered groves will last as long as 200 years or more (until the trees die of old age) (Foster, unpublished; Hartshorn, 1978). In forests of nutrient-poor areas and in forest less than 100 years old, openings created by the fall of a tree are likely to be smaller and less significant in providing successional resources (Foster and Brokaw, unpublished).

River bank erosion and deposition is the most discretely structured of succession-inducing disturbances. With flooding, deposits of silt are laid down on and next to the previous year's deposits and each river bend becomes a fan of different successional stages (Figure 5). Ox-bow lakes become nutrient traps when the river is flooding and are a source of successional aquatic vegetation. The productivity of these strips will depend on the geological source of the sediment carried by the river. It is hard to obtain average rates of meander disturbance for tropical floodplains but a crude estimate is that a given point on a floodplain is completely eroded at a frequency of between once in every 100 to 1,000 years.

Volcanic disturbance is most important in areas ringing the Pacific Ocean and in parts of East Africa. The area covered with volcanic debris from any one "event" may vary from one to more than a hundred square kilometers. Colonization may, at first, be slow due to the sterile and po-

FIGURE 4. Estimates of frequency and area affected by different kinds of disturbance events in tropical forests. The frequency—or, turnover time—is average time between occurrences of a disturbance of a given kind at one site. Axes are logarithmic. The temporal and spatial range of disturbances are deliberately vague because of the enormous variation in disturbance regime for different parts of the tropics. The same parameters used here for all tropical forests could be estimated for specific biological reserves and used in planning. Except for climatic extremes, large area disturbances are usually less frequent than small ones for a given region. A large reserve can contain more kinds of disturbances without being destroyed. Some parameters (for example, climatic extremes) should be measured by severity as well as by area affected. The more severe climatic extremes will be rarer than weak ones, but their effect on a reserve need not be a function of its size.

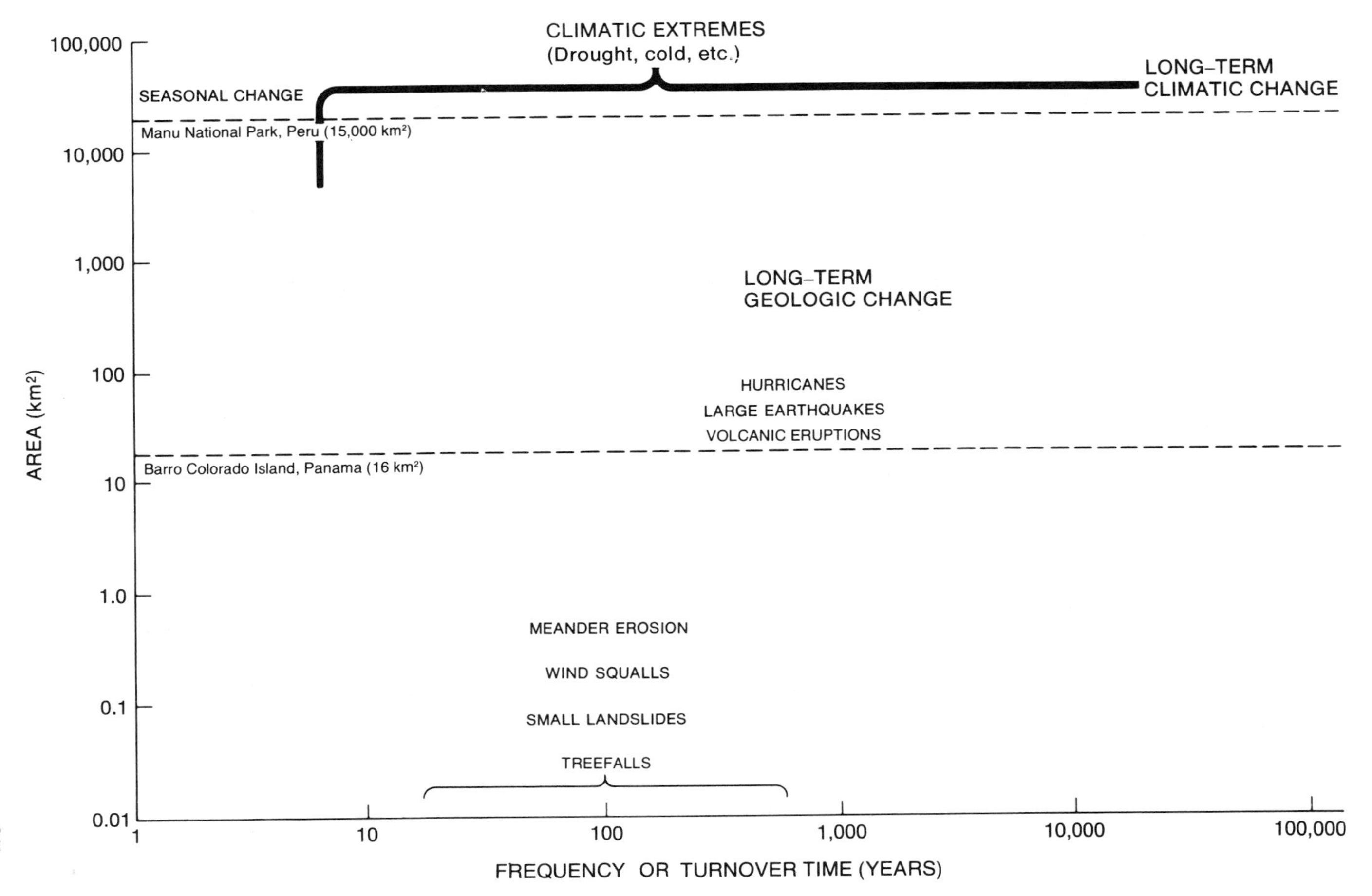
CLIMATIC EXTREMES
(Drought, cold, etc.)
LONG-TERM
CLIMATIC CHANGE
SEASONAL CHANGE
Manu National Park, Peru (15,000 km²)
LONG-TERM
GEOLOGIC CHANGE
HURRICANES
LARGE EARTHQUAKES
VOLCANIC ERUPTIONS
Barro Colorado Island, Panama (16 km²)
MEANDER EROSION
WIND SQUALLS
SMALL LANDSLIDES
TREEFALLS
AREA (km²)
100,000
10,000
1,000
100
10
1.0
0.1
0.01
1
10
100
1,000
10,000
100,000
FREQUENCY OR TURNOVER TIME (YEARS)

FIGURE 5. River meander succession and ox-bow lakes, Rio Manu, Peru. Annual deposits at the river bend and in the lakes create a perpetual array of successional sequences. Mature forest is constantly being destroyed.

rous conditions of the soil. Eventually, however, productivity can be very high due to the nutrient richness of the ash deposits. Depending on geologic conditions, eruptions vary in frequency from every 25 to every 10,000 years, with small eruptions as often as every 4 years (Macdonald and Abbott, 1970).

Landslides are particularly important in mountainous areas, but even in the lowlands, the smaller streams banked by steep slopes are regularly subject to slides. Slides would occur at predictable yearly frequency were it not for the occasional very heavy rains and earthquakes (Figure 6). An earthquake can put areas of 100–200 square kilometers into a highly successional state (Garwood et al., in press). The successional turnover rate for vegetation subject to small slides would be between 50 and 500 years and a function of the geographical relief. Areas subject to large, earthquake-induced slides might have them every 200 to 1000 years. The nutrient status of the cleared area can range from the loss of the entire nutrient capital built up over several centuries to an actual gain in nutrients on slopes in which layers of rich sediment are newly exposed.

Windstorms range from local wind squalls which bring down a few trees, to massive cyclones (hurricanes) affecting hundreds of square kilometers (Watts, 1954; Watt-Smith, 1950). The turnover rate due to wind disturbances can vary from fifty to several hundred years. What starts as

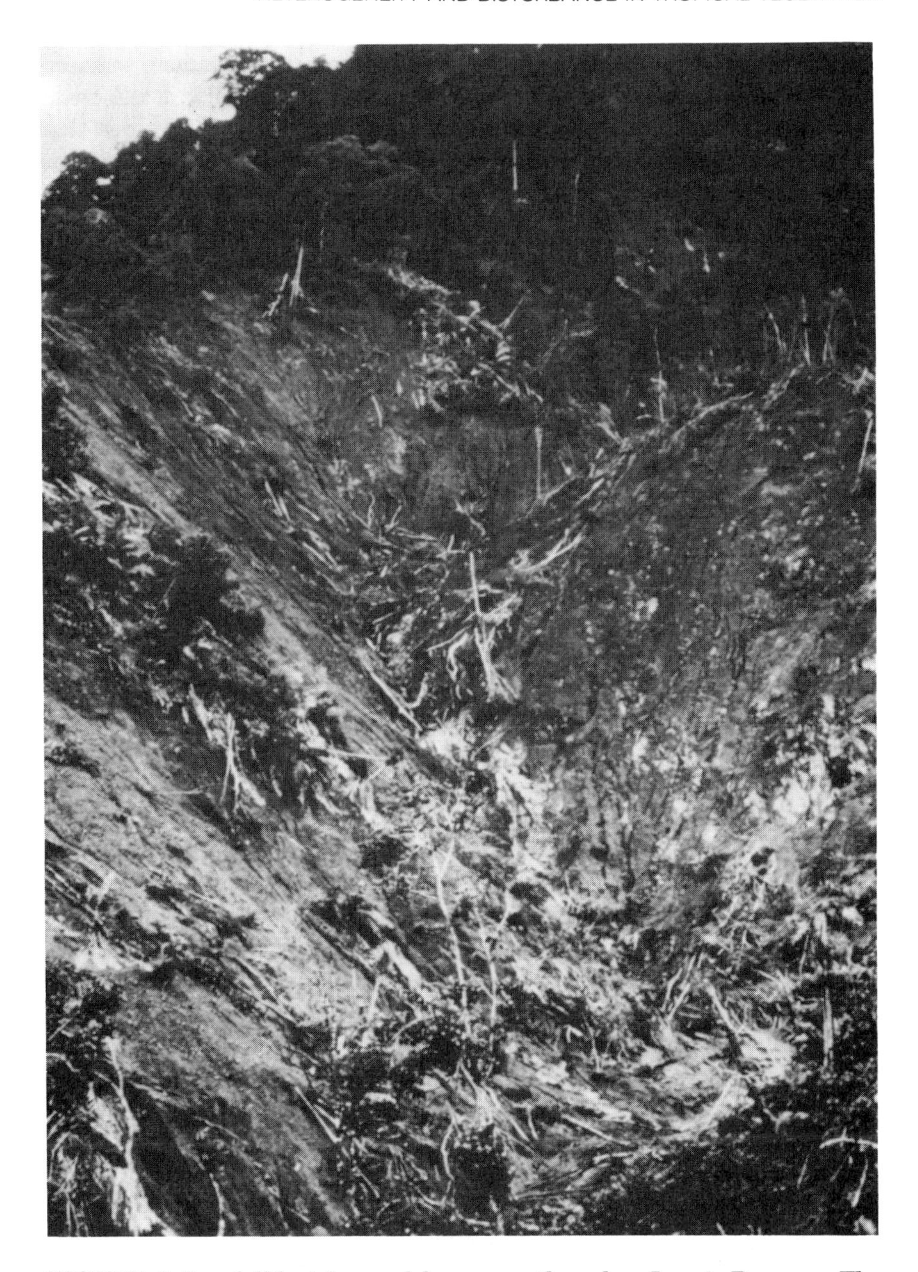

FIGURE 6. Landslide triggered by an earthquake, Jaqué, Panama. The slide leaves a variety of soil conditions. It may require several hundred years for the area to return to a forest similar to the one destroyed. (Photo by N. Garwood)

a single treefall gap in a continuous-canopy forest can be greatly enlarged by wind "erosion" of the forest at the edge of the gap (Whitmore, 1978; R. Lawton, unpublished). Small forest reserves (less than a square kilometer) when surrounded by shorter vegetation are particularly vulnerable to this edge effect on their windward side. Eventually, a tall-forest remnant can be reduced to scrub forest. Large areas of tropical forest that are repeatedly struck by hurricanes develop a distinctive vegetation type —"cyclone scrub" in Queensland, Australia (Webb, 1958) and "hurricane forest" in the West Indies (Beard, 1945). A blown down forest is a mess, but the soil structure beneath it is not greatly altered. There is a great pulse of nutrients and insects from the rotting material, and the tangle provides shelter for many vertebrates. There are several other nonhuman disturbances of tropical forests—insect outbreaks, large mammal damage, fire—but they are usually insignificant relative to those mentioned above.

CONSERVATION IMPLICATIONS

Disturbance and Extinction

Natural disturbance in a community is a two-edged sword. It may often increase the probability of species extinction by either eliminating a small population outright, or by significantly reducing population size (Pickett and Thompson, 1978). In temperate regions, animals that rely heavily on successional conditions are the species most capable of tolerating habitat change (Leopold, 1966; Whitcomb et al., 1976). If this rule applies to tropical forest animals and plants as well, then a high frequency of disturbance may enhance populations of species least vulnerable to extinction at the expense of those most in need of protection (Diamond, 1976).

The importance of disturbance and successional processes (patch dynamics) to conservation is indeed ambiguous. In part, this is because of a great ignorance about the behavior of populations of tropical species. Nevertheless, it is clear that for some species the threat of extinction is minimized by disturbance and succession while for other species it is aggravated. This ambiguity does not necessitate abandoning conservation strategy. Instead, it requires more precision in defining conservation goals. The decisions to be made can then be based on the scale of disturbance relative to the size of the reserve and the capacity for migration and population growth (Pickett and Thompson, 1978).

Should areas with high rates of disturbance be preferentially selected for tropical reserves? The answer should usually be no. Large reserves will almost always have enough disturbance within their boundaries (rivers that flood, windstorms, disturbance-prone mountains) to provide sufficient successional resources. Small reserves require some successional processes but are vulnerable to destruction by large scale natural disturbances. For example, the forest on Barro Colorado Island could suffer a total wipeout from all the disturbances shown in Figure 4 that are on or above the line at 16 square kilometers. Whatever the size of the reserve, high rates of disturbance will favor "weedy" species which are usually not of primary concern to conservationists (but see Chapter 2).

Management Options for Different Taxa

Blanket managerial prescriptions are sometimes unsatisfactory, as specific programs are required for specific taxa. If plants are the major concern, mature forest species that require successional habitats could probably be maintained solely with the succession provided by natural treefall gaps. Great landslide- or hurricane-sized disturbances, whether natural or man made, are not likely to ameliorate the extinction probabilities of plant species. A sudden increase in abundance of one rare plant species as a result of a large disturbance would probably be accompanied by the endangerment of many other species. In small reserves, managed disturbance–perhaps combined with artificial seeding–would be necessary if the floristic composition is to be maintained.

If vertebrates are the primary concern, very large areas, again, will not require induced succession. In very small areas succession will often be a detriment to desirable "deep forest" species because it allows the less desirable forest-edge species to displace them (Whitcomb et al., 1976). Part of the conservation strategy for vertebrates in small reserves would be to actively protect them from anything but low frequency, small scale disturbance.

Insects may be impossible to manage in a very small reserve except on a species by species basis. But artificially created clearings in intermediate sized tropical reserves could greatly enhance the populations of many vulnerable species as well as provide a more stable resource for insect-feeding vertebrates (and, perhaps, insect-feeding plants).

Management Options in Relation to Reserve Size

Where it is not possible to obtain a large, relatively undisturbed area for a reserve, it may be possible to obtain a large area of degraded habitat that contains one or more pockets of the original vegetation (Figure 7). The entire site would grow back rapidly (Opler et al., 1977) if it was naturally or artificially colonized by the more desirable species from the core areas or from nearby sources. It seems entirely reasonable to take large areas of cheap land that are currently of no biological significance and set them aside as reserves which, after succession and with some manipula-

FIGURE 7. A small patch of forest (lower right) surrounded by abandoned clearings. Preservation of the old forest species could be enhanced if they were allowed to gradually invade a protected belt of regenerating forest. (Photo by N. Brokaw)

tion, will eventually harbor a rough approximation of the original community. Conservation efforts do need to be focused on the remaining undisturbed areas, but should not be limited to them. Where protection efforts have already failed or are in danger of failing, it is time to get future reserves growing (Wilson and Willis, 1975).

To a large degree, the size of a reserve will dictate both the frequency and intensity of necessary interference by the humans that manage it. Small reserves have less built-in homeostasis and so require more management effort per unit area, and more human meddling to maintain vulnerable species.

Reserves less than a square kilometer in area might best be considered as botanical gardens or zoos. Active protection of these areas from disturbance by such means as the construction of windbreaks, flood control, and watering during drought, is necessary for long term survival. Each species will eventually require a specific management program to keep it from extinction.

Slightly larger reserves may be drastically affected by windstorms, earthquakes and volcanoes. Barro Colorado Island in Panama (16 square kilometers) is becoming well known for its gradual loss of species, but its size also makes it vulnerable to sudden large losses. While the eruption of a volcano on or near the island in the next 1,000 years is improbable, a severe windstorm or earthquake within the next 50 to 500 years is likely. The flattening or elimination of most of the island's forest would certainly eliminate outright many of the island's plant species, and certainly endanger much of the fauna. Little can be done other than to insure that other patches of similar forest exist, and to restock the areas after such an event. There is also the possibility that insufficient natural disturbance will occur on the island to provide the needed resources of various animal species. In this case some artificial disturbance would be desirable, if it is not at the expense of some of the island's plant species whose entire population consists of a few individuals in a clump.

It is these intermediate-sized reserves that require more difficult decisions because they create more options. They are vulnerable to occasional large scale disturbances but are too large to be protected from them. Induced succession may enhance the stability of resources during times of stress for vulnerable species but may endanger these species at other times. The safest policy is probably to be certain that a low level of succession (less than ten percent of the area) is maintained, and that the size of the successional patches are adequate to include a variety of successional resources. Clearly, if the conservation effort is focused only on one or two species, such a general policy is not needed.

Large reserves provide the best insurance against the deleterious effects of disturbance as well as being the most economical way to provide a healthy mix of disturbance-induced habitats. For an area the size of Manu Park in Amazonian Peru (greater than 15,000 square kilometers) the only significant disturbances are likely to be broad scale climatic variation and, eventually, long-term climatic and geological changes (Figure 4). The latter are probably not relevant given the current crisis of species survival. The effects of seasonality and year to year variation in climate are potentially buffered by inclusion of mountain slopes and river systems within the park. Some species, especially birds, will continue to make seasonal or "bad-year" migrations outside the park. These species will be among the most vulnerable when the upper Amazon Basin is no longer continuous forest. Some less vagile species with very small populations will go extinct within the park due to local disturbance or extreme years (Chapters 6 and 7). Special protection given to these rarest species might be the only biological manipulation needed in a park this size.

SUMMARY

To insure minimal extinction of plant, insect and a few vertebrate species in tropical forests, biological reserves should be selected that represent the true variety of tropical habitats, not just random parcels of "tropical rain forest." To minimize the extinction of most vertebrates and probably most insects, reserves should include areas of high productivity and areas with differing climate or differing response to climate. Choosing areas with high levels of disturbance might be desirable for some conservation goals but not for minimizing species extinctions.

Once a reserve is chosen, the scale of biological manipulation necessary to minimize extinctions is a function of the size of the reserve and the scale of possible disturbance. If a rational plan of minimizing species loss is to be effected, managers of reserves must obtain at least crude estimates of the size and long-term frequency of disturbances and must continuously monitor these disturbances. Small reserves require more manipulation than large reserves. Large, highly disturbed areas that are currently of neither commercial or biological significance could be protected, then allowed or assisted to grow into a condition valuable to conservation in the future.

SUGGESTED READINGS

Holling, C. S., 1973, Resilience and stability of ecological systems, *Ann. Rev. Ecol. Syst.,* 4, 1–23. Emphasizes the importance of heterogeneity to management of ecological systems.

Horn, H. S., 1976, Succession, in *Theoretical Ecology,* May, R. M. (ed.), Blackwell Scientific Publications, Oxford. Discusses the community consequences of adaptation to disturbances.

Pickett, S. T. A. and J. N. Thompson, 1978, Patch dynamics and the design of nature reserves, *Biol. Conserv.,* 13, 27–37. The most important review of the disturbance problem, with data on temperate disturbances.

Prance, G. T. and T. S. Elias (eds.), 1977, *Extinction Is Forever,* New York Botanical Garden, Bronx, New York. Includes several papers on the relation of plant species distribution to their vulnerability and preservation in the American tropics.

Whitmore, T. C., 1975, *Tropical Rain Forests of the Far East,* Oxford, London. The best text covering the dynamics and variability of tropical forests.

Wiens, J. A., 1976, Population responses to patchy environments, *Ann. Rev. Ecol. Syst.,* 7, 81–120. Discusses many aspects of patchy systems not considered here.

PART TWO

THE CONSEQUENCES OF INSULARIZATION

CHAPTER 6

INSULAR ECOLOGY AND CONSERVATION

Bruce A. Wilcox

One of the most profound developments in the application of ecology to biological conservation has been the recognition that virtually all natural habitats or reserves are destined to resemble islands, in that they will eventually become small isolated fragments of formerly much larger continuous natural habitat. Hence, beginning with the seminal monographs by Preston (1962) and MacArthur and Wilson (1967), many ecologists and biogeographers have come to recognize the potential importance of studies of islands and other ecological isolates to conservation.

Typically, the term "isolate" has been used to connote any discrete ecological unit which is insulated from other similar units. In addition to a true island, an isolate can be a stand of trees or other distinct vegetation type, or even a pond or lake. On a smaller scale, an individual tree, shrub or portion thereof can be an "island" from the perspective of very small organisms. For these reasons, "isolate" as it is used here is an ecological community with more or less distinct boundaries. The study of isolate ecosystems, most commonly called "island biogeography," also implies a narrower scope than is actually encompassed. Because of its generality, the term "insular ecology" is preferred here to "island biogeography."

The attractiveness of insular ecology is due, in large part, to the quantitative manageability of discrete ecological entities which have definable physical and biological properties. In particular, because they have definite boundaries, isolates can be described by the amount of area encompassed, proximity to other habitats, as well as taxonomic composition, which is known to be largely related to the former two properties. These properties are those of primary interest to the conservation biologist since they are affected by habitat loss and insularization. Reduction in the total amount of area encompassed by natural habitat and fragmentation into disjunct insular parcels obviously have negative effects on natural ecosystems. So far, insular ecology has only begun to rigorously define the effects and only in a very general way. A much more thorough development of this field is necessary to implement long range conservation strategies. It is hoped this discussion will provide some additional stimuli in this direction. To this end, this chapter begins by reviewing insular ecological theory, then discusses its implications to conservation and considers differences among the major vertebrate taxa in the effects of habitat loss and insularization.

THE SPECIES-AREA RELATION

The study of insular ecology began with the species-area relation (Preston, 1962; MacArthur and Wilson, 1963; 1967). It has become axiomatic in ecology that if isolates or sample quadrats are censused, those of greater area will have more species. This relationship occurs, primarily, because larger areas have more habitat and greater habitat diversity (which includes numerous factors contributing to the stability of individual species populations). The quantitative relationship between species and area can be shown to follow from properties of ecological communities that have practical consequences to conservation.

A typical species-area relation is plotted in Figure 1. Assuming an adequate range of areas are used, species-area plots result in curves which approximate this form. That is, the addition of equal amounts of area contributes fewer species not yet observed. This relation is often expressed by $S = CA^z$, where S and A are the number of species and area, and C and z are dimensionless parameters whose values need to be fitted for each set of species-area data. C is not particularly meaningful here; z is of more interest and its interpretation will occupy much of this discussion.

Since the above expression for the species-area relation is a power function, logarithmic transformation of the variates, species and area, or the axes upon which they are plotted produces a linear relationship (Fig-

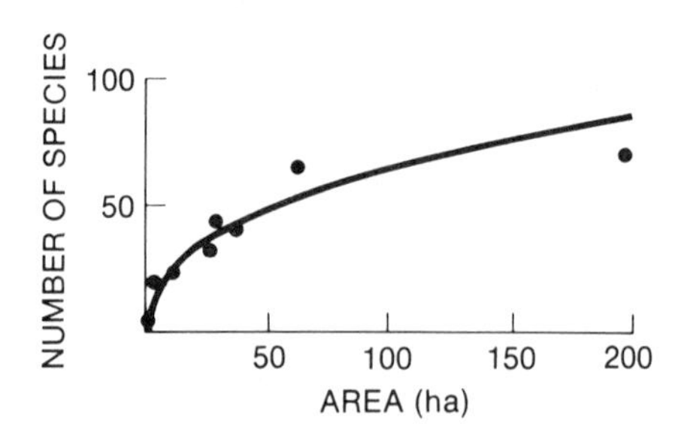

FIGURE 1. An example of a species-area relation: the number of breeding bird species in different size plots of North American deciduous forest. (Data from Preston, 1960)

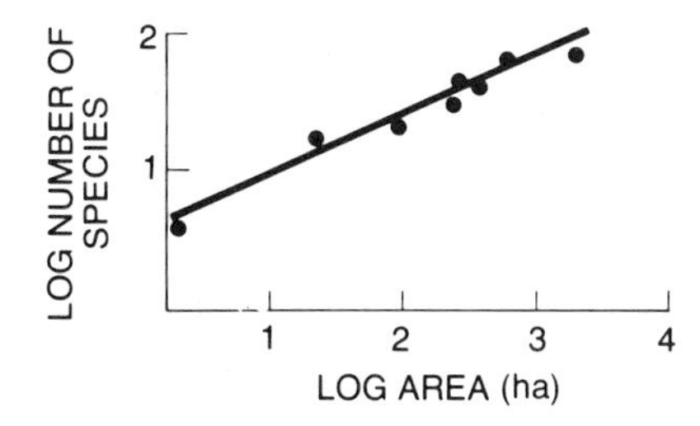

FIGURE 2. The species-area relation in Figure 1 plotted on logarithmic axes.

ure 2). This manipulation of the data allows the values of C and z to be readily calculated by applying linear regression analysis where $\log S = z \log A + \text{constant}$; note that z is now the slope of the line.

The power function can adequately describe many, but not all species-area relations (May, 1975; Diamond and Mayr, 1976; Schoener, 1976). When the power function is appropriate, that is when the log-log transformation produces an approximately linear relation, the resulting z-values (slopes) vary depending on whether the observations are isolates or samples—terms used to describe the nature of the entities being sampled (Preston, 1962). Isolate in this sense is analogous to "universe" in that it refers to the total collection of individuals comprising an ecological community. A sample, on the other hand, is some fraction of such a community. Thus a random transect or quadrat is a sample and successively larger samples together eventually become an isolate (a continent could be considered the largest possible isolate). Aside from the form of the species-area curve, a basic empirical rule is that z-values for samples are lower than those for isolates. Samples range between approximately 0.12 and 0.17 while isolates tend to vary between approximately 0.18 and 0.35 (Preston, 1962; MacArthur and Wilson, 1967), although some values may actually be more extreme than this (Diamond and May, 1976). To understand the reasons for this dichotomy it is important to consider the theoretical basis of the species-area relation as it has been recently clarified by May (1975).

The species-area relation and the apparently narrow range of observed z-values are the result of two underlying quantitative properties of ecological communities. The first is the approximately linear relationship between the number of individuals and area. The second is the relationship between the total number of individuals and the number of species, or the distribution of species abundance as it is often called. The form of this distribution is approximately lognormal for intact communities. As emphasized by May, the lognormal distribution is merely a statistical phenomenon of large, heterogeneous communities and probably has no other underlying biological significance. Also, the apparently narrow range encompassing observed z-values results from mathematical properties of the lognormal distribution.

The divergence of z-values for samples from those for isolates can be explained as a result of the failure of samples to represent a complete lognormal ensemble of species (Preston, 1962). That is, relative to an intact community, a sample is characterized by fewer total species and a higher species-individual ratio. In addition, some species are represented by only one to several individuals. The reason for this can be understood as follows. The smallest possible sample consists of one species represented by one individual. At first, successively larger samples will only contain individuals representing new species. This is because even the most abundant species has far fewer individuals than the rest of the species combined. However, as species accumulate, new species will be recorded less frequently and individuals will be added more frequently. As the size of the sample approaches that which encompasses the entire community, only the least abundant species in the community remain unrecorded. The higher species-individuals ratio of smaller samples inflates the species count. This does not occur in smaller isolates since the species-individual ratio should be more equitable regardless of size.

The key feature that differentiates a sample from an isolate is that the latter is a self-contained ecosystem insulated by barriers that tend to restrict normal movement or dispersal of organisms in or out. All of the immediate ecological requirements of a population are contained within an isolate. However, since vagility varies greatly among species, a habitat which produces isolates for some species may act as samples for others.

Isolates Which Behave As Samples

Depending on the degree of habitat insularity or the vagility of organisms, a system may behave either as a sample or an isolate. As an example, in their studies of bird faunas on the Solomon Archipelago, Diamond and Mayr (1976) calculated a z-value of 0.025 for the "highly vagile" species, conforming to that of extreme samples. However, the species with "low vagility" on the very same islands produced a z-value of 0.28, conforming to that of isolates. The same effect can be seen in another example as a result of distance to an archipelago from the source of colonizers. By assigning the archipelagos to "near" and "far" groups (Figure 3) it is seen that the "near" group behaves like samples and the "far" group like isolates. If z-values are calculated for individual archipelagos (within rather than among archipelagos), however, the opposite trend results: more distant archipelagos have lower z-values. This apparent anomaly is actually consistent with the above since distant archipelagos tend to be inhabited by a proportionately greater number of highly vagile species

(Diamond and Mayr, 1976) or have a larger "effective" species pool (Schoener, 1976).

In summary, the loss and fragmentation of natural habitat will have different consequences depending on the vagility of organisms and the proximity of isolated fragments. Before discussing these consequences in more detail, let us first consider the most far-reaching theory in insular ecology.

THE EQUILIBRIUM THEORY

In addition to the foregoing explanation of the species-area relation, the autonomous nature of island ecosystems led Preston (1962) and MacArthur and Wilson (1963), independently, to postulate another. First,

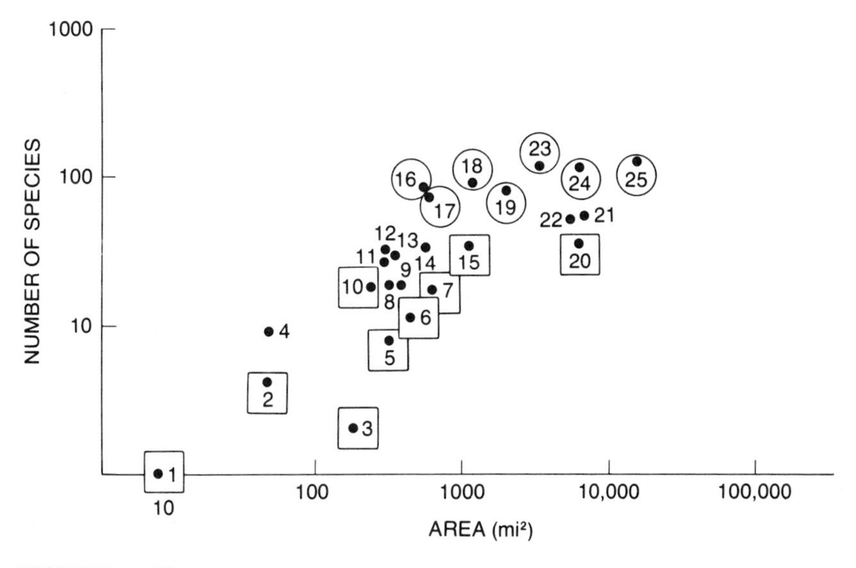

FIGURE 3. The number of land and freshwater bird species on various islands and archipelagos in the Pacific Ocean. "Near" islands (open circles) are those less than 500 miles from a colonization source (New Guinea). "Far" islands (squares) are those greater than 2,000 miles from a colonization source. Dots without circles or squares are islands of intermediate distance. (1) Wake, (2) Henderson, (3) Line, (4) Kusaie, (5) Tua Motu, (6) Marquesas, (7) Society, (8) Ponape, (9) Marianas, (10) Tonga, (11) Carolines, (12) Palau, (13) Santa Cruz, (14) Renell, (15) Samoa, (16) Kei, (17) Louisiade, (18) D'Entrecasteaux, (19) Tanimbar, (20) Hawaii, (21) Fiji, (22) New Hebrides, (23) Buru, (24) Ceram, (25) Solomons. (From MacArthur and Wilson, 1963)

consider that area sets an upper limit on the number of individuals. Second, that the probability of a species becoming extinct will increase with smaller population sizes. This is because small populations are more subject to demographic instability (Chapter 10), inbreeding depression (Chapters 8, 9 and 12) and inclement environmental conditions. Hence, it follows that area acts through extinction to limit the number of species, despite propagules from other islands or the mainland which provide a constant source of new species. This loss of species through extinction can be viewed as being balanced by the gain from colonization.* Thus, the outcome can be expressed as an equilibrium between colonization and extinction.

The equilibrium theory can explain the effects on species number due to variation in area and insularization independently of any assumptions about underlying species abundance distributions. This is demonstrated by the graphical analysis in Figure 4. The curves represent the rates of colonization and extinction for isolates differing in both their distance from a source of colonists and in size, as a function of the number of species present. The curves show that the rate of colonization is highest on "near" isolates and when there are fewer species. This is because the likelihood that a propagule will successfully reach an isolate is greater for shorter distances, and the likelihood that any such propagule will represent a species not yet present is greater when few species exist in an isolate. In a similar manner, the rate of extinction is highest in "small" isolates and when there are more species. This is because extinction probability is greater with less area (smaller population size) and when there are more species to go extinct. At the point of intersection of a colonization and extinction curve, the rates of both processes are equal and the equilibrium number of species, $\hat{s}$, is defined. Figure 4 shows how $\hat{s}$ varies with isolate size and distance; for example, "far" and "small" isolates equilibrate with the least number of species. It also shows (although the curves are purposely drawn in this manner) that the species-area relation can be affected by distance (colonization rates), since the disparity in $\hat{s}$ is greater between the two hypothetical "far" isolates than between the two "near" islands, which would result in a lower z-value for the latter. Figure 3 also illustrates the "distance" effect.

Species Turnover

The most striking feature of the equilibrium theory is its dynamic view of isolates. It proposes that species are constantly being lost and gained, one species being exchanged for another, such that turnover in

*MacArthur and Wilson (1963, 1967) use "immigration" rather than "colonization" as the addition of new species. Both are to be distinguished from the arrival of new individuals representing species already present.

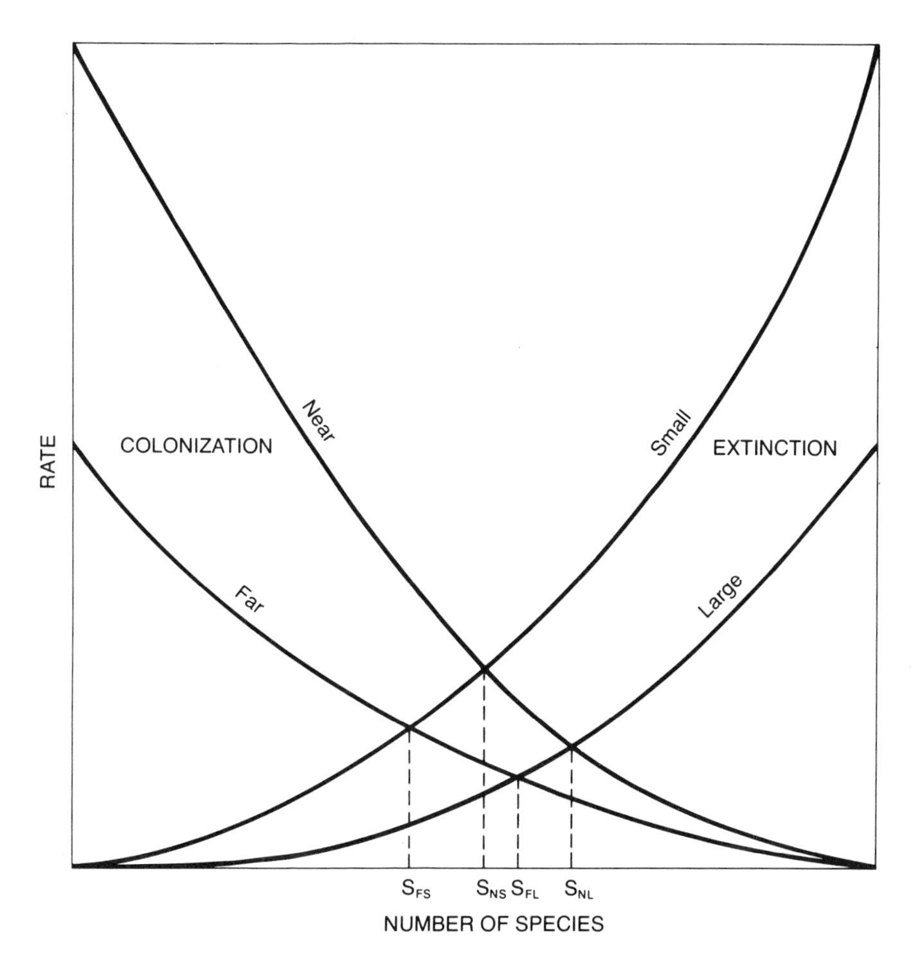

FIGURE 4. Graphic model of the equilibrium theory of island biogeography demonstates the effect of isolation and area on the number of species maintained at equilibrium. The colonization curves represent the rate of addition of new species to an isolate as a function of the number of species already present. This rate should be highest for "near" isolates since proximity to a source of colonists will increase the probability of receiving propagules. The extinction curves represent the rate of extinction as a function of the number of species on an isolate. This rate should be highest on small islands since population sizes are smallest. Concavity of the curves is predicted from heterogeneity in species colonization ability and negative species interactions with increasing species number. For any combination of colonization and extinction curves an equilibrium number of species, $\hat{s}$, is defined at their intersection. (For example, $\hat{s}_{FS}$ is the equilibrium number for an island that is both "far" and "small.")

species composition occurs. Until recently, evidence for turnover, though suggestive, was minimal and based primarily on island bird faunas where census intervals were too infrequent for accurate measurement (Diamond, 1969; 1971; Terborgh and Faaborg, 1973). These early estimates of average annual turnover rates ranged from 0.2 percent to 1.7 percent of the total bird fauna, although total compositional change over a 50 year period was as high as 44.0 percent (Diamond, 1969). Since longer census intervals tend to mask repetitive short-term turnover events, most of these estimates are being modified dramatically as data from short-term censuses become available. Average annual turnover rates for bird faunas are now ranging between 0.9 percent (Jones and Diamond, 1976) and 13.0 percent (Diamond and May, 1976) for islands. Estimates for bird species turnover in mainland habitats are showing similar values. Annual turnover rates of 13.6 percent were calculated for Illinois woods (Whitcomb et al., 1976) and 10.0 percent for a Finnish valley (Järvinen, in press). Where there are a sufficient number of estimates to make comparisons, the turnover rates for smaller islands are highest (Jones and Diamond, 1976). This accords with the prediction of the equilibrium theory for higher per species extinction rates on smaller islands.

These relative turnover rates should be distinguished from absolute turnover rates. According to theory, both vary inversely with area. Distance, however, may not detectably effect relative turnover because the absolute rate and number of species vary in the same direction (Williamson, 1978). Further, absolute turnover may not vary with distance as predicted by the equilibrium theory because of the rescue effect (Brown and Kodric-Brown, 1977).

So far, significant species turnover within the time span of modern biological exploration has not been recorded for vertebrates other than birds. This is not for a lack of data, however, since in theory the time scale of turnover for other groups should be much longer. For birds, at least, the evidence does suggest an important role for colonization-extinction equilibria, particularly in smaller isolates.

NATURE RESERVES

The Sample Effect

Let us now consider insular ecology in terms of nature reserves. Most nature reserves, at the time they are designated as such, are effectively samples in that they represent only a portion of a much larger community. Thus, Serengeti National Park or Manu in the Amazon are samples of African savanna and Amazon rain forest, respectively. They contain fewer species, fewer individuals within each species and more species represented by only several individuals than would similar but larger reserves. A single reserve would have to include virtually the entire Amazon to contain all of its species (albeit a vast system of numerous small reserves could initially accomplish this). On the average though, extrapolating from observed z-values for samples, a reserve will initially exclude roughly 30 percent of the species of a community for each 10-fold decrement in area. For example, a reserve of 10 km^2 will be missing three out of every 10 species found in 100 km^2 of surrounding habitat. We will refer to this initial exclusion as the *sample effect.*

Short-term Insularization Effects

Typically then, a single area designated to preserve any community that does not take in the entire area of the community, will fail to include some of its species. This will occur because of a lack of appropriate habitat or due to species that are rare (Chapter 7) or patchily distributed in spite of habitat (Chapter 4). On the other hand, some species will be included even though their habitat requirements are not met. Under many circumstances, the boundaries of a reserve will not coincide with those of ecological systems, and the requirements for each species cannot be met within its confines. The migration of herds of wildebeest and elephant across the boundaries of large African national parks attest to this fact most dramatically. On any scale, however, the boundaries of a reserve will sever the resources of some species. As long as adjacent habitat remains, these species will be counted among those "protected" within the reserve. However, as the neighboring habitat is removed for agriculture and other forms of land use even more devastating, they will disappear unless intensively managed.

The above implies that dispersal across reserve boundaries as well as the availability of exogenous resources are required for the survival of some species in reserves. Two other conditions can be distinguished where dispersal is responsible for the occurrence of a species. Neighboring habitat may contribute species as non-breeding transients, which are nonetheless recorded as resident species (Simberloff, 1976). Alternatively, neighboring habitat may contribute recruits for minimally stable populations in suboptimal habitats. Since this amounts to preventing extinction in established but marginal populations, it has been dubbed the "rescue effect" (Brown and Kodric-Brown, 1977). The loss of species sustained directly by exogenous resources or dispersal will immediately follow the disappearance of adjacent habitat. These can be considered *short-term insularization effects.*

Long-term Insularization Effects

Colonization-extinction equilibria must, no doubt, play a role in the maintenance of species diversity in samples as well as isolates, although it is much less generally appreciated in habitats without well prescribed boundaries. Here we are simply extending the above notion that individuals dispersing from adjacent or contiguous habitat can shore up a faltering population. It is equally plausible that dispersing individuals may represent new colonists replacing populations which have failed altogether.

As a natural area becomes increasingly insular and colonization sources become increasingly fewer and more distant, colonization rates will, of course, diminish. Since, over the long-term, extinction events are inevitable in even the largest reserves, the reduced colonization rate will result in a net loss of species over time which will continue until the colonization and extinction rates are once again balanced. Of paramount interest, then, is the magnitude and time scale of this putative collapse to a new equilibrium.

Fortunately we need not depend entirely on theoretical conjecture to anticipate *long-term insularization effects.* For example, the frequently cited case of Barro Colorado Island provides a striking illustration (Chapters 2, 3, 4 and 7). Formed around 1914 with the flooding of the Panama Canal, Barro Colorado was designated a reserve soon thereafter. Since that time, however, many of its original species have become extinct, including 48 of its 208 breeding birds. About one-third of the extinctions appear to be the result of a colonization-extinction disequilibrium (Willis, 1974). Similar consequences of habitat fragmentation are documented in Chapter 7.

There is another matter of concern regarding the long-term effects of habitat insularization, particularly in nonflying taxa. These are the shifting climatic zones associated with glacial-interglacial cycles. The Pleistocene Epoch was marked by global temperature oscillations during which fairly extensive latitudinal (and altitudinal) shifts in climatic zones occurred. Normally, as climatic zones shift, so do the associated biotas. Insular regions, however, typically do not span a sufficient latitude to provide refugia for species ill-adapted to a novel climatic regime—and extinction results.

As an example, many species presently adapted to arid southwestern North America probably had the limits of their northernmost ranges pushed several hundred kilometers southward at the peak of the last glacial period. The adjacent islands apparently bear witness to this. Those of higher latitude have fewer nonflying vertebrate species than expected given the present mainland distributions of species. Notably, the arid-adapted reptile species are disproportionately under-represented on the more northern California islands, even on those with apparently suitable habitat (Wilcox, in press).

Barring catastrophic global temperate modification caused by man, the temperature oscillations of the past will be repeated. Therefore, unless long-term climatic shifts are taken into account, another factor can be added to the long-term effects of insularization.

FAUNAL COLLAPSE

Pleistocene Land-Bridge Islands

The combined short- and long-term insularization effects discussed above (even omitting climatic shifts) can potentially result in the eventual loss of most of the vertebrate species in even the largest reserves. This faunal collapse is documented by studies of land-bridge islands, whose mainland connections were severed by the rising sea level around the close of the Pleistocene (Diamond, 1972*b*, 1973; Terborgh, 1974; Chapter 7). The last episode of rising sea level began around 18,000 years ago and ended some 6,500 years ago (Bloom, 1971). During this time the sea rose over 100 meters, inundating lowlands and leaving isolated fragments of such formerly vast continental regions as the Sunda Shelf in southern Asia (Figure 5). In this example, based on the present ocean depths of this region and the rate of sea level rise (Wilcox, 1978), the islands of Borneo, Sumatra and Java, and their fringing archipelagos were probably completely isolated by 10,000 years ago.

By estimating the number of species lost on these and other islands with a similar history, we can gain at least some idea of the fate of our reserves which are presently undergoing the insularization process. As shown in the following example, this is typically done by assuming that each island, before falling to its current level, originally had the same number of species that a mainland sample of the same size has at present.

The faunal composition and distribution of species on the Sunda Shelf prior to inundation cannot be known with certainty, though present faunal resemblance and fossil evidence strongly argues for its similarity to the Malay Peninsula. For example, the islands of Borneo, Sumatra and Java combined have virtually a complete Malaysian mainland mammal fauna though each is deficient in a number of species (Darlington, 1957; Terborgh, 1975). That is, a species missing on one island is often present on another (see Table I in Chapter 4). Evidence that these islands supported larger faunas at the close of the Pleistocene is also provided by fossil remains (Terborgh, 1975). Figure 6 shows the current level of diver-

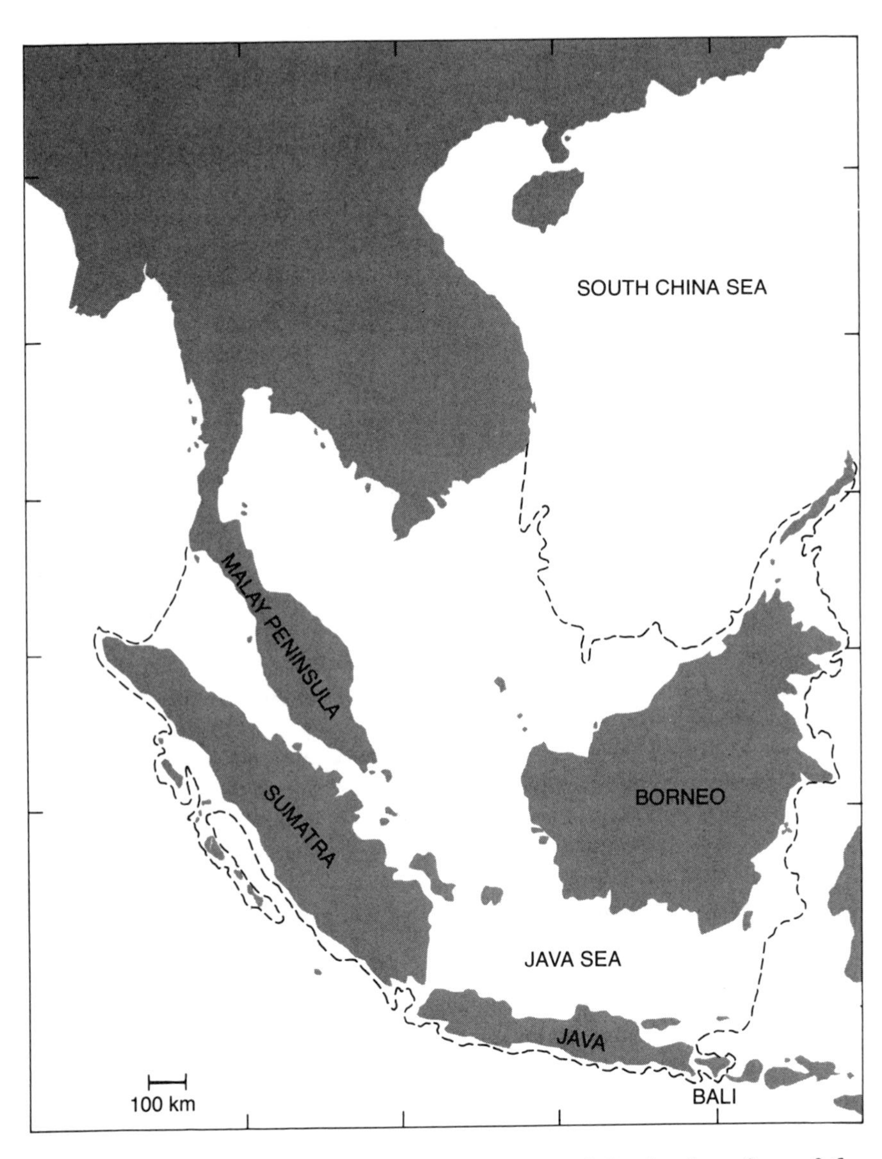

FIGURE 5. The Sunda Islands and the extent of the land surface of the Sunda Shelf during the last glacial period. The dashed line is the present 100 fathom contour and approximates the coastline of 18,000 years ago.

gence between the species area relation for the land mammal faunas of the Malaysian mainland and the Sunda Islands. Although the mainland curve is tenuously based on only two points, its z-value of 0.17 and that of the islands, 0.30, agree with theory and previous findings. Thus, the presumption is that island faunas all began near the upper curve and have since dropped to their present positions. The estimated species loss for

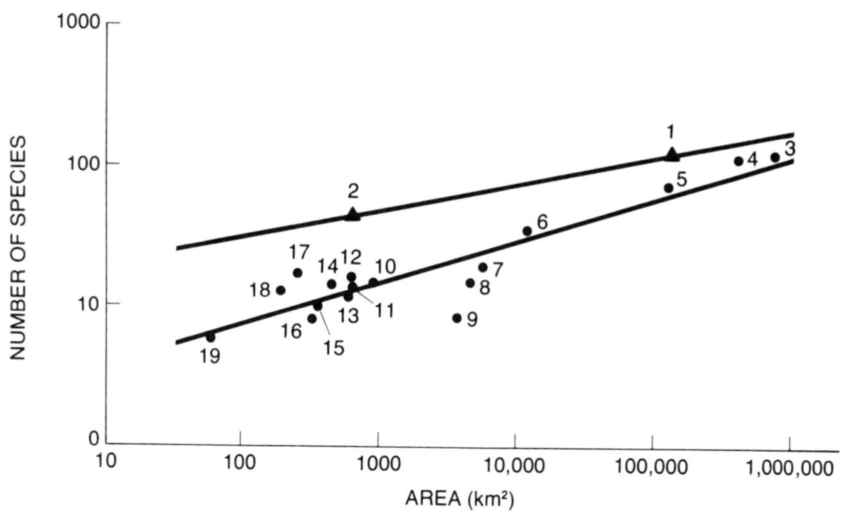

FIGURE 6. Species-area relations for the land mammals (excluding bats) of the Malaysian mainland (upper curve) and Sunda Islands (lower curve). The upper curve ($S = 15.4\ A^{.17}$) is presumed to approximate the species-area relation of the Sunda Shelf prior to fragmentation by the rising sea level at the close of the Pleistocene. The lower curve ($S = 1.8A^{.30}$ by linear regression) differs as a result of fragmentation and collapse of isolate faunas. (1) Malay Peninsula, (2) Krau Game Reserve, Pehang, Western Malaysia, (3) Borneo, (4) Sumatra, (5) Java, (6) Banka, (7) Bali, (8) Billiton, (9) Siberut, (10) S. Pagi, (11) N. Pagi, (12) Sipora, (13) Singapore, (14) Tanabala, (15) Tanamasa, (16) Pini, (17) Penang, (18) Tuangku, (19) Bangkaru. (Data from Medway and Wells, 1971; Chasen, 1940)

each island can thus be calculated as the difference between the projected initial number of species, S_0, and the present number of species, S_p. Table I shows the results of such calculations for the Greater Sunda Islands.

The species losses for these islands, the majority of which are species of large body size or high trophic level, are quite remarkable, especially because most of the islands are an order of magnitude larger than the largest existing nature reserves. Nevertheless, these figures alone probably provide only a crude view of the collapse process. If we could have directly monitored the loss of species from the time of initial isolation, we would, according to theory, observe that the rate of loss would have been highest at first, decreasing exponentially to approach asymptotically a new equilibrium.

The prevailing evidence suggests that the loss of species is approximated by an exponential function, such as $-kS^n$, where k is the relaxation

TABLE I. Estimated number of extinctions of land mammal species (excluding bats) since the formation of the Greater Sunda Islands.

Island	Area (km^2)	Initial number of species* (S_o)	Present number of species (S_p)	Extinctions** Number ($S - S_p$)	Extinctions** Percent $\left(\frac{S_o - S_p}{S_o}\right)$
Borneo	751,709	153	123	30	20
Sumatra	425,485	139	117	22	16
Java	126,806	113	74	39	35
Bali	5,443	66	19	47	71

* Initial number of species, S_0, are estimated from the species-area relation of mainland Malaysia.

** Total species loss is assumed to closely approximate or even underestimate extinctions. See text for more discussion.

parameter and n is an integer that might reasonably take on values from one to four, although two or three seems most probable on the basis of very limited data (Soulé et al., 1979). By using models of this general form, relaxation parameters or comparable measures of the magnitude of species loss with time have been estimated for a number of geographically and taxonomically divergent faunas (Diamond, 1972*b*; 1973; Terborgh, 1974*a*; 1974*b*; 1975; Case, 1975; Soulé et al., 1979). All of these findings show a tendency for smaller islands to have higher extinction rates. Also, the rate of species loss, as a proportion of the initial number of species and as a function of area and time, vary consistently in the manner predicted from intrinsic taxonomic differences. This will be discussed in more detail below.

Collapsing Reserves

The apparent universality in the general properties of the collapse process should allow extrapolation of these findings to nature reserves. One such attempt (Soulé et al., 1979) has been made using the large mammal species of the Sunda Islands as a model to predict the fate of similar species in East African reserves. Application of collapse theory to these reserves is particularly appropriate as insularization is nearly complete. In this study, relaxation parameters were calculated for several Sunda Island faunas to determine their relationship to island size. The size specific values were then applied to 19 East African national parks and game reserves to estimate species loss as a function of area and time. The results predict that the average reserve, which presently has 48 large mammal species and an area of about 4,000 km^2, will lose 11 percent of these in 50 years, 44 percent in 500 years and 77 percent in 5,000 years. Figure 7 shows the probable range of species loss trajectories for three of the national parks. This illustrates that as isolates, even the largest reserves in the world may be incapable of preserving most of their large mammal species without intensive management.

Extinction estimates have been calculated for other taxa in reserves using a similar approach. From his studies of bird faunas on tropical land-bridge islands Diamond (1972) predicted a loss of 51 percent for a hypothetical reserve of 7,800 km^2 in 10,000 years. This is smaller than predicted losses for the large mammals and probably results from smaller body size (allowing higher population densities) and higher colonization rates of birds, since collapse has both a colonization as well as an extinction component.

The theoretically predicted shapes of the colonization and extinction curves (Figure 4) as well as empirical evidence for their shapes approximated from actual data (Gilpin and Diamond, 1976) suggest a rather insignificant role for colonization (even where dispersal is possible) until species number nears equilibrium (Wilcox, 1978). For this reason, the net

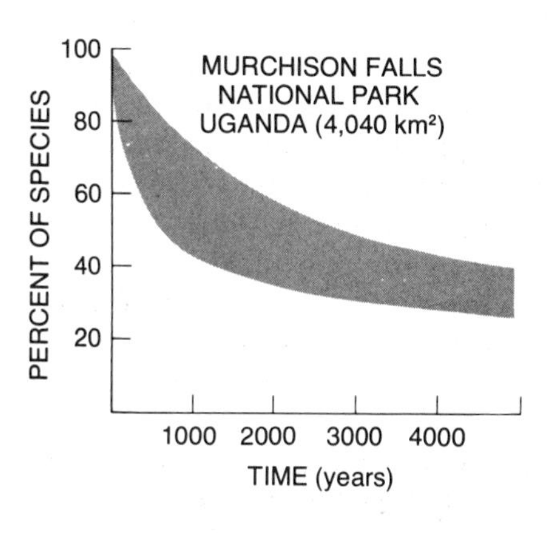

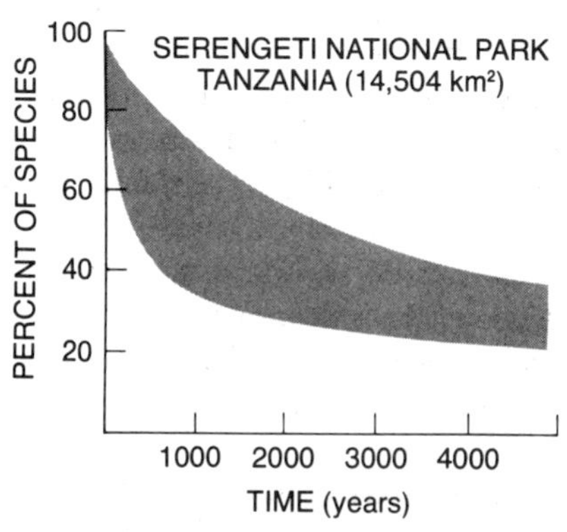

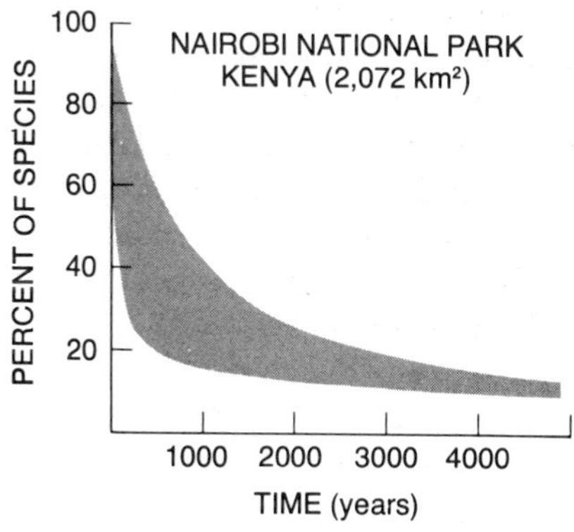

FIGURE 7. Examples of faunal collapse trajectories for large mammals in three East African national parks. Shaded area includes the probable range within which the trajectory will fall for each reserve. The severity of collapse depends on the initial number of species as well as reserve size. (From Soulé et al., 1979)

loss rates, particularly for the mammals shown above, approximate extinction rates. This theoretical detail may be largely irrelevant to collapsing reserves, however, since inter-reserve dispersal of many, if not most, vertebrates will be wholly curtailed. This will certainly be the case for the large mammals of East African reserves, unless heroic measures are taken.

TAXON-SPECIFIC CONSIDERATIONS

A misleading, but necessary, oversimplification of the MacArthur-Wilson equilibrium theory is the assumption of homogeneity in colonization and extinction rates among different species. This problem is partly ameliorated by confining biogeographical analyses to within taxa. However, even within a taxon such as birds, species exhibit a wide range of dispersal ability and extinction vulnerability (Diamond, 1972*b*; 1976; Jones and Diamond, 1976; Chapter 7).

Some generalizations can be made, however, that should bear on conservation decisions. Among the major vertebrate taxa there are two reasons why colonization and extinction rates are expected to differ. First, dispersal is greatly enhanced in the flying taxa (most birds and bats) over nonflying taxa (most mammals, reptiles and amphibians). It should be noted, however, that adaptations for passive dispersal as well as behavioral differences (willingness to cross inhospitable terrain) may vary independently of flight. Second, the lower metabolic rates of ectotherms (reptiles and amphibians) allows higher population densities, thus less vulnerability to extinction than in endotherms (mammals and birds). Among the endotherms, however, the generally larger body size of mammals (resulting in higher absolute metabolic demands and, thus, lower densities), should cause them to be more vulnerable to extinction than birds.

There are two major consequences of these generalizations. First, mammal faunas should collapse more rapidly than bird faunas, while reptile and amphibian faunas should be least prone to rapid collapse. In support of this, in addition to the evidence already cited above from the studies of Diamond (1972*b*) and Soulé et al. (1979), Case (1975) found that area dependent "relaxation times" were greater for lizard faunas than those calculated for bird and mammal faunas (Diamond, 1972*b*). Additional evidence was obtained by comparing relaxation parameters of these same temperate lizard faunas to those of tropical bird faunas on land-bridge islands (Wilcox, unpublished). The results shown in Figure 8 demonstrate an interesting concordance between the change in collapse rate with area, but an almost two orders of magnitude difference in the actual area associated with a given rate of collapse. This follows from the expectation that extinction rates are inversely proportional to isolate size in general, but that ectotherms have lower extinction rates per number of

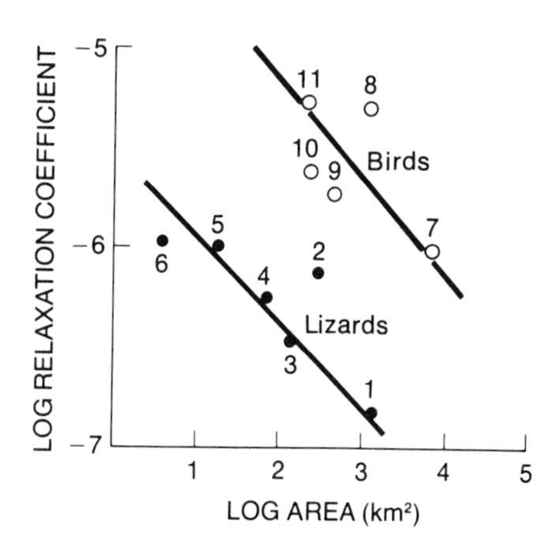

FIGURE 8. Comparison of relaxation parameters as a function of island area for lizard and bird faunas. The curves demonstrate the dependence of extinction rates on area, also the relatively lower collapse rates for lizards. Relaxation parameters are calculated as in Terborgh (1974*a*), except isolation times that are estimated as in Wilcox (1978). Lizard faunas: (1) Tiburon, (2) Cedros, (3) San José, (4) Espiritu Santo, (5) San Marcos, (6) Coronados (data from Soulé and Sloan, 1966). Bird faunas: (7) Trinidad, (8) Margarita, (9) Coiba, (10) Tobago, (11) Rey (data from Terborgh, 1974*a*).

species on an isolate. More convincing, however, is a comparison of different faunas on the same islands in the Gulf of California (Wilcox, in preparation). Reptile faunas of the land-bridge islands appear extremely supersaturated compared to their deep-water counterparts (Soulé and Sloan, 1966; Case, 1975; Wilcox, 1978), while the mammal faunas also appear supersaturated, but less so. Finally, the bird faunas show no evidence of supersaturation, presumably having already collapsed to their equilibria in the 10,000 or so years since the islands were formed (M. Cody, personal communication).

A second major consequence of heterogeneity in colonization and extinction rates among the vertebrate taxa are differences in the relative number of species maintained at equilibrium. Because of a combination of low colonization and high extinction rates, nonflying mammals will be depauperate in relation to other vertebrates in isolates. Indeed, the absence or small numbers of mammal species on most islands without a history of recent land-bridge connections is well known. The great dispersal advantage held by the birds and bats, however, may more than compensate for their high extinction rates in comparison with the reptiles and amphibians, particularly where dispersal barriers are barely surmountable by non-flying organisms. At equilibrium, therefore, bird and bat faunas will be relatively richer than those of reptiles and amphibians in extreme isolates.

These predicted differences in the relative number of species maintained at equilibrium are reflected in species-area relations. For example, it was mentioned earlier that the more highly vagile species among birds produced lower z-values. A similar effect is seen by comparing the flying and nonflying vertebrate taxa of the West Indies. Figure 9 shows that the nonflying taxa are more severely affected by decreased area, that is, they have higher z-values. The birds and bats share the lowest z-value of 0.24, while the highest z-value, 0.48 for the nonflying mammals, follows from their double disadvantage of both poorer dispersal along with lower population densities. Brown (1971) also found a high z-value (0.43) for nonflying mammal faunas which he suggested resulted from a nonequilibrium condition where extinction but no colonization had occurred since the isolates formed. The reptiles and amphibians of the West Indies also produce a z-value (0.38) which is higher than expected for most isolates. This is more likely a consequence of a high rate of *in situ* speciation events on the larger islands, particularly Cuba and Hispaniola.

In general, nonflying mammals, because of poor dispersal, high metabolic rates and large body size are the most vulnerable to extinction due to habitat loss and insularization relative to other vertebrate taxa. Since extinction is the major component of species loss throughout most of the collapse process, reptiles and amphibians will, in time, fare best among the vertebrates—particularly under conditions of extreme habitat loss. Only those taxa that are capable dispersers, such as some birds and bats, will persist indefinitely without careful monitoring and intense management.

FUTURE RESEARCH AND APPLICATION

Applied insular ecology is still very much in its infancy. Although nearly two decades have past since Preston (1962) alluded to the conservation implications of island theory, its potential for application has only recently become generally known—and even then only among limited quarters. Research in this area, with few exceptions, has been basic, not problem-oriented or applied. However, future research must address specific practical problems if insular ecology is to contribute further to conservation.

From a practical viewpoint, there are two primary application approaches of insular ecology. The first involves insular dynamics within a reserve, or "patch dynamics." In a timely and provocative article, Pickett and Thompson (1978) have focused attention on the conservation implications of colonization-extinction dynamics internal to an isolate. They point out that since few external colonization sources are likely to exist for any future reserves, conservation biologists should consider colonization and extinction of species in habitat patches within reserves. With knowledge of species-specific colonization and extinction rates, in combination with information on the spatial and temporal distribution of habitat patches, "minimum dynamic areas" could be defined. This is the

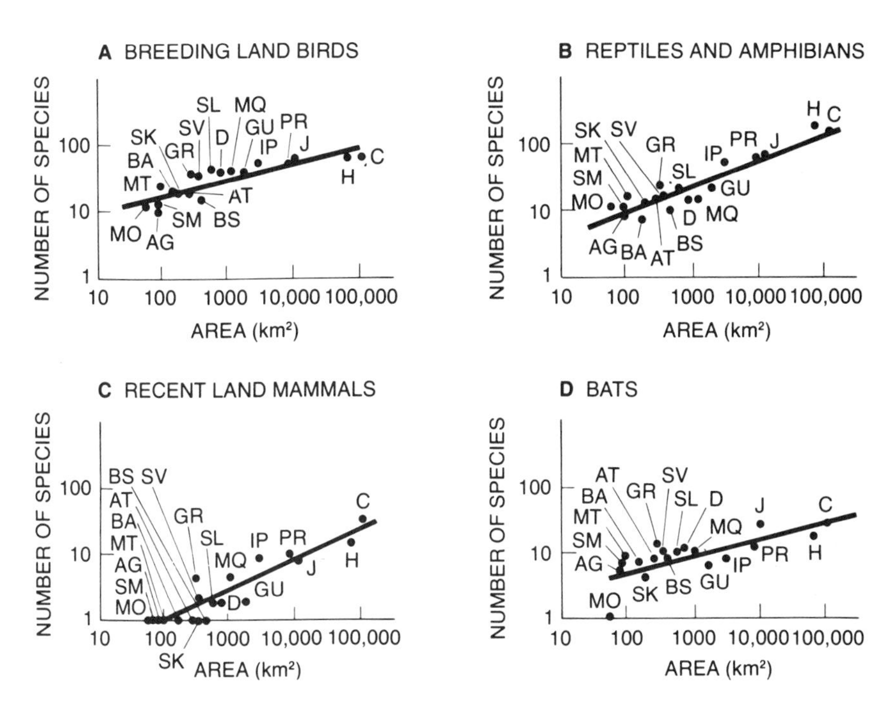

FIGURE 9. Species-area relations of four major vertebrate taxa on the same West Indian islands. The relative differences in the observed z-values follow from taxon-specific colonization and extinction rates. They suggest that consequences of habitat loss and insularization depend on metabolic rate, body size and dispersal ability. Species-area curves fitted by linear regression are as follows: (a) $S = 6.1A^{.24}$, (b) $S = 1.6A^{.38}$, (c) $S = 0.1A^{.48}$, (d) $S = 1.6A^{.24}$. (C) Cuba, (H) Hispaniola, (J) Jamaica, (PR) Puerto Rico, (IP) Isle of Pines, (GU) Guadeloupe, (MQ) Martinique, (D) Dominica, (SL) St. Lucia, (SV) St. Vincent, (BS) Barbados, (GR) Grenada, (AT) Antigua, (SK) St. Kitts, (BA) Barbuda, (MT) Montserrat, (AG) Anguilla, (SM) St. Martin, (MO) Mona. Recent Land Mammals include living and recently extinct native mammals. Many of the present mammal faunas include exotic species which have replaced native species. (Bird data from Bond, 1956, 1961 as compiled in Lack, 1976; reptile and amphibian data from Swartz and Thomas, 1975; mammal data from Varona, 1964)

size of a reserve necessary to insure the occurrence of a sufficient number of habitat patches occupied by a species, such that the probability of simultaneous extinction of the species in all patches is diminishingly small.

This unquestionably should be the approach used where minimum dynamic areas will be of realistic dimensions, given inevitable conflicting land use options. For example, communities of small vertebrates or arthropods adapted to patches of successional habitat would probably require minimum areas no larger than moderate sized national parks. For many species, however, particularly the larger vertebrates, minimum dynamic areas will be considered excessive given the myopic economic demands of most societies. Under these circumstances, the second approach will be required. Multiple reserves or fragments of undisturbed habitat will have to be manipulated as "artificial archipelagos" in order to minimize the probability of simultaneous extinction in each isolate. The second approach will, of course, not be successful in preserving those species incapable of inter-isolate dispersal, or those that cannot be transported successfully or introduced by man. It should be considered a last resort.

Detailed autecological studies are required to determine minimum dynamic areas and optimal design of artificial archipelagos. Here, biogeography will be most useful. The geographical distribution of a species reveals much about its ecological requirements. Diamond's (1976*a*) "incidence functions" are an efficient biogeographic method of defining the survival requirements of species among isolates. The studies of Moore and Hooper (1975), Forman et al. (1976) and those described by Terborgh (Chapter 7) on bird species distributions among disjunct habitat fragments should serve as models for determining preservation requirements of specific insular communities.

Non-avian species have received far too little attention in insular ecology. Perhaps the ease of study and exuberance of ornithologists have contributed to a disproportionate share of attention being focused on birds. Nevertheless, studies of a wider range of taxa are needed, both to test current theory and provide a basis for intelligent conservation decisions.

Of all the biological approaches to conservation, insular ecology is probably of the most immediate importance. The extent to which its lessons are heeded and its successful applications developed in the next few years will play an important role in determining the ultimate fate of this planet's biological diversity.

SUMMARY

The relevance of insular ecology to conservation stems from the similarities between natural isolates, such as islands, and the disjunct fragments of natural habitat resulting from expanding human exploitation of the earth's land surface. Habitat fragmentation can be dissected into two components: habitat loss and insularization. Each contributes to a reduction in the number of species supported, although different mechanisms are involved.

Studies suggest that habitat loss contributes in two ways: first, by excluding a portion of a fauna, particularly the rare or patchily distributed species; second, by increasing the extinction rate of the remaining species as a result of lower population sizes. Habitat insularization also contributes in two ways: first, by extinguishing species "protected" within an area through the removal of required resources outside the area; second, by reducing accessibility for, and sources of, colonists necessary to offset extinction events.

If a reserve were observed from its inception, three classes of these effects could be discriminated on the basis of time-scale. First, the decision to set aside any single area for preservation immediately excludes a number of species characteristic of the greater region as a whole. The proportion excluded is described by the species-area relation for samples. This *sample effect* amounts to the exclusion of about 30 percent of the regional fauna for each 10-fold decrement in area. The sample effect can be minimized by a system of strategically placed smaller reserves. However, other consequences must be heeded.

Short-term insularization effects will closely track encroachment of habitat contiguous to reserves resulting in the loss of self-sufficiency for species dependent on resources outside a reserve. Such species that might be mistakenly thought to be protected by their mere presence in a reserve, include non-breeding transients and populations that are breeding but marginally sustained by dispersal (rescue effect).

Long-term insularization effects describe the consequences of colonization-extinction disequilibria and climatic changes. Extinction is inevitable for any population, though the probability rises sharply with decreasing population size, or reserve area. The decreased colonization rate imposed by artifical dispersal barriers and fewer, more distant colonization sources will add to the collapse towards a new equilibrium with fewer species. Shifting climatic zones associated with the glacial-interglacial cycle will contribute further to extinctions unless reserves are large enough to include refugia.

Using land-bridge islands as models of faunal collapse suggests that large mammal faunas even in reserves of several thousand km^2 will begin to lose a very measurable number of species almost immediately. These losses will amount to as much as half of the initial reserve census of species in several hundred years and three-quarters in several thousand years. Even these estimates may be optimistic since they assume that the reserves will be left to themselves and not subject to poaching and other types of incursion or encroachment.

The rate of faunal collapse and the eventual number of species maintained at equilibrium in a reserve will depend not only on size and insularity of reserves, but also on taxonomic differences. Among the vertebrates, for example, mammals (excluding bats) have combined poor dispersal and high metabolic requirements (lower population densities on average). As a result, they are most vulnerable to extinction, manifesting the highest rates of faunal collapse and fewest numbers of species at equilibrium. Birds and bats, however, because of lower metabolic requirements and superior dispersal ability, are slightly less vulnerable to extinction, manifesting lower collapse rates and greater numbers of species at equilibrium. Reptiles and amphibians generally have much lower metabolic requirements than birds and mammals. As a result, they are, by far, the least vulnerable to extinction, and manifest the lowest collapse rates. In some natural isolates, reptiles and amphibians may maintain equilibrium faunas with nearly as many species as birds or bats. Under the conditions of extreme insularity often produced when humans impose dispersal barriers, however, nonflying vertebrate faunas in general will ultimately be thoroughly extirpated (barring human intervention).

Two primary approaches in insular ecology are suggested for future research and application. First, "minimum dynamic areas" should be defined, whose colonization-extinction equilibria are maintained internally. Second, since such area requirements will be excessive for many vertebrate communities, multiple disjunct reserves should be set aside and maintained as artificial archipelagos. For both approaches more autecological studies in insular systems are needed, especially of non-avian species.

The manageability of future reserves as well as the ultimate fate of biological diversity will depend to no small degree on the extent to which insular ecology is successfully applied in the next few years.

SUGGESTED READINGS

Diamond, J.M., 1975, Assembly of species communities, in *Ecology and Evolution of Communities,* M.L. Cody and J.M. Diamond (eds.), Harvard University Press, Cambridge, Massachusetts, pp. 342–444. A unique and important study of island bird faunas of the Pacific that aims to elucidate the determinants of species' distributions and composition. Introduces "incidence functions" and is the first comprehensive theoretical treatment of an area which requires a thorough understanding for the formulation of sound conservation strategies, particularly concerning natural or artificial archipelagos.

Hooper, M.D., 1971, The size and surroundings of nature reserves, in *The Scientific Management of Animal and Plant Communities for Conservation,* E. Duffey and A.S. Watt (eds.), Blackwell Scientific Publications Ltd., London, pp. 555–561. An excellent, not excessively theoretical, discussion of some of the topics central to this chapter.

MacArthur, R.H. and E.O. Wilson, 1967, *The Theory of Island Biogeography,* Princeton University Press, Princeton, New Jersey. A pioneering study of the ecology and evolution of island communities that provides the theoretical framework for much of insular ecology.

May, R.M., 1975, Patterns of species abundance and diversity, in *Ecology and Evolution of Communities,* M.L. Cody and J.M. Diamond (eds.), Harvard University Press, Cambridge, Massachusetts, pp. 81–120. Discusses the theoretical basis of species abundance distributions and implications to the species-area relationship.

Moore, N.W. and M.D. Hooper, 1975, On the number of bird species in British woods, *Biol. Conserv.,* 8, 239–250. A good example of a biogeographic study of manmade isolates, with conservation implications.

Simberloff, D.S. and L.G. Abele, 1975, Island biogeography theory and conservation practice, *Science,* 191, 285–286. This paper and a series of rebuttals entitled "Island biogeography and conservation: strategy and limitations" (J.M. Diamond, *Science,* 193, 1027–1029; J.W. Terborgh, *Science,* 193, 1029–1030; R.F. Whitcomb et al., *Science,* 193, 1030–1032; D.S. Simberloff and L.G. Abele, *Science,* 193, 1032) should be read together.

Soulé, M.E., B.A. Wilcox and C. Holtby, 1979, Benign neglect: a model of faunal collapse in the game reserves of East Africa, *Biol. Conserv.,* 15, 259–272. Describes, hypothetically, the fate of large mammal species where there is a lack of conservation strategy which considers insular theory.

CHAPTER 7

SOME CAUSES OF EXTINCTION

John Terborgh and Blair Winter

If we are to be successful in preserving examples of our biological heritage for the appreciation and enlightenment of future generations, it is essential that we learn as much as we can about the causes of extinction, particularly as they pertain to fragmented remnants of the natural landscape.

Some of the most instructive lessons can be derived from the study of land-bridge islands and habitat islands. Land-bridge islands are true islands which possess the distinction of having once been connected to the adjacent mainland. Many of them lie on continental shelves, having been isolated by rising sea levels at the close of the Pleistocene about 10,000 years ago. Habitat islands, as the name implies, are disjunct patches of one kind of vegetation surrounded by vegetation of a different kind. The patches may be natural ones, such as bogs or mountaintop forests, or they may be man made, as when a woodlot is left amidst fields and pastures. Many parks and wildlife preserves are destined to become habitat islands as we continue to intensify our use of the earth's land resources.

The faunas of land-bridge and habitat islands generally contain fewer species than those of the adjacent mainland or of larger tracts of equivalent habitat. One can readily ascertain which species are missing from a particular land-bridge or habitat island by comparing its fauna with that of an equivalent segment of the mainland, or a control block of habitat. Here we follow the now well-established practice of assuming that the missing species have gone extinct during the interval of isolation (J. H. Brown, 1971; Diamond, 1972; Willis, in press; Terborgh, 1974*a;* Case, 1975; Wilcox, 1978; Soulé et al., 1979). If, in examining a number of such island-mainland comparisons, we find that the missing species shared certain common characteristics, it might then be possible, by extension, to identify in advance the most vulnerable species contained within future parks and reserves. This is our objective.

We shall begin by reviewing the principal generalities that have emerged from previous studies of land-bridge and habitat islands. Then we shall present some new results which enable us to answer a number of further questions about the biological traits that contribute to susceptibility or resistance to extinction. And finally, we will draw some inferences about why extinctions result from fragmentation of the natural landscape.

Area Dependence

The most far-reaching result of past studies is the demonstration that extinction is strongly area dependent, as anticipated by island biogeographic theory (MacArthur and Wilson, 1967). This is implied by the unusually steep slopes of species-area regressions of land-bridge islands, a result that can be explained as follows (J. H. Brown, 1971; Diamond, 1972; Terborgh, 1974*a*). At the time of their isolation, land-bridge islands contain a full complement of mainland species. The total number of species involved is generally far greater than would be expected on an oceanic island of the same size. This implies that a land-bridge island initially holds far more species than can be maintained by colonization at equilibrium. The "extra" species gradually drop out, until after a sufficient time, which may be measured in tens or hundreds or thousands of years (Diamond, 1972), the island approaches an equilibrium condition. How fast an island loses the initial excess and achieves equilibrium is a function of its size; small land-bridge islands (<100 km^2) may be indistinguishable from oceanic islands after a few thousand years, while large ones ($>10{,}000$ km^2) continue to carry a considerable species excess even after 10,000 years (Diamond, 1972; Terborgh, 1974*a;* Wilcox, 1978). Because of this, the increase of species number with area is far greater for land-bridge islands than for oceanic islands. Since habitat islands and dismembered reserves are analogous to land-bridge islands, the numbers of species they can be expected to hold after a period of isolation will be strongly area dependent. This conclusion argues for preserving habitats in large blocks.

Though the steep area dependence of land-bridge island faunas is inherently transitory (it disappears when the largest island attains equilibrium), it holds for both large areas ($>1{,}000$ km^2) over the post-Pleistocene interval (around 10,000 years), and for small areas (1 to 100 km^2) over relatively short intervals (for example, 50 to 100 years). Both of these situations are relevant to the design of nature preserves, and will be considered later.

Rates of Extinction

The problem of estimating extinction rates on land-bridge islands has been studied by Diamond (1972), Terborgh (1975) and Soulé et al. (1979). The models tested so far have related the extinction rate to the number of species momentarily present. A better fit is obtained when the pace of extinction is held proportional to the square or cube of the number of species present than when a simple linear proportionality is used. This suggests that extinction probabilities vary considerably from one species to another, possibly because of competitive interactions or other reasons to be discussed later.

Another important finding is that the greater the excess of species over the expected equilibrium number, the more extinctions there will be per unit time, both absolutely and relatively. This implies again that large reserves will be most effective because they will initially contain smaller excesses of species and can be expected to sustain lower rates of loss.

Patterns of Extinction

Here we come to a crucial issue, for if it were possible to identify extinction prone species in advance, it might be practical to engage in certain types of preventative management. The only attempts to do this so far are those of Brown (1971), with temperate mammals and Willis (1974; in press), with tropical and subtropical birds, respectively. The two concur in their judgements that large size and habitat specialization are both detrimental to long-term survival. Moreover, Willis (in press) identified several avian guilds that appear especially extinction prone. Though both recognize the jeopardy contained in small population size, neither has attempted to discover whether the added risks associated with large body size and habitat specialization are anything more than consequences of low population levels. In the following sections this question will be examined more closely.

PATTERNS OF EXTINCTION IN A LONG TERM/LARGE SCALE SETTING

Procedure

Previous analyses of extinction on land-bridge islands have purposely employed sets of islands that were homogeneous with respect to their faunas and geography. Here we do quite the opposite on the premise that, if there are universal generalities in the patterning of extinctions, they should be largely independent of taxonomy and geography.

To examine this proposition we used the land bird faunas of five major land-bridge islands representing four biogeographic regions (Table I). Four of the islands are at tropical or near tropical latitudes, and one is temperate (Tasmania). In size, they range from ones that are smaller than a good many existing parks (<5,000 km^2) to ones that are considerably larger (>60,000 km^2).

TABLE I. Present, past and presumptively extinct land bird faunas of five major land-bridge islands.

Island	Area (km^2)	Postulated number of species at isolation	Present number of species	Number of presumed extinctions	Percent extinct
Fernando Po	2,036	360	128	232	64
Trinidad	4,834	350	220	130	37
Hainan	33,710	198	123	75	38
Ceylon	65,688	239	171	68	28
Tasmania	67,978	180	88	92	51
Total				597	

Lists of breeding land birds for the islands and equivalent mainland control areas were compiled from the following sources: Trinidad and Venezuela: French, 1973; Meyer de Schaunesee and Phelps, 1978. Fernando Po and Cameroon: Amadon, 1953; Hall and Moreau, 1970; Searle, 1950; 1954; 1965; Searle and Morel, 1977. Ceylon and India: Ali and Ripley, 1968; Henry, 1971. Hainan and China: Cheng, 1976. Tasmania and Australia: Ridpath and Moreau, 1966; Sclater, 1970; 1974.

Lists of the present and presumptive past land bird faunas of the islands were compiled from recent field guides and handbooks (see footnote to Table I). Migratory and vagrant species were not included, nor were those of unconfirmed breeding status or aquatic forms such as ducks, herons and rails. Presumptive faunas at the time of separation (about 10,000 years ago) were assumed to be identical to the present faunas of mainland control areas. These were matched to their respective islands as closely as possible with regard to area, rainfall, habitat distribution and topography. The differences between the mainland and island lists were then assumed to represent extinctions which had occurred over the post-Pleistocene interval.

In following this procedure, we encountered two kinds of problematical species: insular endemics and nonendemics which are absent from the designated mainland control areas. To avoid having to make a series of arbitrary judgements, we deleted both categories of species from the comparisons. In all cases, such species constituted less than five percent of the insular totals except in Ceylon where they made up nine percent. Deleting the insular endemics results in an overestimation of extinctions because some of the insulars are derived from closely allied mainland forms which are counted as having died out on the islands. Range shifts on the part of mainland populations present another procedural quandary. Subsequent to insularization some species may have deserted the mainland control areas while others may have invaded. The latter will be scored incorrectly as ones which have expired on the corresponding islands. The method also assumes that insular extinctions are terminal, that is, that no recolonizations have occurred during the term of isolation, even though recent evidence indicates that such recolonizations are frequent on small islands at equilibrium (Diamond and May, 1976*a*). In spite of these unavoidable flaws in the procedure, we pursue the analysis in the hope that the net errors are tolerably small. Taken together the five islands provide a sample of 597 presumptive extinctions, quite enough for the resolution of any major patterns.

Quantitative Patterns

The results agree with those of previous studies in indicating that extinction is inversely area dependent. The quantitative trend is obscured by a considerable scatter in the values, the causes of which are mostly unknown. Fernando Po has suffered much heavier apparent losses than Trinidad even though the size difference between them is relatively small. Trinidad's greater proximity to the mainland and the existence of a series of intervening stepping stone islands may have facilitated an appreciable number of recolonizations. It is also possible that Trinidad may have been connected to the mainland more recently than Fernando Po (Wilcox, 1978). At the opposite end of the size scale, Tasmania shows an unexpectedly high incidence of extinctions. This could be a consequence of Tasmania's temperate location, or it could be an artifact, if Tasmania lacks certain vegetation types that are represented in southern Victoria. There is nothing to be gained in belaboring these questions because the available information is insufficient to resolve them.

Qualitative Patterns

To screen data for qualitative patterns we prepared a family by family itemization of the numbers of surviving and extinct species for all the islands. The numbers for each island were then compared with the overall survival (or extinction) rates for that island. Our initial expectation was that certain families would consistently exhibit outstanding survival across all the islands, while others would prove to be poor survivors. Such outcomes were highly exceptional. Most families showed mixed or near average performance. Only a few ran consistently above or below the mean.

Pigeons, cuckoos and swifts showed outstanding survival ability, while kingfishers, thrushes (Turdidae) and sylviid warblers were somewhat less decisively above average (Table II). The first five of these families are composed of strong flyers, and are disproportionately represented in the faunas of oceanic islands. Their good showing could simply be due to frequent recolonizations following extinctions or it could be real (Brown and Kodric-Brown, 1977). The presence of these families on oceanic islands carries a dispersal prerequisite, but exceptional survivorship can also be implied. In their trophic status, foraging tactics and nesting habits these families show no obvious common denominator.

TABLE II. Taxa showing unusual resistance or susceptibility to extinction on land-bridge islands.

Apparently resistant	Apparently susceptible
Columbidae (pigeons)	Falconidae (falcons)
Cuculidae (cuckoos)	Phasianidae (pheasants)
Apodidae (swifts)	Picidae (woodpeckers)
Alcedinidae (kingfishers)	Timaliidae (babblers)
Turdidae (thrushes)	Tinamidae (tinamous)*
Sylviidae (O.W. warblers)	Cracidae (guans)*
	Bucerotidae (hornbills)*
	Ramphastidae (toucans)*

* Uncertain because of small sample size.

Several families appear to be particularly extinction prone. These are the falcons, pheasants, woodpeckers and babblers. Tinamous, guans, hornbills and toucans may also be vulnerable, but they occur in only one or two of the comparisons, so no clear trend can be claimed. The latter families, however, are all composed of large bodied forest frugivores, suggesting that these may be especially susceptible to local extinction. The same result emerges from the short-term, small-scale study that will be reviewed in the next section.

Extinction vs. Body Size

J. H. Brown (1971), Willis (1974) and Wilcox (Chapter 6) have all remarked that large species were among the first to drop out of land-bridge island faunas. Among the apparently extinction prone families listed above, all but one (the babblers) are of greater than mean body size. On the other hand, a number of families of relatively large birds display normal to good survival. These are the hawks, vultures, pigeons, parrots, owls, nightjars, drongos, crows and cuckoo-shrikes. Comparisons between families, in any case, are hardly appropriate, as many factors may be involved. A better test would be to compare the incidence of extinction in large vs. small members of the same family. To do this we drew families from the comparisons in which the mainland list included at least six (and usually more than 10) species. Then, we scored the species which had presumptively gone extinct according to whether they were larger or smaller than the median size for the mainland species in the family. Summing all of these, we found that out of 179 presumptively extinct species, 85 were smaller and 94 larger than the median size for their families. Thus there does appear to be a slight trend in the expected direction, but it is too small to be statistically significant. We shall show later why this is so.

Extinction vs. Competition

Another question which can be answered with the data is whether a species' survival chances are lessened by the presence of close competitors in the same fauna. We asked simply whether species that are alone in their genera (in the mainland control faunas) survive with a higher frequency than those which co-occur with one or more congeners. Using the Venezuela-Trinidad comparison, we found no discernible difference; 63 percent of the species that were alone in their genera survived and 62 percent of the species which at least initially shared the island with congeners survived.

So far we have failed to uncover any consistencies that could serve to predict extinctions with a useful degree of precision. Either there are no such patterns (in other words, extinction is random) or they reside in biological properties that we have not yet examined.

PATTERNS OF EXTINCTION IN A SHORT TERM/SMALL SCALE SETTING

More immediately germane to the design of parks are short term studies of land-bridge islands, or, better yet, remnant islands of natural habitat that have been left intact within a sea of pasture and hedgerows or otherwise modified vegetation. Breeding bird censuses of isolated woodlots in North America clearly reveal their degenerate state, not only with respect to the total numbers of species present, but more markedly with respect to the altered species composition of their communities (Forman et al., 1976; Galli et al., 1976; MacClintock et al., 1977).

Even more dramatic changes have recently been documented by Willis (in press) in a largely nonmigratory subtropical fauna in the state of Sao Paulo, Brazil. Unbroken forest once covered the entire region, but today the landscape is dominated by cropland, cattle pasture and coffee plantations. Only a few small and isolated tracts of forest habitat remain. Willis provides a carefully supported account of the breeding birds of three of the remnant forest tracts. We are most grateful for permission to quote his valuable results here.

With areas of 21, 250 and 1,400 hectares, the tracts cover a size range of nearly two orders of magnitude. It is not known when they became detached from the formerly continuous woodlands of the Sao Paulo plateau, as deforestation of the region has been progressing over the last 150 years. Elapsed times since isolation are all probably a few decades. Faunal depletion may have begun before any of the tracts reached their present dimensions because a number of large and conspicuous birds disappeared long ago from the entire region (for example curassows, large eagles, macaws).

Quantitative Patterns

The effect of habitat fragmentation on this subtropical avifauna is truly drastic (Table III; Figure 1). Even the 1,400 hectare "control" forest has lost an appreciable number of species, as judged from the records of early collectors, while the communities of the smaller plots bear little resemblance to their former selves. The steepness of the relationship between extinction and area once again proclaims the quintessential role of space in furthering the long term survival of isolated populations.

TABLE III. Breeding birds of three remnant forests in Sao Paulo State, Brazil.

		Number of breeding species			
Forest	Area (hectare)	Formerly	Now	Number of presumed extinctions	Percent extinct
Barreiro Rico	1,400	203	175	28	14
Santa Genebra	250	203	119	84	41
Uni-camp	21	203	76	127	62

(From Willis, in press)

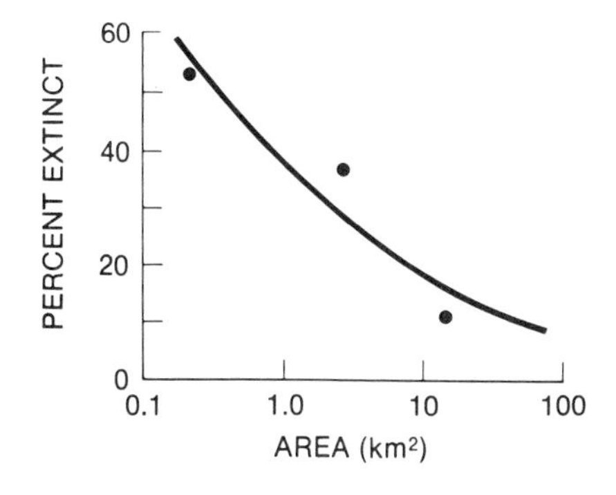

FIGURE 1. Percent of apparent extinctions during the last century based on presumed initial bird faunas for three Brazilian woodlots. (Data from Willis, in press)

Qualitative Patterns

Willis tabulated persistence vs. disappearance by guilds, and concluded that certain guilds are especially extinction prone, notably large raptors, large canopy frugivores (parrots, toucans, cotingids), large terrestrial or near-terrestrial insectivores and small insectivores of bamboo thickets and forest tangles. Trophic status and nesting habits, as in the earlier long-term, large scale analysis, do not correlate in any obvious way with disappearance.

There is at one level, however, a conspicuous orderliness in the results. It is perceived in the fact that the remnant faunas form an almost perfect nested set. That is, virtually all the species that have persisted in the smallest forest island are also present in the medium sized one, a pattern that continues through the largest. This could happen only if species were dropping out in a consistent order. In the Sao Paulo woodlots at least, extinction is not a random process. Definite rules are being followed; it remains for us to discover what they are. Further scrutiny of Willis' data suggests that the key factor is population density.

Extinction vs. Initial Rarity

Willis' manuscript lists the numbers of individuals of each species detected in each forest island per 100 hours of observation. Our analysis of these data rests on two assumptions: (1) the current abundances of species in the control (1,400 hectare) plot reflect initial abundances at the time of fragmentation and (2) the initial abundance of each species was similar in the three tracts, notwithstanding the differences in soils, drainage and vegetation mentioned by Willis. Although both assumptions are crude approximations at best and, if not met, should obscure the results, the trend is abundantly clear (Figure 2).

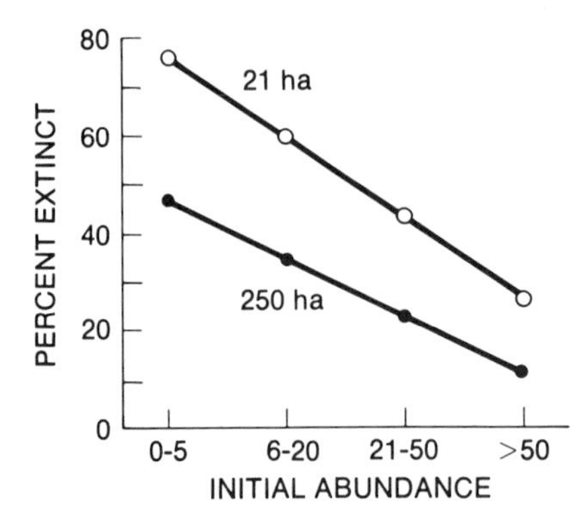

FIGURE 2. Percent extinction among birds having different initial levels of abundance in two Brazilian woodlots. Abundance values are numbers of individuals detected per 100 observer hours. (Data from Willis, in press)

Now that rarity emerges as a reasonably good predictor of vulnerability to extinction, it would help us to know more about what kinds of species are rare. Everyday experience tells us that nearly all rare species conform to one of the following patterns: (1) species that are near the limits of their geographical ranges or habitat spectra, (2) species that are constitutively rare, that is, that exist at low population densities wherever they occur and (3) species that specialize on certain types of patchily distributed habitats. Familiar avian examples of these types of rarity are, respectively: (1) the Common Black Hawk and Coppery-tailed Trogon in the southwestern United States, (2) California Condor and Ivory-billed Woodpecker and (3) Henslow's Sparrow and Kirtland's Warbler in the eastern United States.

It should now be apparent why there was such a poor correlation between large body size and disappearance from the faunas of land-bridge islands. Not all large birds are rare, and by no means are all rare birds large. Indeed, of the three types of rarity, only constitutive rarity is commonly associated with large body size. Habitat specialization, to the contrary, seems most frequent in small species.

Many of the birds which have disappeared from the Sao Paulo woodlots show type 2 or type 3 rarity. Virtually all of the species which were missing in the 1,400 hectare control plot exist at inherently low population densities and are large (for example, several eagles, macaws, large parrots, toucans, tinamous, a wood-quail, a pigeon and a fruit crow). (The fates of many of these were undoubtedly hastened by hunting pressure.) Habitat specialists (birds of treefall openings and tangles in the forest interior) had survived in the large plot but had all but vanished (eight out of nine) from the small one. Their places in the small plot had apparently been taken by widespread species characteristic of edge and second growth.

The only group of species which went extinct with high frequency, and which did not conform to the pattern of low initial population density, was comprised of forms that subsist on fruit and nectar. These are resources which can fail altogether following unusual climatic perturbations (frost, drought). In such instances dependent populations may be obliged to emigrate to avoid heavy mortality. In the tropics, many hummingbirds are known, even in normal years, to vary drastically in abundance at a given site (Feinsinger, 1976), while parrots and macaws may commonly be seen flying from horizon to horizon between their feeding and roosting areas. For such species, even 1,400 hectares does not provide an adequate minimum amount of living space.

THE CAUSES OF EXTINCTION

While the proximal cause of a species' extinction can be described as the failure of recruitment to offset mortality (for example, Mertz, 1971), the ultimate cause is generally the advent of a stressful change or a novel additional element in its environment. While many ultimate causes have played known or suspected roles in extinction, only a minority of these are likely to be important in undisturbed land-bridge or habitat islands (Table IV). These deserve a more detailed consideration.

TABLE IV. Causes of extinction in habitat islands.

1. Primary: stochastic extinction consequent to reductions in population size (mainly large or nomadic species, habitat specialists).
   a. Fragmentation.
   b. Negative population growth.
   c. Species dependent on irregular resources.
2. Secondary: ecological imbalances resulting from 1a and b above.
   a. Alteration of interaction webs, especially predator-prey and pollinator-disperser relations.
   b. Secondary or tertiary loss of food plants, prey species, mutualisms and so on, due to combined effects of all of the above.

Fragmentation: Its Effects on Population Size and Dispersal

While attempts to model extinction mathematically have nearly always focused on stochastic population death (MacArthur and Wilson, 1967; Leigh, 1975), it is, for our purposes, the least interesting mechanism of extinction. This is because the final sputter of almost any collapsing population can be viewed as a stochastic event. A population becomes vulnerable to random fluctuations in its environment only when it has already reached dangerously low levels. The same can be said of genetic deterioration via inbreeding. Our goal as conservationists should be to prevent such dangerous situations from arising; our goal as scientists should be to understand how they arise. The following offers some speculative answers to this question.

A population may be immediately threatened with stochastic extinction if fragmented into unnaturally small population units. This is a trivial case but the minimum area necessary to secure a viable population of certain species may not be trivial. For example, the entire 1,500,000 hectare Manu National Park in Peru may contain only eight or 10 family groups of the giant otter *(Pteroneura brasiliensis)*. For most species this would be considered a precarious situation.

A healthy species is normally at equilibrium with its environment, that is, overall its numbers are neither growing nor declining. At a local level, however, single demes or clusters of demes may deviate from the equilibrium condition. "Good" habitat will produce excess individuals that will be obliged to defer reproduction until vacancies open up in the local territorial mosaic, or to disperse to "poor" habitat to try their luck at beating the odds in the "reproductive sweepstakes."

A population that becomes isolated through habitat fragmentation may be located in a "poor" region within which its long term growth rate is negative. This being so, its numbers will gradually dwindle until only the best territories are occupied. A new equilibrium may or may not be established, but even if it is, the mean number of individuals may be dangerously small—within the range in which stochastic extinction is a likelihood. Although as yet purely hypothetical, this mechanism seems highly plausible and is especially relevant to fragmented populations. The type of population dynamics it anticipates should be looked for in future studies of habitat islands.

The vanished frugivores and nectarivores of the Sao Paulo woodlots are an exceptional case of sensitivity to fragmentation because of their dependence on irregular resources. Northern finches and boreal raptors are more familiar examples of species which are periodically obliged to emigrate from their normal haunts (Bock and Lepthien, 1976). Although such populations can legitimately be viewed as being subject to wide stochastic fluctuations, we treat them as a distinct category because their mean or initial numbers in a disjunct piece of habitat may appear adequately large. When essential food supplies fail, however, their continued survival depends on the availability of alternative resources in other, often distant, locations. Certain tropical mammals also show this type of behavior (for example, some fruit bats, nectar bats and forest pigs).

Secondary Extinctions Resulting from Ecological Imbalances

Herein lies a potentially important area of research that has scarcely been breached (Chapter 2). Paine (1966) and others have clearly demonstrated that the loss of a top predator in a rocky marine intertidal ecosystem can lead to a drastic simplification of the residual community. Yet we know next to nothing about what consequences follow the loss of top predators in terrestrial ecosystems.

At this point we would like to recount a set of causal observations which suggests, in a tentative way, that the presence of top predators may exert a stabilizing influence on the structure of tropical forest communities. The Smithsonian Institution maintains a biological research station on Barro Colorado Island, a former hilltop that became an island when the Chagres river was dammed in connection with the construction of the Panama Canal. Completely protected for more than 50 years, Barro Colorado has become a veritable zoo without cages. Medium and large mammals are unwary and remarkably abundant. These include the collared peccary, agouti, coati mundi, three-toed and two-toed sloths, tamandua, armadillo and howler monkey.

Terborgh has travelled extensively in the Neotropics, and nowhere, even in areas protected from hunting, has he found any of these mammals to be as abundant as on Barro Colorado. One could think of several possible reasons for this, but one stands out because its logic is so simple. In most places mammal populations are subject to varying degrees of predation pressure, from man, from natural predators or from both. But this is not so on Barro Colorado which is too small (17 km^2) to sustain large predators. The mammals mentioned above constitute the bulk of the prey of jaguars, pumas and harpy eagles, none of which have bred on Barro Colorado for decades. It seems reasonable to infer that large mammals are abundant on Barro Colorado, at least in part because they live in a predator free environment.

Since its isolation from the mainland, Barro Colorado has been losing species at a rate commensurate with those calculated for other land-bridge islands (Terborgh, 1974*a*). Birds, reptiles and mammals are included in the list of species that have disappeared. Because of Barro Colorado's status as a research preserve, the extinctions have been carefully documented, not just presumed (Willis, 1974). In spite of this, the underlying causes remain a mystery.

Some 15 to 18 forest-dwelling birds have vanished. Although these represent a wide range of families and ways of life, many of them share the trait of nesting on or near the ground (Willis, 1974). It could be that excessive densities of partially predatory terrestrial mammals, especially peccaries and coatis, were instrumental in the disappearance of these birds. Both of these animals opportunistically eat eggs and nestlings. In an environment where more than 75 percent of nesting attempts result in failure, an increased incidence of nest predation could easily lower an affected population's recruitment rate below the maintenance level. All this is merely speculation, but it has a persuasive ring.

Complex ecosystems, particularly tropical ones, are built upon elaborate webs of mutualism. Frugivorous birds and mammals depend on a seasonal succession of resources for the stability of their populations. Removal of one or more of the fruit species in the sequence (for example, via selective logging) could lead to a devastating gap in the annual food budget. Plants, in turn, are dependent on their animal pollinators and dispersers. The disappearance of an important frugivore through overhunting could be just as destabilizing to its mutualistically dependent food plants, though the reaction would necessarily be slower since it would be measured in the longer units of tree generations (Temple, 1977).

Insectivores are certainly more immune to the loss of individual resource species, though even insectivores evolve mutualistic dependencies. In the Neotropical forest there are birds which habitually profit from other animals, using them as beaters to flush out hidden insect prey. Some follow monkey troops, others follow peccary herds and still others, in the most obligate of all these commensualisms, follow army ants. The ant followers may not be able to survive in the absence of their beaters, for they have all but vanished from the two smaller Sao Paulo woodlots where the major army ant species are no longer found (Willis, in press).

SUMMARY

Comparisons of the faunas of land-bridge and habitat islands with control areas indicate that the long-term survival of isolated populations is strongly area dependent. Time and area play inverse roles in extinction; what happens over centuries in a large area occurs within decades in a small one. Extinction is impartial in choosing its victims; species of all sizes, trophic levels and taxonomic groups fall prey. Rarity proves to be the best index of vulnerability. Rare species include top predators, and frequently other large species, as well as habitat specialists and species near the limits of their ranges. Frugivores and nectarivores, though they may be common, are also vulnerable because they are occasionally obliged to emigrate temporarily during periods of resource scarcity.

Fragmentation of habitats appears to result in a deteriorative ecological chain reaction which begins with the stochastic loss of rare species, among which top predators have disproportionate importance because of their key roles in regulating prey populations. Their loss may trigger a cascade of unexpected secondary extinctions consequent to the disruption of evolved predator-prey relationships. The culmination of this process is the eventual attainment of a new equilibrium, one that is far less complex than the original one in its diversity of species and interactions.

If we are to prevent these reactions from running their course in our parks and reserves, we must first take pains to preserve complete ecosystems. But having done this, a policy of benign neglect will not be sufficient (Soulé et al., 1979). We will have to keep a watchful eye on the health of the community to be sure that dangerous imbalances do not develop. Corrective and preventative management, practices that today are in their infancy, will have to become the bywords of the conservation movement the world round.

SUGGESTED READINGS

This field is so new that there are no general references. However, the following supplementary readings are suggested.

Diamond, J. M., 1974, The island dilemma: lessons of modern biogeographic studies for the design of nature reserves, *Biol. Conserv.* 7, 129–146.

Terborgh, J., 1974, Preservation of natural diversity: the problem of extinction prone species, *Bioscience* 24, 153–169.

Willis, E. O., 1974, Populations and local extinctions of birds on Barro Colorado Island, Panama, *Ecol. Monogr.* 44, 153–169.

CHAPTER 8

EVOLUTIONARY CHANGE IN SMALL POPULATIONS

Ian Robert Franklin

The arguments in favor of conservation have usually been based on ecological, economic or even ethical grounds. There are, however, important genetic considerations. Any strategy for the conservation of a particular species should, in part, be determined by knowledge (or inference) about the genetic structure of that species. The most stringent requirements arise when we consider the future needs of plant and animal breeders. If we are to feed the human species and its domesticates, it is necessary that a diversity of species be maintained—each in sufficient numbers to assure that genetic variability is adequate for continuing selection. It is not easy to predict man's future needs in this area. There are many plants which have not been evaluated as sources of oil, food and fiber, and the possibility of using single cell cultures, including the tissue culture of animal cells, has only barely been examined. Nevertheless, seed banks have been established (Harlan, 1975) and there is limited storage for the semen of some domesticated animals, but these efforts are just a beginning. We have already lost potentially valuable commercial strains of plant and animal species; the wild relatives of cultivars, which are a source of disease resistance and other useful genetic variation, are in peril.

Important as these issues are, I do not wish to dwell on them, but rather to concentrate on the effect of small population size on evolutionary change and survival. The case for preservation of genetic resources has been presented in detail before (Frankel, 1974).

We have already seen (Chapter 6) that endangered species can be maintained in parks and wildlife sanctuaries in limited numbers, and that restriction of population size has in itself a number of genetic consequences. Perhaps the most important of these, certainly in the short-term, is inbreeding depression, which for many species places an immediate lower bound on the population size which is compatible with survival. However, even in a population which is large enough to escape serious inbreeding, there may be a gradual loss of genetic variability which limits future evolutionary change. I suggest in this chapter that in random mating populations, such as are found in most mammals and birds, inbreeding considerations alone require that population numbers should be not less than fifty individuals. In the long-term, genetic variability will be maintained only if population sizes are an order of magnitude higher. The latter argument is based on the assumption that: (1) continued, and often rapid, evolutionary change is necessary for survival and (2) response to natural selection is limited by small population size.

Before developing this argument in detail, we should ask if evolutionary change is what we want. Do we wish to conserve the elephant, or ensure the survival of its elephant-like descendants? This is an important issue. If we are concerned with preserving the precise phenotype of a species, rather than a phylogenetic line in which we allow continued evolutionary change, our strategies will be very different. In captively propagated populations, it is possible to maximize evolutionary change, and in some circumstances to almost halt it entirely.

In the following discussion, I will be concerned largely with random mating populations, and with phyletic evolution entirely. Speciation is discussed in Chapter 9.

ADAPTATION

Adaptation and Extinction

The historical record shows many examples of episodic extinction and rapid evolutionary change, the most recent of which began in the late Pleistocene with the disappearance of many species of large mammals. There is a controversy over the cause of the Pleistocene extinctions; some attribute them to climatic change, others to the rise of the ecological dominance of man (Chapter 16). There is no doubt, however, about the cause of the current crisis. In addition to direct competition and predation by man, there has been a massive shuffling of the world's biota. In Australia, for example, introduced placental mammals, ornamental plants and pasture species are in direct competition with the native flora and fauna. On top of all this, we have the well documented large-scale habitat destruction (see other chapters).

In response to all of this ecological disruption, we expect, and have already seen, a great burst of evolutionary change. In a matter of years, we have observed the emergence of industrial melanism, insects which are resistant to insecticides, rabbits immune to myxomatosis and grasses which can grow on the tailings of lead and copper mines. Some species will adapt readily to their new environments; others, for reasons which are largely unknown, will not and face extinction. The immediate cause of extinction may be a capricious event, such as fire or an epidemic, but a species which is vulnerable to these chance events is likely to be so because it has, in the past, adapted less well than its competitors to a changing environment. Ultimately, a population disappears if it cannot keep up with its physical or biotic environment. Adaptation is a crucial concept in any understanding of population survival.

Adaptation and Natural Selection

Adaptation, the acquisition or modification of traits which 'fit' the organism more perfectly to its environment, is a process known only by inference. The products of adaptation (for example, the structure of teeth for grazing or browsing, protective coloration and mimicry, migratory and territorial behavior, pollination and seed dispersal mechanisms) are recognized from the relationship between structure and function. Rarely, however, do we understand adaptation in terms of changes in gene frequency, since in general we are a long way from translating physiology and behavior into instantaneous birth and death rates for known genotypes.

Notwithstanding our ignorance, it is likely that the most important adaptive traits are continuously varying (quantitative) characteristics. Since adaptation is the result of natural selection (although not all natural selection is adaptive) we can draw on considerable experience with selection in laboratory and domesticated animals to make some inferences about population size and evolutionary change.

We can distinguish between three modes of selection for quantitative traits. These are: (a) stabilizing selection, or selection against extreme values; (b) directional selection, in which one extreme is at an advantage; and (c) disruptive selection, or selection for extreme values, and against intermediates.

There is no doubt that stabilizing selection is the predominant mode of selection in natural populations. For many traits the greatest fitness, as measured by survival or number of offspring, has been found to be very near the mean for the population. Examples are shell size in snails (Weldon, 1901), size of duck eggs (Rendel, 1943), clutch size in swifts (Lack, 1954) and adult height and birth weight in man (Cavalli-Sforza and Bodmer, 1971). Conversely, directional selection, while common in animal or plant breeding, is probably rare in nature. It is thought that the best adapted individuals "exhibit a harmonious combination of all characters" (Lerner, 1954), so that extreme individuals, which may be "fit" for one trait, will be unbalanced with respect to others. Consequently, fitness often declines under directional selection for a single trait.

The third mode, disruptive selection, is not well understood; its importance in natural populations is controversial. It is, perhaps, operative in patchy environments and may play a role in speciation.

Natural selection, then, appears to be a complex process in which a slowly changing species accompanies a slowly moving niche, "always slightly behind, slightly ill-adapted, eventually becoming extinct as it fails to keep up with the changing environment because it runs out of genetic variation on which natural selection can operate," (Lewontin, 1978). For any trait changing under selection, the rate of response is a function of the selection intensity and the heritability of the trait. Both of these quantities are affected by genetic drift. It is well known that populations which have been through a bottleneck or which have been maintained at a small population size do not show as great a response to artificial selection as do large populations.

GENETIC DRIFT AND EFFECTIVE POPULATION SIZE

In a finite population, the array of genotypes in any generation is formed by sampling gametes from the previous generation; virtually all of the genetic effects which arise in small populations are an unrepresentative consequence of sampling, a process known as genetic drift. In small populations, gene frequencies change from generation to generation, even in the absence of selection, mutation or migration. These random changes lead to a gradual increase in homozygosity, and, if they go unchecked, will ultimately lead to fixation at all loci and hence a complete loss of genetic variability. At the level of the phenotype, we can commonly observe three effects of genetic drift: (a) inbreeding depression, due either to loss of heterozygosity or to fixation of deleterious genes; (b) random change in the phenotype, especially changes in the means of quantitative characters; and (c) a decrease in genetic variance (and hence a lowering in heritability).

For a given population size, the rate of genetic drift varies depending on the mating structure and the distribution of offspring number (Chapter 12), so the theoretical consequences of drift are usually calculated for an idealized population in which each individual contributes gametes equally to a pool from which the next generation is formed.

Any population of N individuals will have a sampling variance in gene frequency which is equivalent to the sampling variance of an ideal population of N_e individuals. N_e is then the effective population size (more strictly called the variance effective number) of the population whose census number is N. This is an important concept. In much of the following discussion concerning inbreeding depression or loss of genetic variance the population size will be the effective rather than the census number.

Kimura and Crow (1963) present a comprehensive account of this theory, but we can illustrate the relationship between actual and effective population number with a few important examples.

1. *The effect of variance in progeny number.* Let the variance in progeny number be σ^2, and the census number N. Then

$$N_e = \frac{4N}{2 + \sigma^2}\text{, approximately.}$$

Suppose, for example, that the mean family size is two, and the variance is four. Then $N_e = 2/3\,N$. On the other hand if every mating pair contributes exactly two offspring to the next generation, that is, there is no variation in family size, then $N_e = 2N$. Effective number is decreased by increased variation in progeny number and, conversely, N_e is maximized when all families contribute equally to the next generation. Crow and Morton (1955) calculated from observed distributions of progeny number in a variety of species that the effective size ranged from 0.6 to 0.85 of the census number.

2. *Unequal numbers of the two sexes.* Whenever the numbers of each sex contributing to the next generation are not equal,

$$N_e = 1 \Big/ \left(\frac{1}{4N_m} + \frac{1}{4N_f}\right), \tag{1}$$

where N_m and N_f are the number of males and females respectively. This is particularly important in species which maintain stable harems (for example, elephant, seals, zebras, vicunas, and some bats). A breeding population in which there are 90 effective females but only 10 effective males has a total effective size of 36, not 100.

3. *Fluctuation in population size.* If population size varies from generation to generation, the effective number is the harmonic mean:

$$\frac{1}{N_e} = \frac{1}{t}\left(\frac{1}{N_1} + \frac{1}{N_2} + \ldots . + \frac{1}{N_t}\right), \tag{2}$$

where N_t is the effective size at the tth generation. Suppose, for example, that a population which normally maintains an effective size of 1000 drops for one generation to 50. Then, over a ten generation interval,

$$N_e = 10 \Big/ \left(\frac{1}{50} + \frac{9}{1000}\right) = 345. \tag{3}$$

A rare crash in size, as frequently occurs through drought or disease, clearly has an important effect on effective number.

This treatment of effective size assumes that each individual mates essentially at random with the other members of the group. Such an assumption may be appropriate for some highly vagile animal species, but is inadequate for many plants and many nonflying animals. In these latter species, inbreeding is very common; often a population is sharply subdivided into many smaller interbreeding groups. When the degree of relationship between mates is a function of geographical separation, a genetical analysis of population structure is much more complicated.

SHORT-TERM EFFECTS: INBREEDING DEPRESSION

This subject is covered in some detail in Chapters 9 and 12, and will only be briefly discussed here. Empirically, it is clear that species differ markedly in their resistance to inbreeding depression. There are numerous plants which reproduce naturally by self-fertilization. Of all the animal species, mammals and birds are probably the most sensitive, perhaps because of their low inherent reproductive rates. In the latter species inbreeding depression is the most important consequence of reduced population size.

The degree of inbreeding in a population is measured by the inbreeding coefficient, f; if the effective size is N_e, f increases by $1/2N_e$ per generation. In small populations, therefore, inbreeding depression accumulates gradually over many generations. A single instance of close inbreeding through, for example, the founding of a population with a small number of individuals, does not necessarily mean that the descendants will be highly inbred. Immigration of unrelated individuals into an inbred population reduces the level of inbreeding dramatically.

Inbreeding has deleterious effects on survival and reproduction, and affects such characters as growth rate and adult size. For many traits inbreeding depression, as a proportion of the mean, is of the order of one third of the inbreeding coefficient, and perhaps twice as great for fecundity. Overall, the effect of inbreeding on fitness is probably much greater than this. Latter and Robertson (1962) estimated that the effect in *Drosophila* of inbreeding on a competitive index (which for this purpose we will equate to fitness) was $e^{-2.7f}$. That is, for a small f, the reduction in fitness is 2.7 times the inbreeding coefficient. Sved and Ayala (1970) showed that homozygosity in *Drosphila pseudoobscura* for one of the autosomes representing about one fifth of the genome, reduced fitness by about 20 percent. Despite these deleterious effects, a small amount of inbreeding can be tolerated. Animal breeders accept inbreeding coefficients as high as a one percent increase per generation (that is, $N_e = 50$) in domestic animals without great concern. Natural selection will, of course, tend to counter the immediate deleterious effects of genetic drift. There will be a minimum population size, depending on the species, at which the population will be able to cope with the inbreeding effects, albeit with some cost in survival and reproductive rate. In the absence of regular introduction of unrelated stock, I suggest an effective size of at least 50 for large mammals.

Finally, it is important to emphasize that inbreeding accumulates at a rate which is directly related to the generation interval. A population of 50 randomly breeding elephants will take several hundred years to reach a ten percent level of inbreeding; mice will attain this figure in less than ten years.

LONG-TERM EFFECTS

Even in populations which are unaffected by inbreeding depression, sampling effects cause fluctuations in gene frequency which have important consequences for the future evolution of the species. Some genes will be lost by chance, especially those which are initially rare in the population, and alleles which are maintained by selection at some intermediate frequency will drift away from their equilibrium value. The theory of selection in finite populations is very complicated; the outcome depends very much on assumptions about the mode of natural selection.

At the phenotypic level, we observe changes in the frequency of simply inherited characteristics, and for quantitative traits, drift in the mean and, most often, a reduction in genetic variance. These last two effects will be discussed in detail, but it is the loss of genetic variation which, I believe, is of critical importance.

Random Changes in the Phenotype

Aside from the general effects in inbreeding depression, changes in phenotype are obvious in inbred lines of laboratory animals. Wright (1977*a*), for example, recounts his early experience with inbred lines of guinea pigs, each of which could be recognized as a result of changes in morphology, physiology, color pattern and temperament. In his general view of the evolutionary process, Wright attaches great importance to changes in the mean of small populations, for it is only through genetic drift that new adaptive combinations of traits are created. Hence, Wright suggests, the splitting of species into isolated subpopulations promotes evolutionary change. This view has wide but by no means universal acceptance. Fisher, for example, always argued that evolutionary change is maximized in large populations.

Since for most quantitative traits natural selection tends to favor an intermediate value, any change in an already well adapted organism will be deleterious and will be opposed by selection. On theoretical grounds, the drift in populations of reasonable size (for example, 500) is not great and should be countered by natural selection with little selective death. In fact, the observation that a trait changes in mean value in small populations probably indicates that it is not adaptively important in that environment.

Depletion of Genetic Variance

Phenotypic variance can be partitioned into three components: environmental variance, genotypic variance and a genotype-environment interaction variance. The last term is often overlooked. It arises when the best genotype in one environment is not necessarily the best in another. The genotypic variance can be further partitioned into additive, dominance and epistatic terms. Of these, it is through additive variance (sometimes called the genetic, or genic variance) that a population responds immediately to natural selection. The ratio of the additive to phenotypic variance is called heritability; if the heritability of a trait is zero, there can be little, if any, adaptive change.

Our understanding of the forces underlying the maintenance of genetic variance for quantitative traits is quite superficial. We do know however, that variation is ubiquitous, and that most traits respond readily to selection. The ultimate source of genetic variability is, of course, mutation, but the probability that a new mutation will reach a reasonable frequency depends on the mutation rate, the population size and the current mode of selection for the trait. Under directional selection, any mutation which favors the change but is not deleterious in other respects will tend to increase in frequency. Stabilizing selection, however, allows only genes with a small effect to accumulate (Fisher, 1958).

Let us ignore for the present the effects of natural selection and consider only the rates of gain and loss by mutation and drift. In very small populations, the loss of variability by sampling will be greater than the gain by mutation, and there will be a net loss of variation. Conversely, in very large populations, mutation will dominate the process, and we expect a steady gain. For each trait there will be a population size at which the rates of gain and loss are equal, and there will be no net change in the existing level of variability.

If there is no dominance or epistatic variance, additive variance is lost at the same rate as heterozygosity, that is, at a rate of $1/2N_e$ per generation. This value is probably reasonably accurate for most traits. The rate of gain by mutation is less easily determined. We have some meager information on the rate of production of new variation for bristle characters in homozygous lines of *Drosophila*. These data, reviewed by Lande (1976), suggest that this rate is of the order of one thousandth of the environmental variance. For a number of bristles, therefore, we expect an approximate equilibrium between gain and loss when $1/2N_e = 10^{-3}$, that is, $N_e = 500$. Selection and linkage complicate this simple picture enormously, but I will attempt to show that even strong selection does not result in dramatic reductions of additive variance.

The present level of genetic variation in a population will have been largely determined by the selection history—the time since the last bottleneck—and the effective size of the population. The effects of continued stabilizing selection are most difficult to assess, but there is little evidence that additive variance is affected markedly. In theory, stabilizing selection reduces all components of the phenotypic variance. Reduction in the genotypic component occurs through fixation of genes, the buildup of linkage disequilibrium (Bulmer, 1976) and, ultimately, developmental feedback systems that evolve to stabilize the phenotype (Waddington, 1957). These are, however, very slow processes. Kimura (1965), Latter (1970) and Lande (1976) have shown that considerable genetic variance can be maintained by mutation under stabilizing selection.

Directional selection also reduces additive variance, but again, slowly. Theoretically, genetic variance changes at a rate proportional to the third moment of gene effects, and if, as conventional wisdom suggests, quantitative traits are controlled by a large number of genes each with small effect, the third moment will be close to zero and genetic variance will change very slowly. In addition, there is the possibility that closely linked genes can act as a repository of variability. Under selection pressure this variation can be released slowly by recombination.

In general, experimental evidence supports the theoretical arguments. Reduction in variance through stabilizing selection is difficult to achieve, but positive results have been obtained (Rendel, 1960; Prout, 1962). These experiments, however, have been difficult to interpret, and it is not always clear whether it is the genetic or the environmental component of variance which has been changed. The consequences of directional selection are much clearer. There have been numerous experiments of this kind; one of the longest running is the Illinois corn experiment in which there has been continued response to selection for oil and protein content for 76 generations. Selection began in 1896 with a starting population of 163 open pollinated ears. Twelve to 24 parents were chosen by each generation. The response to selection for oil content is shown in Figure 1 and the pattern for protein was similar. Over the course of the experiment, a response of 20 standard deviations of additive variance for high oil and protein content has been achieved. Less spectacular results have been obtained for selection (in reasonably large populations) in diverse laboratory organisms such as *Drosophila, Tribolium* and the mouse. Examples may be found in Falconer (1960). These experiments demonstrate convincingly that additive genetic variance is not depleted rapidly by selection.

What, then, can we conclude about the effect of natural selection on genetic variance? Strong selection reduces additive variance, but weak directional or stabilizing selection probably has little effect. In some ex-

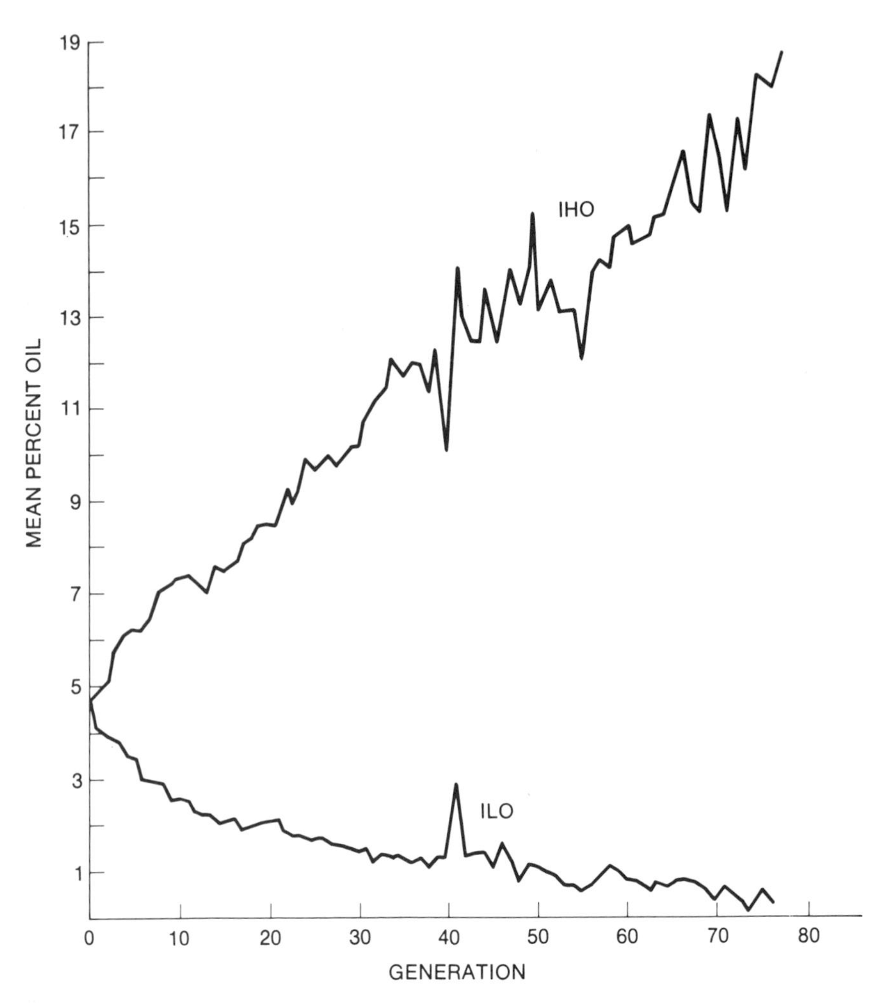

FIGURE 1. Mean percent of oil in ears of *Zea mays*—selected high (IHO) and low (ILO). (From Dudley, 1973)

periments, even with prolonged selection there has been as much additive variance in the selected population as there was in the original population. Therefore, we can tentatively conclude that the major determinant of the level of genetic variance in natural populations is the balance between genetic drift and mutation.

FOUNDER EFFECTS AND SELECTION RESPONSE

The previous section was concerned with the chronic effects of small population size. Here I want to examine the effects of bottlenecks (restriction in size followed by recovery) or of the founding of populations from a small number of individuals. The experimental investigations of response to selection in small populations (Frankham et al., 1968; Hanrahan et al., 1973) show that response to selection is less in small than in large populations, but in general the effects are not great. Even in the most extreme situation, when a population has been founded from a single pair, three quarters of the additive variance remains, which means that, on the average, there is still an opportunity for an appreciable response.

Hammond (1973) collected some data which illustrate the effect of small founder populations very well. He established populations of *Drosophila* from one, 10 and 50 pairs, and then measured the response to selection for abdominal chaetae over 10 generations. These data (Table I) show that the realized heritability (a measure of the observed change) in populations established from 10 pairs differs very little from that in populations established from 50. As expected, populations founded from one pair, but not maintained at this level, showed three quarters of the maximum response.

Founder effects, although they may not greatly restrict immediate selection response, nevertheless introduce major changes in the gene pool so that the pattern of change may be very different from that in the population from which the gene pool was derived. An important theoretical contribution to the theory of bottlenecks is that of Nei et al. (1975). The founding of new populations from small numbers is of considerable evolutionary significance; the subject warrants much more theoretical and experimental attention.

TABLE I. The relationship between founder population size and subsequent selection response.

Founder size	Treatment*	Number of replicates	Realized heritability
1 pair	1-1	28	.051 ± .018
	1-10	21	.157 ± .007
	1-20	18	.152 ± .007
10 pairs	10-10	24	.185 ± .004
	10-20	21	.177 ± .006
50 pairs	50-50	17	.202 ± .006
	50-100	9	.205 ± .006

*The treatments (for example, 1–10) indicate that the population size was reduced to a certain number of pairs (for example, 1) for a single generation, and was subsequently maintained at a higher number (for example, 10 pairs per generation).

(From Hammond, 1973)

In the preceding discussion I considered the effect of small population size on evolutionary potential—assuming no human interference. I now wish to briefly consider the consequences of intervention. In some circumstances, particularly when a breeding population is maintained in a zoo or reserve, decisions which relate to breeding have evolutionary consequences. It must be remembered that programs that minimize inbreeding and genetic drift are not conducive to simultaneous evolutionary change. For example, one way to maintain effective population size is to ensure that each family contributes equally to the next generation by culling only within families. Another option, if management is sufficiently intense, is to employ mating schemes which lead to maximum avoidance of inbreeding. All such methods reduce the response to natural selection and postpone the elimination of especially abnormal genotypes. Given a choice, however, it is perhaps best to sacrifice evolutionary change at the present, if by so doing one is conserving genetic variation for future evolutionary change. This is especially true in captive breeding situations where the return of stock to the wild is anticipated. In such cases we wish to avoid domestication.

Though domestication may be no substitute for conservation—such a step is essentially irreversible—it is, I feel, inevitable, at least in the long run. I do not wish to enter this arena, other than to point out that many of the behavioral traits which accompany domestication appear to have a high heritability and can be selected for relatively easily (Fuller, 1969). Far too little is known about the changes which occur in the process of domestication; we need to identify the traits involved, understand their inheritance and perhaps catalog those species which are preadapted for commensalism with man. Hale (1969) gives an excellent review of the evolution of domestication.

Finally there is the question of whether to maintain a single large population or to split the species into a number of smaller breeding units. Such decisions will be primarily made on political or ecological grounds, but the latter course seems to have distinct genetical advantages. If a species is maintained in a number of small populations, not only is the danger of accidental extinction (for example, by disease) reduced, but an opportunity for local adaptation exists which may increase the chance of ultimate survival. Genetic drift can be countered by allowing occasional migration. In the absence of selection we know that the subpopulations would have the properties of a single intermating group if only one or two migrants were exchanged each generation. There is little information on the joint effects of selection and migration in small subdivided populations, but some theorizing by Avery (1978) suggests that migration rates of the order of one per generation increase the overall genetic variability, both for heterotic loci and for loci selected in opposite directions in the subpopulations.

CONCLUSIONS

We know very little about the selective forces involved in the maintenance of polymorphisms for single genes or adapted gene complexes, and I have not considered the effects of genetic drift at this level. I have instead treated the problem of the conservation of genetic variability from the point of view of a quantitative geneticist, whose observations are the means, variances and covariances of continuously varying traits rather than gene frequencies, mean heterozygosity or the proportion of polymorphic loci. Based on our current knowledge of quantitative traits, I have suggested that the evolutionary potential of populations maintained in reasonably effective numbers (that is, in terms of hundreds and not tens of individuals) is not seriously impaired. I think that most quantitative geneticists working with animal populations would come to the same conclusion.

The subject of the quantitative geneticist has been the laboratory bred organism or the domesticated species, rather than the natural population. Nevertheless, most important adaptive changes which occur in evolution are changes in continuously variable traits, and there is every reason to believe that the single gene theory is inadequate for our understanding of the dynamics of such changes. Extrapolation from laboratory to natural populations is greatly complicated by the differences in complexity of the two kinds of environments. The reduction in size or disappearance of one species has repercussions on others (Chapter 2). I think it is fair to say that we have an incomplete understanding of these interspecific interactions. Another unknown is the importance of density dependent selection. A trait which is favored at high density may be disadvantageous at low density, and hence a reduction in number would add to the genetic load. Any change in the ecosystem, therefore, has the potential to alter the patterns of adaptation, and a species may face a serious crisis if it has to make too many changes at once. Negative genetic correlations make the situation worse, for if we try to select for two antagonistic traits simultaneously, we can expect little change in either.

I have already suggested that in the short-term the effective population size should not be less than 50. I tentatively propose that in the long-term the minimum effective size should be 500. Below this latter value, it is likely that genetic variance for complex traits is lost at a significantly faster rate than it is renewed by mutation. In reaching this conclusion I have adopted the view that it is important to maintain a pool of variation upon which future selection may operate. The number I have chosen is based on extremely meager evidence, and I stress that much more research is needed before we can answer such questions with confidence. There is no doubt that a species which maintains an effective size ten times this value would be in less danger of genetic deterioration.

Some species can recover from very small populations. The elephant seal has suffered extreme reductions in size and has apparently lost a great deal of its heterozygosity (Bonnell and Selander, 1974). Nevertheless, its ability to respond to selection is unknown. Other species may be in danger at a much greater effective size. In populations with a density dependent birth rate, there may be a critical size (well above 500) below which the population is doomed to extinction on purely ecological grounds.

Although I have argued that reduction in size does not seriously limit selection response, this does not mean that a population will be able to adapt to new conditions. The increase in genetic load due to a change in the ecosystem may be so great that the necessary genetic change, no matter whether the population size is 10 or 10 million, is beyond the reproductive capacity of the species.

Finally, it should be pointed out that these conclusions are concerned with evolutionary changes in identifiable, random-mating subpopulations, and hence are biased towards highly vagile animal species. From the point of view of conserving the genetic variation of the species, one elephant may be as good as another. This is certainly not true of most plants where microdifferentiation and ecotypic variation need to be taken into account. If you have seen one redwood you have not seen them all.

SUGGESTED READINGS

Quantitative Genetics

Falconer, D., 1960, *Introduction to Quantitative Genetics,* Oliver and Boyd, London.

Pollak, E., O. Kempthorne and T. Bailey (eds.), 1977, *Quantitative Genetics,* Iowa State University Press, Ames, Iowa. This recent review of the field contains some very useful summary papers.

Hill, W. G., 1972, Estimation of genetic change (I. General theory and design of control populations), *Animal Breeding Abstracts,* 40, 1–15. This excellent review of the design of control populations is relevant to the management of species in zoos or nature reserves.

Conservation of Genetic Resources

Frankel, O. H., 1974, Genetic conservation: our evolutionary responsibility, *Genetics,* 78, 53–65.

Harlan, J. R., 1975, Our vanishing genetic resources, *Science* 188, 618–621.

Brown, A. H. D., 1978, Isozymes, plant populations, genetic structure and genetic conservation, *Theoret. Appl. Genet.,* 52, 147–157. The papers by Harlan and Brown review and discuss the problems associated with plant resources.

Also, the September, 1978 issue of *Scientific American* is devoted entirely to evolution. The article by Lewontin on adaptation is particularly relevant.

CHAPTER 9

THRESHOLDS FOR SURVIVAL: MAINTAINING FITNESS AND EVOLUTIONARY POTENTIAL

Michael E. Soulé

The major goal in conservation genetics is the development of criteria for determining the population size (or minimum area) which will provide for the maintenance of fitness and adaptive potential. This is no academic exercise; economic and political forces relentlessly encroach on the land and the budgets given to conservation programs. Unless conservationists produce sound and defensible criteria for minimal population sizes there will be no rational way to counter these attacks, and all of our efforts to salvage samples of our magnificent large plant and animal species will be wasted. This chapter is an attempt to produce such guidelines.

A useful device for considering the relevance of population and evolutionary genetics to conservation is the "time scale of survival." Employing this scale, one can see, somewhat arbitrarily, three survival problems or issues: (1) the short-term issue is immediate fitness—the maintenance of vigor and fecundity during an interim holding operation, usually in an artificial environment; (2) the long-term issue is adaptation—the persistence of the vigor and evolutionary adaptation of a population in the face of a changing natural environment; (3) the third issue is evolution in the broadest sense—the continuing creation of evolutionary novelty during and by the process of speciation.

I will discuss the first and third of these issues (fitness and speciation). The issue of gradual evolutionary adaptation to a changing environment is taken up in Chapter 8. It will be mentioned here only briefly. The following section is, in part, abstracted from a more comprehensive account of conservation genetics (Frankel and Soulé, in press, Chapter 3).

SHORT-TERM FITNESS AND SURVIVAL

Many conservation programs are holding actions, particularly those involving small numbers of individuals in captive or controlled environments. Intensive projects of this type are often necessary until permanent survival programs can be established. Such programs, though, have a built-in hazard—genetic drift (random genetic change and the fixation of deleterious genes or gene combinations). Breeders of plants and animals have learned through experience that vigor, fecundity and other aspects of fitness decline at a rate proportional to the degree of random genetic change, and that this in turn is inversely related to population size. Therefore, the goals of such captive breeding programs must be to (1) minimize genetic and phenotypic deterioration and change and (2) to minimize the loss of genetic variation so that future adaptive options are retained. The cost of ignoring these objectives is almost certain failure (extinction).

The central problem of short-term conservation genetics is the relationship between population size and fitness. Two biological disciplines converge on this problem and contribute to its solution; they are empirical population genetics and quantitative genetics. Quantitative genetics, the oldest of the two, provides a wealth of data on the relationship between inbreeding and fitness in domesticated or laboratory lines. These data permit us to generalize about the effect on fitness produced by different rates and amounts of inbreeding. As discussed below, the key to our problem is found in this literature. To proceed on this basis alone, however, and extrapolate from the inbreeding effects produced in the laboratory or on the farm to the consequences of genetic drift and inbreeding in organisms fresh from nature, is to be nagged by certain doubts. A more satisfying approach, the one used here, is to demonstrate that the fitness of individuals in both captive and natural populations is related to their relative levels of heterozygosity (proportion of heterozygous loci). The following section explores the qualitative evidence for a relationship between population size and fitness in nature.

Heterozygosity and Fitness in Natural Populations

Hundreds of electrophoretic studies and surveys have been performed, but only a fraction are useful in addressing whether a decrease in heterozygosity (or genetic variance) in a natural population will lead to a diminution of fitness. Nevertheless, there now exist several such studies, and they provide us with a nearly unanimous answer (notwithstanding that the causal or mechanistic basis of the relationship remains obscure). These studies can be broken down into two principal categories: comparisons among individuals within populations and comparisons between populations. Many of the former involve age- or size-class comparisons, whereas the latter depend on differences in mean heterozygosity and fitness between populations of the same species. Some of the technical and theoretical issues raised by these studies are discussed elsewhere (Frankel and Soulé, in press).

Table I summarizes the results of 13 studies. In 11 of these studies the relationship between various measures of fitness and heterozygosity was or could be tested within populations. Data in 10 out of these 11 reports could be interpreted as supporting such a relationship, although the authors of some of these studies advance alternative hypotheses. One of the most complete and convincing investigations was that of Schaal and Levin (1976) on the herb *Liatris cylindracea*. *Liatris* is a perennial composite of the dry prairies. Its perennating organ is a corm which can be aged in some populations by counting the rings of pigmented cells deposited periodically. Survival is related to heterozygosity as shown in Figure 1. Among young individuals there is a large excess of homozygotes, probably because of inbreeding, but the mortality of the most homozygous plants is disproportionately high. Greenhouse tests performed on seedlings by Schaal and Levin established that age to sexual maturity as well as reproductive output and vegetative output were all significantly correlated with individual heterozygosity.

Table I also refers to a somewhat similar study in animals, one on

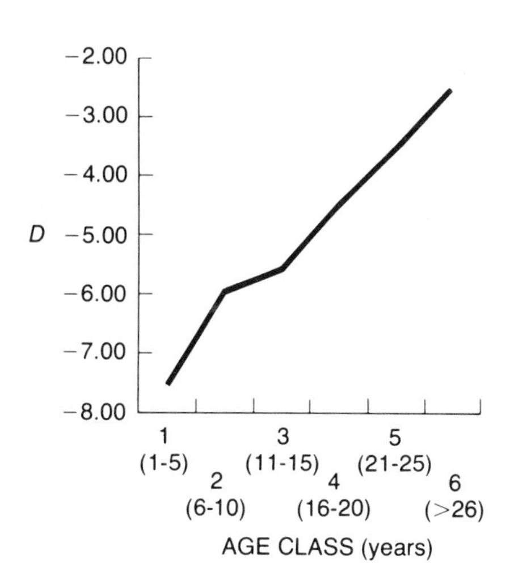

FIGURE 1. The decrease in heterozygote deficiency with age in a natural population of the herb *Liatris cylindracea*. Ordinate *(D)* is deviation from expectation assuming random mating. (Data from Schaal and Levin, 1976)

TABLE I. Summary of studies evaluating the relationship between heterozygosity and fitness in natural populations.

Organism	Number of polymorphic loci	Fitness criterion	Results	Reference
A. skipjack tuna (*Katsuwonus pelamis*)	1	survival to size	inconclusive	Fujino and Kang, 1968
B. ribbed mussel (*Modiolus*)	1	survival to size	decrease in homozygote excess with age	Chaisson et al., 1976
C. ribbed mussel (*Modiolus*)	1	survival to size	decrease in homozygote excess with age	Koehn et al., 1973
D. California mussel (*Mytilus*)	3	survival to size	decrease in homozygote excess with age*	Tracey et al., 1975
E. edible mussel (*Mytilus*)	6	survival to size	heterozygote excess in 1 of 6 loci	Koehn et al., 1976
F. lizard (*Sceloporus*)	7	survival to size	adults more heterozygous than juveniles for the esterase locus**	Tinkle and Selander, 1973
G. *Colias* butterfly	1	survival to age	heterozygote excess increasing with age	Watt, 1977
H. American oyster (*Crassostrea*)	5	growth rate	correlation of heterozygosity and body weight for 4 of 5 loci	Singh and Zouros, 1978
I. *Liatris* (perennial herb)	14	survival to age	decrease in homozygote excess with age	Schaal and Levin, 1976

oysters by Singh and Zouros (1978). They found that the fastest growing individuals were significantly more heterozygous than the slowest growing individuals. Among the other studies in Table I, B through K all contain evidence for the selective superiority of relatively heterozygous in-

J. killifish *(Fundulus)*	5	morpho- logical variabil- ity	heterozygotes less variable than homozygotes	Mitton, 1978
K. monarch butterfly *(Danaus)*	6	morpho- logical variabil- ity	heterozygotes less variable than homozygotes	Eanes (personal communication)
L. old field mouse *(Peromyscus)*	18	aggres- siveness	aggressiveness correlated with mean heterozygosity	Garten, 1976
M. side-blotched lizard *(Uta)*	18	develop- mental stability	morphological symmetry correlated with mean heterozygosity	Soulé, 1979

*Tracey et al., 1975, attribute this result to population structure, not selection.
**Only one of the seven loci was strongly polymorphic.

dividuals in natural populations, but in some cases the relationship is probably fortuitous.

Two obvious caveats to Table I are (1) that negative results often go unreported and (2) that general surveys of electrophoretic variation usually fail to demonstrate an excess of heterozygotes. On the other hand, most general surveys do not distinguish among age or size classes, and the opportunity of observing differential survival or fitness is obviously decreased when such classes are pooled. In addition, the small sample sizes of many studies preclude the observation of statistically significant departures from Hardy-Weinberg (expected genotypic) frequencies.

The other approach to the relationship between genetic variance and fitness in natural populations is to compare populations known to differ in mean heterozygosity. Such studies exploit the differences in genetic variation that often occur among populations of species with poor powers of dispersal. For example, nonflying vertebrates tend to have more genetic variation near the center of their range or where the species is most dense (Soulé, 1973; 1976) and some predominantly inbreeding plants have single, peaked clines in allozyme diversity (Clegg and Allard, 1972; Rick et al., 1977; Nevo et al., in press).

The last two items in Table I refer to investigations of this type. Garten (1976) observed that body size and success in aggressive encounters in old field mice *(Peromyscus polionotus)* were correlated with the mean heterozygosity in the population from which the mice were collected. Soulé (1979) observed in island populations of the side-blotched lizard *(Uta stansburiana)* that fluctuating asymmetry, a putative inverse measure of developmental homeostasis, is negatively correlated with mean heterozygosity among the populations. (One disadvantage of such studies is the hazard of spurious correlation, but this danger exists in *within* population work as well.)

Surveys of the genetic structure of predominantly inbreeding plants (for example, Brown, 1978) yield a vast range of allelic polymorphism. Even within a single species one may observe some populations with no observable polymorphism (evidence that heterozygosity is not a necessary condition for survival), while other populations of the same species have levels approaching those of outbreeding species. Apparently, some kind of heterozygote advantage is a factor contributing to survival in some polymorphic plant populations, since these plants maintain twice the expected level of heterozygosity in the face of considerable inbreeding (Clegg and Allard, 1972; Rick et al., 1977).

In summary, heterozygote advantage is a reasonable interpretation of much of the data from natural populations. On the other hand, the data do not allow us to determine what kind of selection model (overdominance, frequency dependent, marginal overdominance) is most appropriate. In any case, the prudent strategy is to preserve genetic variability and to avoid, as far as possible, unnecessary decreases in population size, since the rate of genetic erosion depends on population size as discussed in Chapters 8 and 12.

Inbreeding Depression in Captive Populations

The literature on inbreeding is replete with examples of the deleterious effects of homozygosity. One example reproduced in Table II is from Wright (1977*a*). He recounts an experiment on Poland China swine (McPhee et al., 1931) that may typify the scenario for other species which lack a history of intense inbreeding and selection. The experiment was designed to study the effects of sib (brother-sister) mating. With sib mating, a line loses 25 percent of the remaining heterozygosity in each generation, at least in theory. At such a rate of inbreeding, many deleterious (usually recessive) genes in the line will be fixed by chance, because under these conditions the rate of fixation is greater than the rate at which selection against homozygotes can eliminate them. The swine experiment was discontinued after two generations following a precipitous drop in fitness. Table II shows that the mean number of pigs per litter went from 7.15 in the noninbred herd to 4.26 in the second generation of inbreeding. In addition, the survival of newborn pigs dropped by over 50 percent. Multiplying the size of the litter by the percent of young raised to 70 days gives the number of surviving offspring per litter; this value dropped from four to one. Accompanying this decrease in fecundity and viability went a change in sex ratio favoring males (Chapter 12 explains why this effect is expected), thus producing a shortage of sows. The symbol f in Table II is Wright's inbreeding coefficient, and can be thought of as the amount that heterozygosity has decreased compared with the base population (general herd).

TABLE II. Vital statistics of a herd of Poland China swine and the progeny of two generations of sib mating.

	No.	f Dam	f Litter	Size of litter	Percent born alive	Percent raised to 70 days	Sex ratio
General herd	694	0+	0+	7.15	97.0	58.1	109.7
F_1 inbred	189	0.09	0.33	6.75	93.7	41.2	126.1
F_2 inbred	64	0.33	0.42	4.26	90.6	26.6	156.0

(From McPhee, Russel and Zeller, 1931. After Wright, 1977.)

Harmful changes in reproductive characters typically occur upon inbreeding; a large part of inbreeding depression is made up of such changes. A survey of inbreeding experiments led to the generalization that increasing the inbreeding coefficient by 10 percent induces a 5 to 10 percent decline in a particular reproductive trait (Frankel and Soulé, in press). Note that an f equal to 10 percent approximates the amount of inbreeding that would theoretically occur in a population of five adults breeding at random for a single generation, or—using the formula $f = 1 - (1 - 1/2N_e)^t$ where t is the number of generations—in a population of 25 adults breeding at random for 5 nonoverlapping generations. A 5 or 10 percent decline in fecundity might not appear to be very serious, but if the effects of inbreeding depression on the other traits (such as viability) are also considered, this amount of inbreeding can lower reproductive performance as a whole by 25 percent (Abplanalp, 1974, for example). Inbreeding depression can be very potent indeed.

Another way of looking at the effects of close inbreeding is in the survival of inbred lines. Table III summarizes data from a variety of organisms. *A priori,* it is expected that survival will depend on and be inversely related to the load of deleterious mutations in the stock population. Everything else being equal, this load should be greatest in stocks lacking a history of inbreeding or artificial selection, because such lines have not been purged of any of their deleterious genes. It is not surprising, therefore, that the Japanese quail *(Coturnix coturnix),* the least domesticated of the birds in Table III, show the most inbreeding depression.

Considering the above results, it is obvious that conservationists ought to view inbreeding as anathema. Although inbreeding will purge a stock of some of its deleterious genes, most lines do not survive a bout of inbreeding of this intensity. Those that do survive are typically handicapped by lower fitness and by random changes in morphological, physiological and behavioral traits. It rarely, if ever, happens that an inbred line is equal in overall vigor or fecundity to its outbred control, particularly in a natural setting. Therefore, to purposely inbreed one of the last stocks of an endangered species, on the off chance that it will be "cleansed" of its genetic load, is the height of folly.

This fiat does not apply with equal force to species which normally inbreed. Young and Murray (1966) observed that the degree of inbreeding depression in domesticated plants is correlated with the usual amount of cross-pollination; in order of increasing cross-pollination and inbreeding depression (genetic load) these plants were barley, cotton, tomato and corn.

The Basic Rule of Short-Term Survival

The data in the preceding two sections may motivate one to ask how to best and safely manage populations in order to minimize the loss of fitness, but such data do not provide a basis for specific recommendations. It is as if we knew that a particular substance was poisonous but

TABLE III. Survival of inbred lines following intense inbreeding.

Organism	Number of lines at beginning	Number of lines surviving	Percent of lines surviving	Length of survival and remarks	References
Drosophila	10	1	10.0	10 or 20 generations	Clayton et al., 1957
Japanese quail	388	3	0.8	3 generations	Wallace and Madden, 1965
Japanese quail	388	0	0.0	4 generations	Sittman et al., 1966
Poultry	279	30	10.8	3 generations (short hatch period)	Abplanalp, 1974
Poultry	279	8	2.9	15 generations	Abplanalp, 1974
Mice	20	1	5.0	12 generations	Bowman and Falconer, 1960
Mice	14	2	14.3	6 generations	Lynch, 1977
Guinea pigs	35	5	14.3	12 generations	Wright, 1977*a*

were ignorant of the lethal dose. That is, enough is known about inbreeding to justify concern that any increase in homozygosity reduces absolute vigor and fecundity, but so far we have no way of deciding how much is tolerable. We can not even pinpoint with any certainty the cause of inbreeding depression, although we suspect that this loss of fitness is mostly attributable to a load of deleterious recessive genes and to altered gene interactions rather than to a loss of single gene heterosis concomitant with decreasing heterozygosity (Eberhart, 1977; Comstock, 1977; Falconer, 1977; Singh and Zouros, 1978). Finally, we know that a partial antidote for the accumulation of such deleterious genes during inbreeding is the winnowing effect of natural selection. Our problem is that we lack data on the relative strengths of the poison (genetic load) and the antidote (natural selection).

We cannot look to population genetic theory and models for help because the models for finite populations require several assumptions (about mutation rates, gene frequencies, selection coefficients) about which we can only guess. In addition, one must decide the mode of selection (dominance, overdominance, or even neutrality, a paradoxical choice). Clearly, some more down to earth approach is needed. Perhaps the best way to proceed is to side-step theory and resort to empiricism.

By trial and error, animal breeders have discovered how much inbreeding can be tolerated by domestic animals before the lines begin to decline in performance and fertility. Their rule of thumb is that the per generation rate of inbreeding should not be higher than two or three percent (Dickerson et al., 1954; Stephenson et al., 1953). Higher rates of inbreeding fix deleterious genes too rapidly for selection to eliminate them from the line. I prefer a slightly more conservative value, an upper limit of one percent inbreeding per generation. The reasons are (1) domesticated stocks have been partially purged of deleterious genes over the millenia, so they can tolerate higher rates of inbreeding than can outbreeding species fresh from the wild, and (2) animal breeders can safely ignore some classes of phenotypic change resulting from inbreeding and genetic drift, whereas conservationists wish to preserve the "wild-type". That is, conservationists may have to practice selection on more traits than do animal breeders in order to preserve the original phenotype, and the rate of inbreeding must be decreased if the number of traits being selected increases.

How does this basic rule (the one percent rule) translate into population size? The rate of loss of heterozygosity (increase in homozygosity) or f per generation is equal to $1/2N_e$, where N_e is the effective population size. Thus, N_e must equal 50 or more if the inbreeding rate is to be kept below the one percent threshold.

Even at the rate of one percent, however, the loss of genetic variation is appreciable after a few generations. That is, even though natural selection may be able to nullify the most harmful inbreeding effects at this

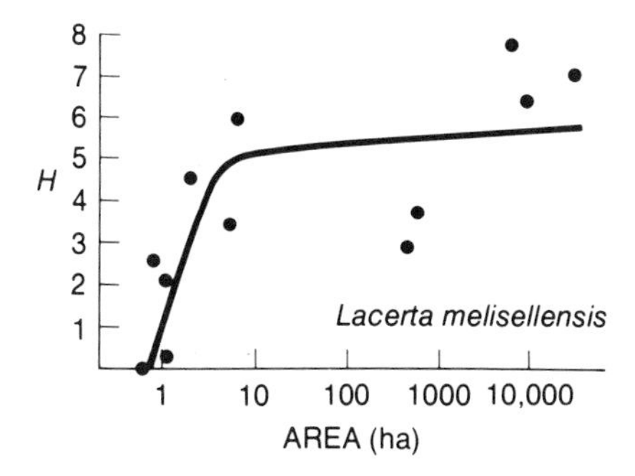

FIGURE 2. Relationship between area and percent heterozygosity *(H)* among island populations of the lizard *Lacerta melisellensis.* (Data from Gorman et al., 1975)

rate, it cannot prevent a gradual attrition of genetic variation. Sooner or later, depending on N_e, the population will become virtually homozygous, although there is evidence for an increase in heterozygote advantage at higher values of f (Frankel and Soulé, in press, Chapter 3). This is why the basic rule is a short-term criterion. A population held in check at $N_e = 50$ for 20 to 30 generations will have lost about one-fourth of its genetic variation, and along with it, much of its capacity to adapt to changing conditions.

The effects of a chronically small population on genetic variation seem to be indicated by the "small island effect" shown in Figures 2 and 3. These figures suggest that decreasing island size has little effect on heterozygosity until a critical area is reached—about five hectares for lizards. Below this size the populations appear to lose heterozygosity. For the species shown, a five hectare island has somewhere between 250 and 2000 individuals. These particular islands were formed by rising eustatic sea levels between eight and ten thousand years ago. Some or all of the populations might be that old; some or all might be much younger, having been established by one (gravid) or more founders at a later date. In any case, the founder effect alone is not a sufficient or likely explanation

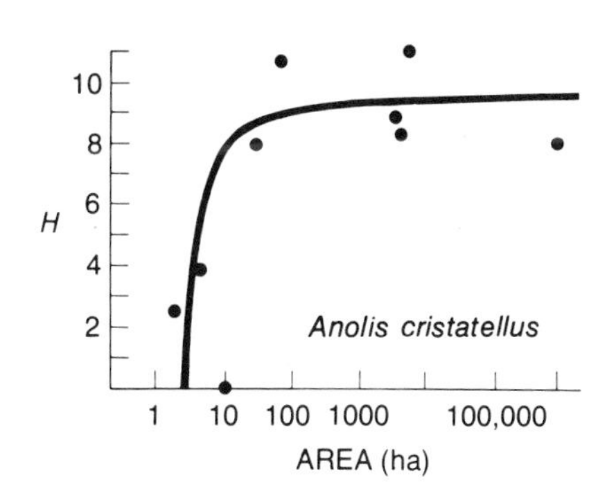

FIGURE 3. Relationship between area and percent heterozygosity *(H)* among island populations of the lizard *Anolis cristatellus.* (Data from Gorman et al., 1979)

for the near absence of heterozygosity in a few of these populations; genetic drift is almost certainly a contributing factor. The reason for presenting these data is simply to illustrate that the depletion of variation can and does happen in nature, given small enough populations.

Incidentally, several workers have pointed out that the number of founders in a colony, so long as it is greater than about five individuals, is not nearly as important as the long-term maintenance size of the colony (Chapter 12; Nei et al., 1975; Denniston, 1978). The reason is that a single bottleneck event has much less impact on heterozygosity than does genetic drift in a perennially small population.

Over how many generations can a small population persist before it has a high probability of going extinct? A possibly useful empirical rule is suggested by the experience of animal breeders. Animal breeders have noticed that there is an obvious effect on fecundity in small populations as the inbreeding coefficient approaches 0.5 or 0.6. Using these numbers as arbitrary thresholds, we can estimate the number of generations it will take at various values of N_e before the group becomes susceptible to extinction from genetic causes. Setting Δf equal to 0.6 (or 0.5) in the formula

$$\Delta f = 1 - \left(1 - \frac{1}{2N_e}\right)^t$$

and solving for t, the number of generations, we obtain the rough approximation $t \cong 1.5\ N_e$, or in other words, the expected number of generations to the extinction threshold is about 1.5 times the effective population size.

Effective Population Size and the Basic Rule

The effective application of the basic rule requires more than merely maintaining a stock of 50 or more breeding adults. The reason is that the actual (census) number of adults may not be even close to the effective number. Franklin (Chapter 8) discusses this distinction in some detail, but some further comments and examples will be given here.

Breeding structure is absolutely critical. In species with harems, leks or other polygynous systems, the effect of a relatively small number of breeding males is a significant increase in the minimum census number, N, required to reach a desired effective size. It may be quite common for the ratio N/N_e to be 3 or 4, meaning that the minimum size of the colony is 150 to 200 individuals.

In species where social inhibition or other factors prevent many adults from breeding, the ratio is even higher. In canids, for example, it is typical for only the dominant pair in the pack to breed (Chapter 14). This applies to the cape hunting dog, *Lycaon pictus*. In the Serengeti, one of the largest national parks in Africa, the population of hunting dogs numbers 30 or so adults. Assuming that these animals are divided among five packs, this means that only about 10 animals are breeding in any given year. Thus, the rate of inbreeding is about five percent. Situations of this kind call for immediate and continuous preventative management (that is, artificial gene flow).

Fluctuations in population size are another fly in the ointment, as Franklin points out. We once believed that tropical species, especially vertebrates, hardly fluctuated at all, but the discussion by Foster (Chapter 5) puts this pleasant fantasy to rest.

Overlapping generations, as occur in perennial plants and in most vertebrates, may also shrink the effective number of breeding adults. Hill (1977) states that "with overlapping generations, even if there are no fertility differences among survivors and there is random death of breeding individuals, the inbreeding rate will be higher than [expected] since the distribution of lifetime family size is not Poisson. With an exponential distribution of deaths the rate can be nearly three times as high as in the simple formula (Felsenstein, 1971)."

Here, it might be heuristic to estimate the minimum size of a nature reserve dedicated to the short-term conservation of a canid species, for example, the wolf. The necessary parameters might be as follows: the density of wolves is about one adult per 20 square kilometers (Rutter and Pimlott, 1968), but only about one-third of the adults actually breed. Say that the population fluctuates, reaching an ebb of about 10 percent of the carrying capacity on the average of once in 10 years. Employing the basic rule, we would start out with an absolute minimum effective size of 100 adults, since many adults fail to breed in a given season, and probably in a given generation.

Next, it would be advisable to double the minimum tolerable size to at least 200 because of the problem of overlapping generations. Finally, we assume that the population increases by 50 percent each year following a crash. A little experimentation with formula 3 in Chapter 8 will show that the minimum bottleneck size is about 60 (that is, $1/N_1 \approx 0.0167$). Bottleneck sizes below this, given the assumed conditions, prevent N_e from reaching 200. Therefore, the carrying capacity must be 600 or more, and the size of the reserve would have to be at least 12,000 square kilometers, or substantially larger than Yellowstone National Park. In Chapter 8, an effective size of 500 is suggested as the lower limit for the preservation of variation and evolution in a changing environment. If we apply Franklin's estimate of a long-term criterion to the wolf example, the minimum size of the reserve would be an order of magnitude larger.

Genetic considerations, therefore, raise the issue of the minimum sizes for reserves in quite a different light than it exists in the biogeographic literature (Chapter 6). That is, most existing parks are too small to ultimately preserve many large species on purely genetic and evolutionary grounds. This conclusion buttresses the empirically based conclusion of biogeographers that extinction rates are inversely correlated with area.

THE END OF VERTEBRATE EVOLUTION IN THE TROPICS

We now turn from the issues of fitness and evolutionary adaptation to the third level of biological survival—speciation, the capacity of evolutionary lines to generate new species. Speciation is the splitting of evolving lineages. Many current evolutionists consider speciation to be the principal source of biological novelty, relegating the role of gradual phyletic change to fine tuning of adaptations within already established species (Gould and Eldredge, 1977; Stanley, 1975). According to this thesis, speciation is the major, if not the only, process by which significant evolutionary changes come about. As a generator of diversity, therefore, speciation emerges as preeminent.

The question posed here is whether tropical nature reserves will be large enough to permit speciation in higher vertebrates and plants. I emphasize the tropics because probably 90 percent of species are tropical, and also because the rate of destruction of tropical habitats (Chapter 17) means the imminent insularization of tropical floras and faunas.

Fortunately, this question—the minimum area required for speciation—can be solved empirically. All that is necessary is to determine the size of the smallest island on which a particular taxon has speciated autochthonously *(in situ)*—that is, where there is sufficient room for a species to split into two or more species. For example, the occurrence of endemic genera and families of lemurs, chamaeleons, birds and plants on Madagascar is evidence that these taxa have speciated on that island.

It must be understood, however, that most island radiations are not autochthonous in the sense meant here. Most insular radiations occur on archipelagos, water gaps between islands being the barrier to gene flow. The Darwin finches on the Galápagos, for example, did not speciate on a single island. Islands the size of these in the Galápagos archipelago are too small to have physical features large enough to isolate conspecific populations of birds within an island.

Speciation by multiple invasion, like speciation in an archipelago, is not relevant to the problem at hand. This form of speciation requires that an island be invaded at least twice by the same taxon. For example, two endemic species of lizard in the *roquet* group of the genus *Anolis* occur on the island of Grenada in the Caribbean, but this is the result of two separate colonizations (Yang et al., 1975).

Figure 4 portrays the sizes of the smallest islands on which particular taxa are known to have speciated by processes occurring exclusively on the island. In simple terms, it appears that large or vagile organisms require much more space to produce two or more contemporaneous species

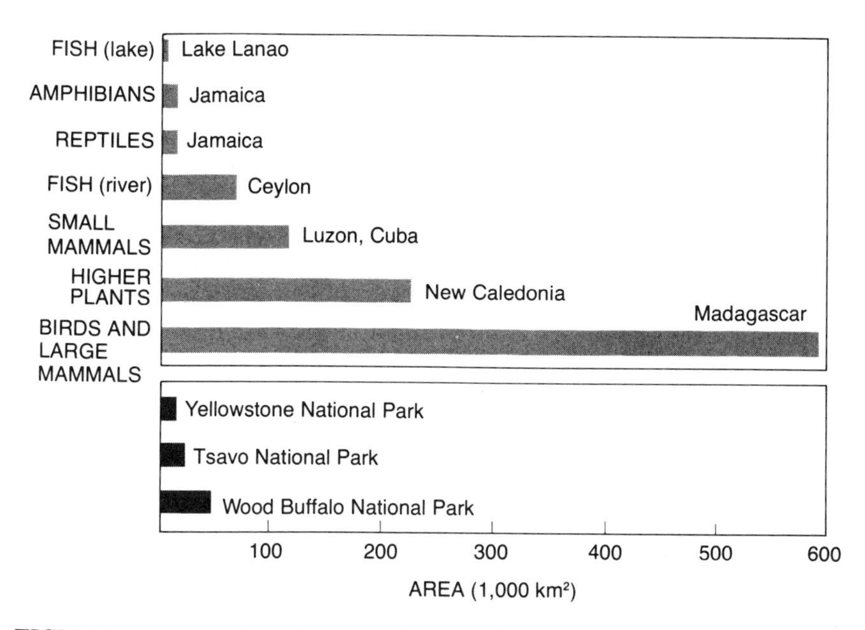

FIGURE 4. Above: Smallest islands on which autochthonous speciation is known to have occurred for various taxa. Below: Three large nature reserves for comparison.

than do small or sedentary organisms. Note that, roughly speaking, as absolute vagility increases, so does the size of the landmass necessary for apparent autochthonous speciation. Also note that the largest national parks in the world would seem to be too small for the speciation of higher plants, birds and mammals.

Not only are the islands in Figure 4 large, they are also old. New Caledonia and Madagascar, for example, are ancient islands, being remnants of the breakup of Gondwanaland. Therefore, it would appear that the element of time is also a factor in the production of endemic radiations.

Although the data are strongly suggestive, one could take exception with the above interpretation. An alternative is that the apparent dependence of speciation on area is spurious, and that the actual area-dependent process is the longevity of newly arisen species. This is to say that speciation in plants, birds and mammals may actually occur on smaller islands than suggested in Figure 4, but that the turnover or extinction rate (Chapter 6) is so high on these smaller islands that at any given time it is unlikely that there will be more than one species surviving per lineage. In other words, extinction rate, not speciation rate, is the area-related factor. (Note that Wilcox in Chapter 6 presents biogeographic evidence that increased body size and metabolic requirements result in higher extinction rates in vertebrates and Terborgh and Winter in Chapter 7 argue that population size or density is the best predictor of extinction in birds; thus on small islands, the probability of extinction for large organisms is very high. Assuming that their generalizations hold up, it means that the apparent thresholds for autochthonous speciation could, at least in part, be attributable to thresholds for speciation-extinction equilibria for numbers of species greater than one per genus.)

In practice it makes little difference whether the area effect on speciation in Figure 4 is a reflection of taxon-specific thresholds for autochthonous speciation, or is a reflection of thresholds for speciation-extinction equilibria. Even if it is the latter, which I think is unlikely, it means that existing nature reserves are still too small to generate more species than they now contain. In fact, they will contain fewer and fewer species with the passage of time because most reserves are now supersaturated and will lose species rapidly over the next few centuries (Chapter 6). It therefore appears inescapable that, for the first time in hundreds of millions of years, significant evolutionary change in most higher organisms is coming to a screeching halt in the tropics.

One might object that this entire argument is misleading on the grounds that reserves can support speciational processes because of the isolation between them (that is, on the model of archipelagos). The minimum requirements for this kind of speciation is that a species occur in two or more reserves, and that the reserves be isolated long enough for the evolution of intrinsic isolating mechanisms such as differences in habitat preferences or post-mating reproductive barriers. Superficially, this model seems reasonable, assuming that reserves are continuously viable for thousands of years.

The flaw in this line of thought is that it ignores the high rate of extinction of large organisms on islands. Figure 7 in Chapter 6 shows that we cannot count on the species surviving long enough to speciate, even in the largest tropical reserves. In fact, the extinction rate is so high that half of the original large mammals species in such a reserve are expected to go extinct in from 500 to 2000 years unless, that is, man intercedes.

The irony is that such intercession itself will prevent speciation of this type from occurring. The prevention of extinction of such animals will require the exchange of individuals between the populations in the different reserves, this being the only available means of preventing genetic deterioration of the largest or least dense species. Large doses of gene flow, of course, are inimical to incipient speciation and the development of locally adapted races because it turns the subdivided species into a genetic omelette. But even if genetic deterioration were not a factor and gene flow was not a necessary preventative therapy, artificial migration would still be necessary in most instances in order to prevent stochastic events such as drought, disease and other catastrophes from driving fragmented populations to extinction.

Thus, it would appear that with or without management, evolution of large, terrestrial organisms in the fragmenting tropics is all but over. Ironically, the very means at our disposal to protect the large life forms in the tropics (artificial migration) is the same means that will impose evolutionary stasis. It is sad that the necessary condition for survival itself contributes to the end of speciation for large organisms. We are witnessing (and causing) a biological crisis without precedent.

CONCLUSIONS

The criteria for population size and area proposed in this and accompanying chapters (Chapters 8 and 12) should force some painful reappraisals of existing and contemplated conservation programs. For example, only a handful of captive breeding programs for birds and mammals meet the minimum criteria for short-term fitness, and only a fraction of national parks and reserves are of sufficient size to sustain evolutionary viable populations of large animals. Thus, planners are faced with three choices: (1) They can ignore the genetic and evolutionary criteria (business as usual); (2) They can accept and attempt to implement the criteria; and (3) They can decide that it is hopeless and change professions. The latter choice is understandable, especially when one considers the economic and political barriers confronting conservation planners.

Perhaps it is some solace that a less than ideal program is better than none at all, assuming it makes the best use of available resources. Unfortunately, many existing captive breeding and wildlife conservation programs are failing to achieve even the latter goal. For example, national and individual ambition and pride often prevent the establishment of breeding and management consortia, the implementation of which would significantly increase the effective population sizes of many captive and endangered species. Realistically, in some cases the only solution will be a legislative one—the penalty for uncooperativeness being removal of preferred tax or licensing status, and the incentive being grants and official recognition.

Finally, an apology and a caveat. Some biologists will be appalled by the blanket prescriptions for survival suggested here, especially in view of the heterogeneity in population structure, genetic variability and, probably, in genetic load that exists among species—even closely related ones. Indeed, criteria such as those recommended here lack precision. The caveat is that the luxuries of confidence limits and certainty are ones that conservation biologists cannot now afford, given the rate of habitat destruction documented in many of the chapters of this book. Constructive criticism is welcome, but to embrace the purist's motto of "insufficient data" is to abandon the bleeding patient on the operating table.

SUMMARY

1. Conservation genetics can be applied to three distinct survival problems occurring on three time scales. In order of increasing temporal duration these problems are (1) immediate fitness; (2) adaptation within a phyletic line; and (3) speciation, the generation of evolutionary novelty by the splitting of phyletic lines.

2. Data from natural populations supports the inference that relatively heterozygous individuals have greater viability and, in some cases, fecundity, than relatively homozygous individuals. Within populations, relatively heterozygous individuals are observed to have high growth rates, high survivorship and greater morphological stability. In general, individuals from relatively polymorphic populations, compared to those from less polymorphic populations, seem to have similar advantages.

3. Inbreeding experiments provide another body of information relevant to conservation genetics. A review of the literature shows that even a small amount of inbreeding typically undermines fecundity and viability; a 10 percent increase in homozygosity may reduce total reproductive performance by as much as 25 percent. Most lines that are inbred become extinct after three to 20 generations, and the chance of surviving a bout of inbreeding may be virtually nil in some species fresh from nature.

4. Based on a rule of thumb employed by animal breeders, the basic rule of conservation genetics is that the maximum allowable rate of inbreeding is one percent, which corresponds to a genetically effective size of 50. In the world of real (in contrast to ideal) populations, several times this number may be needed.

5. The basic rule applies only to the maintenance of short-term fitness, because population sizes of this order of magnitude do not prevent the gradual erosion of genetic variation, and with it the raw material for future evolution.

6. Data from island populations of lizards support the thesis that natural populations suffer attrition of genetic variation when their numbers drop below a few hundred.

7. Even the largest nature reserves are probably too small to guarantee the permanent survival of large herbivores and carnivores.

8. This century will see the end of significant evolution of large plants and terrestrial vertebrates in the tropics. Nature reserves are too small for the necessary isolation required by allopatric speciation. Further, it is

unrealistic to hope that isolation between reserves (using an island model) can substitute for within-reserve isolation, because the extinction rates within reserves are too high to justify expectation of long-term survival. One measure mitigating the extinction problem is artificial gene flow between reserves, but this will impede local adaptation and prevent speciation.

SUGGESTED READINGS

Refer to the *Suggested Readings* given in Chapter 8.

CHAPTER 10

DEMOGRAPHIC INTERVENTION FOR CLOSELY MANAGED POPULATIONS

Daniel Goodman

In much of wildlife research the single greatest challenge is simply to obtain reasonably accurate estimates of population size. Then, mathematical expertise may best be employed in devising and evaluating censusing programs and, perhaps, in investigating the relationship between population growth rate and population density or size. In some instances, however, we are well aware of the population size—most often when the population is very small. In such circumstances, population managers may profitably turn to mathematically sophisticated demographic models, most especially models which take account of age structure, in the effort to stabilize a small population, or perhaps to enhance the rate of recovery to a larger population size.

The data required for a full demographic analysis are not likely to be available except for a closely observed population. The interventions specified by such an analysis often will not be practical to apply except for a closely managed population. Nevertheless, as we shall see in this chapter, these detailed management programs based on detailed analysis can prove appreciably superior to a less specific approach. As the list of species whose population numbers are small continues to grow, and as the intensity of environmental modification increases, there will be more and more populations for which detailed programs of demographic interventions are both feasible and advantageous.

The mathematical foundations of demography were laid by Lotka (1939) and Leslie (1945). A full exposition of modern mathematical demography may be found in Keyfitz (1968). Cole (1954) and Lewontin (1965) pioneered in the use of mathematical demography in ecological analysis. During the past fifteen years a great deal of progress has been made in understanding the mathematics of life history optimization, the evolution of life history strategies and the ecological consequences of various life history patterns. Much of this literature is reviewed in Goodman (1979). Some examples of the application of demographic modelling to specific, managed populations may be found in Mertz (1971), Fowler and Smith (1973), Miller and Botkin (1974) and Foose (1978).

In this chapter I will first follow in the tradition of treating population models in the abstract to see what sorts of mathematical techniques may be brought to bear on the questions that concern us, and to see what general sorts of insight they offer. We will consider three sorts of population management problems where a full demographic analysis would suggest a particular strategy of management. The first deals with a small wild population that is declining, or at any rate failing to grow, even though its food resources do not seem to be limiting. The second deals with a wild population that is pressing the limits of the carrying capacity in its available habitat. The third deals with an endangered captive population from which a surplus must regularly be removed. In the final section of this chapter application of the analyses will be illustrated with the model of an actual population.

AGE STRUCTURE IN DEMOGRAPHIC MODELS

Before proceeding with the specific analysis of the management problems described above, we will briefly review the mathematical foundations of the relevant demographic models. Our concern with age structure leads us to models which distinguish individuals by age and which assign individual parameters, such as probability of survival or fertility rates, also according to age.

Actuarial data necessarily group individuals by age classes of finite duration, so a discrete time model will appear more natural in application. In general, however, continuous time models are mathematically more convenient, but except for noting this fact, discussion will be confined to discrete time models.

We scale time in units corresponding to the span of an age class. Let m_x be the per capita effective fecundity of the xth age class. This is the mean per capita number of offspring born to individuals aged $x-1$ to x during a unit time interval, where these individuals are censused at the beginning of the time interval and the offspring are censused at the end. For the moment we shall consider only the female segment of the population, so fecundity is scaled in female births per female.

Let p_x be the survival rate of the xth age class. This is the fraction of individuals entering age class x which survive to enter age class $x + 1$. Obviously, the age specific death rate is defined simply as

$$q_x = 1 - p_x \tag{1}$$

The survivorship schedule, l_x, as represented in a conventional life table, is related to survival rates according to

$$l_x = \prod_{j=1}^{x-1} p_j \tag{2}$$

The survivorship of age class x is the fraction of individuals surviving to enter age class x, per censused newborn.

The net maternity function ϕ_x for the xth age class is the product of the fecundity and survivorship for that age class

$$\phi_x = l_x m_x \tag{3}$$

The two schedules of vital rates, fecundity and survival (or, alternatively, fecundity and either mortality rates or survivorship), suffice to describe the growth of a population for as long as these schedules remain true. We usually assume that the vital rates will remain constant for a constant environment, but changing population density may well affect the environmental state, and for that matter it may influence the life table parameters directly.

A matrix operator, called the Leslie matrix, $\boldsymbol{A}$, will serve for projection of a general population vector $\mathbf{n}_t$, where $n_{j,t}$, the jth element of the vector at time t, gives the number of individuals in age class j at that time. The projection is obtained according to

$$\boldsymbol{n}_{t+1} = \boldsymbol{A}\boldsymbol{n}_t \tag{4}$$

The Leslie matrix, a square matrix of the same order as the number of age classes in the population vector, is formed from the vector of age specific fecundities as its first row and the vector of survival rates as its principal subdiagonal; it is zero everywhere else.

The ultimate effect of repeated multiplication by the matrix $\boldsymbol{A}$, as the population continues to grow (or decline, or remain stationary) under a constant schedule of vital rates, may be learned from eigenanalysis of the matrix. Generally, the age distribution will converge to a stable set of proportions given by the dominant eigenvector; once the population age distribution has essentially this form, the total population size, N_t, will be growing according to a constant multiplicative factor given by the dominant eigenvalue, λ,

$$N_{t+1} = \lambda N_t \tag{5}$$

There are special sorts of Leslie matrices for which age distribution does not converge to stable form, but these will not be encountered in analysis of long-lived repeatedly breeding animals where the vital rates are organized on a time scale which results in nonzero fecundity in at least two consecutive age classes. Details of the conditions for convergence are treated in Sykes (1969).

The stable age distribution, given by the schedule of fractions each age class represents in the ultimate population, is calculated from

$$c_x = \frac{\lambda^{-x} l_x}{\sum_{y=1}^{w} \lambda^{-y} l_y} \tag{6}$$

where c_x is the fraction of the population aged $x-1$ to x, and w is the oldest age achieved.

The factor of population increase is related to the life table schedules in the equation

$$\sum_{x=1}^{w} \lambda^{-x} \phi_x = 1 \tag{7}$$

This equation is a wth order polynomial in λ, so it will have w roots (solutions), but only one of these is positive and real, and this one has the largest modulus. The rate of convergence to stable age distribution is given approximately by the ratio of the one real positive root to the modulus of the root with the next largest modulus.

A final age specific statistic which will prove of great use is the schedule of reproductive value, V_x, defined as the number of ultimate descendants expected from the future reproduction of an individual now aged x relative to an individual now newborn. Reproductive value is calculated from

$$V_x = \frac{\lambda^{x-1}}{l_x} \sum_{y=x}^{w} \lambda^{-y} \phi_y \tag{8}$$

When too little is known about a particular population to construct a life table model based on annual age classes, a number of alternatives are available which still permit the use of the theory developed in this section. One is simply to use a coarser time scale, grouping individuals in age classes spanning more than one year, and to construct a Leslie matrix which projects the population vector through time intervals equal to the span of an age class. A certain amount of potential detail is lost in such lumping, but if the original data were available only in this form, we never really had the information that is "lost."

A second approach retains an annual time scale, even though the data allow significant statements only about classes of individuals spanning several years in age. This is accomplished by assuming that within each multiyear age class the vital rates expressed on an annual basis do not change from one annual age class to the next. The rates do change when the transition is made to the next multiyear age class. If this assumption of a certain smoothness of the vital rates as functions of age is reasonable, this approach will restore some of the detail that would have been lost by lumping age classes. Of course, if this assumption is radically wrong, as it would be if there were a major change in the real vital rates during the span of a multiyear age class, the "restored" detail would be spurious. Obviously, other smoothing assumptions might be made in order to flesh out an annually based life table from scanty data. In practice, these methods seem to work quite well, at least for long lived organisms. Several examples are illustrated by Goodman (in press).

Finally, it may prove more realistic to group individuals according to size, or some other morphological status or physiological state, without explicit reference to age. This renders it likely that the various classes will not span equivalent time intervals, and that their duration will, in any case, not equal the time unit of the projection matrix. For this reason, the projection matrix will not necessarily have the form of a Leslie matrix. For example, there may be nonzero entries in positions that would be impossible in a Leslie matrix. These might correspond to the probability that an individual will advance two classifications during a particular time unit, or that it will remain in its present class, and so on. Accordingly, the specific equations derived for the Leslie matrix are no longer applicable, but the underlying conceptual framework, and certain formal mathematical properties remain the same. In particular, it is still true that certain of the eigenvalues and eigenvectors of the new projection matrix are the critical quantities. The analogs of some, but not all, of the results reported in this chapter have been worked out for the more general projection matrix where organisms are grouped according to state (for example, Caswell, 1978).

REPOPULATION STRATEGIES FOR SLOWLY REPRODUCING POPULATIONS

Endangered wild populations may linger for some time with small numbers that are slowly declining or haphazardly fluctuating. Typically, these will be species that are long lived as adults, and which reproduce slowly, such as the California condor or Indian rhinoceros. Often it is easy enough to attribute the population's endangered status to habitat loss;

but practically, habitat restoration (in the sense of providing more acreage) may be the least feasible conservation measure. In some instances the population's food supply may, in any case, appear adequate. Then we may wish to investigate the relative payoffs of some specific management tactics intended to enhance the population's growth rate.

A given management program may have effects on the fecundity of, and may affect the survival rates of, particular age classes. The optimal allocation of management effort will be that which maximizes the increase in the population's rate of increase. That is, we are interested in evaluating the impact of modification of age specific vital rates on the ultimate factor of population increase.

It is not possible to write a general solution for equation 7 which will isolate the factor of increase, λ, on one side of an equal sign. Nevertheless, we may explore the sensitivity of λ to specific changes in the life table by implicit differentiation.

For modification of the fecundity of a particular age class, i, we obtain

$$\frac{\partial \lambda}{\partial m_i} = \frac{\lambda^{-i+1} l_x}{\sum_{x=1}^{w} x\lambda^{-x}\phi_x} \tag{9}$$

and for the special case of interest, where the factor of increase is very nearly one (zero growth), we note that

$$\left.\frac{\partial \lambda}{\partial m_i}\right|_{\lambda = 1} = \frac{l_i}{\sum_{x=1}^{w} x\phi_x} \tag{10}$$

Since the survivorship schedule is necessarily a decreasing function of age, equation 10 leads us to the common sense conclusion that an increase in the fecundity rates of a young age class will yield a greater improvement in the population growth rate than will a proportionate increase in the fecundity rates for an older age class.

For modification of the survival rate of a particular age class, we obtain

$$\frac{\partial \lambda}{\partial p_i} = \frac{\sum_{x=i+1}^{w} \lambda^{1-x}\phi_x}{p_i \sum_{x=1}^{w} x\lambda^{-x}\phi_x} \tag{11}$$

and for the special case of zero growth,

$$\left.\frac{\partial \lambda}{\partial p_i}\right|_{\lambda = 1} = \frac{\sum_{x=i+1}^{w} \phi_x}{p_i \sum_{x=1}^{w} x\phi_x} \tag{12}$$

The tail sum

$$\sum_{x=i+1}^{w} \phi_x$$

is insensitive to i for prereproductive ages, and decreases with i for reproductive ages. It is, of course, zero for ages past the oldest reproductive age. So, neglecting p_i in the denominator, the numerator of equation 12 indicates that it is more efficient to increase the survival rates of younger reproductive or prereproductive age classes than to increase those of older reproductive or postreproductive age classes.

Since it is to be expected that survival rates of the juveniles will be lower than those of adults in their prime, the p_i in the denominator of equation 12 makes the sensitivity of λ to changes in the survival rate of juveniles even greater, relative to older age classes, for most life tables.

Equation 12 can be reexpressed, with the help of the definition for reproductive value (equation 8), as

$$\left.\frac{\partial \lambda}{\partial p_i}\right|_{\lambda = 1} = \frac{l_i V_{i+1}}{\sum_{x=1}^{w} x\phi_x} \tag{13}$$

Comparison with equation 10 now reveals that the returns on an increase in survival rates of age class i will be greater than the returns on an increase in fecundity for age class i for any age where the reproductive value of a one age class older individual is greater than one. By definition, the reproductive value of a newborn is one, and for the usual life table reproductive value increases with age until it reaches a maximum near the age of peak reproduction, after which it declines. Thus, for any age class younger than the age of onset of senescent decline, it would be more advantageous to arrange an increase in the age specific survival rates than a proportionate increase in fecundities.

To be sure, these results must be weighed against the possibility that it may be easier to achieve an increase of a given magnitude in fecundities than in survival rates. That is, the analysis presented here calculates the payoff schedule; cost functions depend on the particular case.

It may be unrealistic to envision a management program which orchestrates changes of the life table, age class by age class, in a wild population. Interventions which accomplish an across the board increase or decrease in fecundities of all age classes, or survival rates of all age classes, are more likely.

If survival rates in all age classes are increased by a multiplicative factor D, and if we apply this same factor to the fecundity schedule $\{m_x\}$ on the assumption that the implicit survival term accounting for survival from the instant of birth until the first census behaves accordingly, this amounts to multiplying the entire Leslie matrix by D. Then the new factor of increase relative to λ_0, the pre-intervention factor in increase, is simply

$$\lambda = D\lambda_0 \tag{14}$$

There is no equivalent formula relating the achieved λ to an increase in age specific birth rates (that is, applying only to the instantaneous birth rate component of m_x, and so being applied only to the m_x schedule and not the p_x schedule) in a general life table.

To gain some insight into the relative advantages of changes in fecundity rates or survival rates of all age classes, we must resort to approximating the net maternity function of the actual life table with some special distribution which permits algebraic solution of the summation in equation 7. Consider, for example, a situation where the mortality rates for all the adult age classes are the same, and where the fecundities of all adult age classes are the same. Then we may replace m_x with m, for x greater than or equal to α, the age at first reproduction, and $m_x = 0$ for x less than α; and we may replace p_x with p, for x greater than or equal to α. With these substitutions, the summation in equation 7 becomes a geometric series, which simplifies the relationship between λ and the life table, yielding the equation

$$\lambda^{\alpha} - p\lambda^{\alpha-1} - ml_{\alpha} = 0 \tag{15}$$

where l_α is the survivorship until the age of first reproduction.

Implicit differentiation of equation 15 reveals the sensitivity of λ to modifications affecting the entire life table, for life histories where the actual net maternity function reasonably fits a geometric sequence. Evaluating the derivatives for the relevant case of populations experiencing zero growth, we find

$$\left.\frac{\partial\lambda}{\partial m}\right|_{\lambda=1} = \frac{l_a}{p + \alpha\,(1-p)} \tag{16a}$$

$$\left.\frac{\partial\lambda}{\partial p}\right|_{\lambda=1} = \frac{1}{p + \alpha\,(1-p)} \tag{16b}$$

$$\left.\frac{\partial \lambda}{\partial l_a}\right|_{\lambda = 1} = \frac{m}{p + \alpha(1 - p)} \qquad \textbf{(16c)}$$

$$\left.\frac{\partial \lambda}{\partial \alpha}\right|_{\lambda = 1} = 0 \qquad \textbf{(16d)}$$

Since survivorships are necessarily less than one, equations 16a and b show us that more will be gained from increasing the survival rates of all adult age classes by a certain amount than from increasing the fecundities of all adult age classes by the same amount. The difference in the gain will be greater for life tables where survivorship until the age of first reproduction is low.

Equation 16c indicates that the gain in population growth rate from an increase in the rate of survival until the age of first reproduction will be proportional to fecundity. For life histories with the moderate adult fecundity rate of one female offspring per adult female (bearing in mind that this refers to offspring that survive to be censused at the beginning of the next time interval, so the actual birth rate would have to be higher than this value), the sensitivity of λ to the rate of survival until first reproduction is the same as the sensitivity to the adult survival rate. For low fecundity life histories—for example, one offspring of either sex per adult female per year—the sensitivity to l_α is appreciably less than the sensitivity to p.

The insensitivity of population growth rate to the age at first reproduction (equation 16d) requires further explanation. When α alone is varied, the effect is that of displacing the net maternity function, $\{\phi_x\}$, along the x axis. In fact, a marginal change of this sort will not affect λ for a zero growth life table. If, however, changing the age at first reproduction also affects the survival to the age at first reproduction, there will be a change in the population growth rate in accord with equation 16c.

Most statistics computed on the assumption of a geometric series net maternity function will be quite accurate when applied to data for long lived, slowly reproducing organisms. For life histories which do not fit this pattern, other analytical distributions are available which may be substituted as approximations for the actual net maternity function.

When a great deal of detail is known about the life table of the population in question and about the specific consequences for the schedule of vital rates if the contemplated interventions are instituted, optimization techniques developed by Taylor, Gourley, Lawrence and Kaplan (1974) may be applied to find the particular program of management that will yield the maximum achievable population growth rate. In the absence of such detailed information, the more general results of this section will at least point the way toward economical investment of management efforts in the attempt to increase a population's rate of increase.

STABILIZATION OF CONTROLLED WILD POPULATIONS

When a wild population is too numerous for the available resources, some program of population control must be instituted to avert habitat deterioration and a population collapse. Two considerations weigh heavily in the design of the control program. It is usually desirable that the managed population, so maintained, be as large as is compatible with the habitat, and it is desirable that the ongoing program of removals be as simple as possible to administer. A large population size caters to the esthetic delight of seeing a lot of animals, and it may confer biological benefits through the maintenance of genetic diversity and by lessening the probability of chance extinction. Yet this goal conflicts with the objective of simplifying the control program, for the more closely a population presses upon the carrying capacity, the more precise must be the regulation of its numbers.

A population's ultimate growth rate may be calculated just from its life table. But, unless the population has a stable age distribution, the actual rate of growth in numbers depends on an interaction between the life table and the current population age structure. For this reason, arranging for a managed population to achieve a schedule of fecundities and survival rates which confer an ultimate factor of increase, λ, equal to one, does not guarantee that the population's numbers will be constant.

Imagine, for example, a managed population initiated with a cohort of newborn under a regulated life table with an associated λ of one. At first, the population's numbers will decline. None of the individuals will be of reproductive age, and so the only processes represented in the population will be ageing and death. As the cohort reaches reproductive age, there will be a burst of reproduction, so the population will then increase in numbers temporarily. The offspring which appear during this episode will contribute no further reproduction till their period of maturation has passed, so there will be a plateau, or perhaps another slump, in the birth rate followed by yet another boom. Ultimately, these oscillations will damp out, as the population settles into its stable age distribution and manifests its zero growth rate; but, in the interim, the changes of numbers and the episodes of peculiarly skewed age distribution may have brought some harm either to the population or to its environment.

A continual revision of the harvesting schedule, so as to compensate for the inherent oscillatory dynamics of the population, frustrates our desire to have the control program be simple to implement. In any case it may not succeed, for a changing control program imposes a time-varying life table on the population, driving it even farther from what might have been its stable age distribution. The ideal situation would be a population initiated with an age distribution matching the stable distribution associated with the zero growth life table that will be achieved under a constant management scheme. This would bypass the period of oscillatory behavior which preceeds the attainment of a stable age distribution. Evidently, then, the solution is to trim the population to the desired population size and the appropriate age distribution with a very detailed and carefully calculated culling at the inception of the control program. Thereafter, it will be necessary only to maintain a constant and relatively unselective program of population control.

Formally, the solution is applied as follows. Assume, for example, that a simply applied culling program would consist in annually removing a fixed fraction, h, of individuals in age classes a through b. The ages to be harvested might be selected on the basis of convenience, economic yield or the manager's judgement that certain age classes are relatively expendable. If it is difficult to determine the precise age of a free individual, the ages (a,b) for a practical removal program may have to cover a broad span (for example, obviously mature individuals or all obviously immature individuals, or, even all individuals). The rate of removal must confer zero population growth, so the value of h must be the solution to the equation

$$\sum_{x-1}^{a}\phi_x + \sum_{x=a+1}^{b+1}\phi_x(1-h)^{x-a} + (1-h)^{b-a+1}\sum_{x=b+2}^{w}\phi_x = 1 \tag{17}$$

where the vital rates refer to schedules of fecundity and survivorship achieved prior to the institution of the removal program.

The associated stable age distribution in the controlled population will be given by

$$c_x^* = \begin{cases} \dfrac{l_x}{s} & \text{for } x \leqslant a \\[2ex] \dfrac{l_x(1-h)^{x-a}}{s} & \text{for } a < x \leqslant b+1 \\[2ex] \dfrac{l_x(1-h)^{b-a+1}}{s} & \text{for } b+1 < x \end{cases} \tag{18}$$

where

$$s = \sum_{x=1}^{a} l_x + \sum_{x=a+1}^{b+1} l_x(1 - h)^{x-a} + (1 - h)^{b-a+1} \sum_{x=b+2}^{w} l_x \tag{19}$$

Accordingly, the first year of the control program should consist of tailoring the population's age distribution to match the schedule of equation 18. Thereafter, removal of a fraction h of all individuals in age classes a through b, where a and b were chosen in advance (at least in part for convenience), will suffice to maintain a stable population.

OPTIMAL AGE STRUCTURE FOR A CAPTIVE POPULATION

When a captive population begins to grow to such a size that it threatens to overrun the available facilities, some program of removal must be initiated. Just as in the above section, one of the objectives of the removal schedule will be the attainment of an effective life table which confers zero population growth. However, since in this case the population is so small and so carefully observed that the history, or at least the age, of each individual is likely to be known, a quite detailed age specific schedule of removal may be worth considering for an ongoing plan of management.

Let the removal schedule be designated $\{h_x\}$ where h_x is the fraction of age class x removed during each time interval. The zero growth feature is achieved for any removal schedule which satisfies the relation

$$\sum_{x=1}^{w} \phi_x \prod_{j=1}^{x-1} (1 - h_j) = 1 \tag{20}$$

where ϕ_x refers to the vital rates in the absence of removals.

Equation 20 leaves a great deal of flexibility in the apportionment of removals among the various age classes. The schedule of removals will determine the stable age distribution achieved in the managed population according to the relation

$$c_x^* = \frac{l_x \prod_{j=1}^{x-1} (1 - h_j)}{\sum_{y=1}^{w} l_y \prod_{j=1}^{y-1} (1 - h_j)} \tag{21}$$

Thus we may enquire how to allocate the age specific removal rates so as to obtain the most favorable age distribution.

Generally, the highest priority in the management of a small captive population will be minimization of the probability of extinction. Aside from genetic considerations, we may think of strategies for reduction of extinction rates as having two components: minimizing the population's vulnerability to chance events which reduce the population growth rate, and maximizing the population's capacity for growth during a recovery episode. We will first treat the latter.

An individual's contribution to population growth is measured by its reproductive value. In fact, the summed reproductive value of a population grows exponentially at a rate given by the ultimate rate of population increase, regardless of the age distribution. So, the force of population growth in a given population may be measured as the product of λ and the population's summed reproductive value. Thus, in order to maximize the population growth potential of a certain number of individuals, we would attempt to maximize their total reproductive value. In other words, the age distribution for the managed population which will maximize that population's potential for recovery (growth)—whenever the removal program is suspended—will be the age distribution which maximizes the per capita reproductive value in the population (with reproductive value calculated on the basis of the vital rates in the absence of removals).

We can refer to the schedule of reproductive values, calculated according to equation 8, and find the age at which the highest reproductive value occurs. A population consisting solely of individuals in this age class certainly will, in some sense, have a maximum per capita reproductive value, but of course this is not a sustainable solution. The individuals in the population will age, and they will give birth to individuals which necessarily enter the population through age class one. So the population's per capita reproductive value will decline relative to the initial situation. A supply of individuals of the desired age will require a steady stream of younger individuals which will then enter this age class, and so on. Evidently, our problem is to reconcile the desire for concentrating individuals in certain age classes with the need to have the removal schedule permit the maintenance of the target age distribution: the target age distribution must correspond to some age distribution achievable by manipulating the schedule $\{h_x\}$ in equation 21 subject to the constraint of equation 20.

Beddington and Taylor (1973) have proven that a large class of harvest problems of this sort, where the objective is to maximize some per capita age specific property in either the harvested individuals or in the population that remains, will necessarily be solved by a harvest which removes all the individuals at some particular age and which partially harvests one other age class. This formal result makes possible an efficient computer search for the optimal removal schedule in our problem. The algorithm is as follows:

1. Treating the ages j and k as unknowns, where k is the age of total removal and j is the age of partial removal, sequentially try values of k,

starting from the oldest in the life table.

2. For each trial value of k, test all possible values of j, between one and $k-1$. Satisfying the zero growth constraint (equation 20) now reduces to

$$\sum_{x=1}^{j}\phi_x + (1 - h_j)\sum_{x=j+1}^{k}\phi_x = 1 \tag{22}$$

thus yielding an explicit solution for h_j given any particular value for the ages j and k. The test criterion for each combination is the value of V_r, the equilibrium per capita potential reproductive value associated with each set of ages (j,k) according to the equation

$$V_r = \sum_{x=1}^{w} c_x^* V_x \tag{23}$$

The age distribution c_x^* is calculated with the removals taken into account, whereas the reproductive values are calculated on the basis of the life table in the absence of removals. Thus equation 23 reduces to

$$V_r = \frac{\sum_{x=1}^{j}\sum_{y=x}^{w}\lambda^{x-y-1}\phi_y + (1 - h_j)\sum_{x=j+1}^{k}\sum_{y=x}^{w}\lambda^{x-y-1}\phi_y}{\sum_{x=1}^{j} l_x + (1 - h_j)\sum_{x=j+1}^{k} l_x} \tag{24}$$

where all the parameters refer to the life table in the absence of removals.

In this test, retain the value of j which confers the largest V_r for a given value of k, and retain that value of V_r.

3. Return to step 1 for the next trial value of k.

4. Once step 2 has been performed for all possible values of k, find the largest among the values of V_r retained. The combination of j and k associated with that value of V_r will be the ages of the partial removal and the total removal, respectively, in the optimal culling program, and the optimal partial removal rate will be given by equation 22.

The removal program arrived at in this manner will result in an age distribution which maximizes the potential for growth in the managed population. The second consideration, minimizing the vulnerability of the managed population, is accessible to a similar program of analysis. The only modification of the algorithm is that the life table parameters substituted in the calculation of V_x in equation 23 should be calculated on the basis of the vital rates achieved under the worst expected circumstances (for example, during an episode of disease), in the absence of the removal. This will result in a stable age distribution where the most resistant age classes dominate to the extent possible. The balance between maximizing recovery rates and minimizing vulnerability will be a matter of judgement to be weighed in each case.

A NUMERICAL EXAMPLE: THE PRIBILOFF FUR SEAL LIFE TABLE

To complement the abstract formulations, developed in the preceding sections, with some actual calculations of the sort that are likely to be encountered with a real population, we will consider the case of the Pribiloff fur seal. This is a relatively cheerful matter; at present the population is not in danger of extinction, and seems to be faring well under a program of management negotiated by international treaty.

Early in this century, unregulated harvesting greatly reduced the fur seal population. The governments involved in the commercial take then agreed at first to cease all harvest. Later, as the population began to recover, a controlled harvest, concentrating almost exclusively on prereproductive males, was instituted. Under this program the population continued to increase in numbers—at an average annual rate of about 8 percent—for about two decades, after which the population stabilized at somewhere near half a million adult females. The population remains at that level today.

The history of the fur seal population, and a critical review of the research on its dynamics, are presented in a recent article by Smith and Polacheck (1979). The values for the vital rates employed in the present chapter are adapted from their tables. We will consider only the female segment of the population.

The smoothed schedule of present fecundity rates is shown in Figure 1. This set of values is based on proportions of females observed to be pregnant in a set of pelagic samples taken about twenty years ago.

The available survivorship data are less satisfactory. Observations of numbers of individuals in different age classes over several successive years allowed estimates of rates of survival from one age class to the next, at that time. The estimates for ages twelve and older seem relatively secure. For ages eight to 12 there were some complicating circumstances which render the estimates less reliable. No direct observations were made on survival rates of younger age classes, except for one number bearing on the survivorship from birth to age four. A smoothed schedule of survival rates is presented in Figure 2. Rates for the younger age classes were filled in by first extrapolating and then by adjusting rates in the youngest age classes to be consistent with the assumption of zero population growth. That is, the values were adjusted until the net maternity

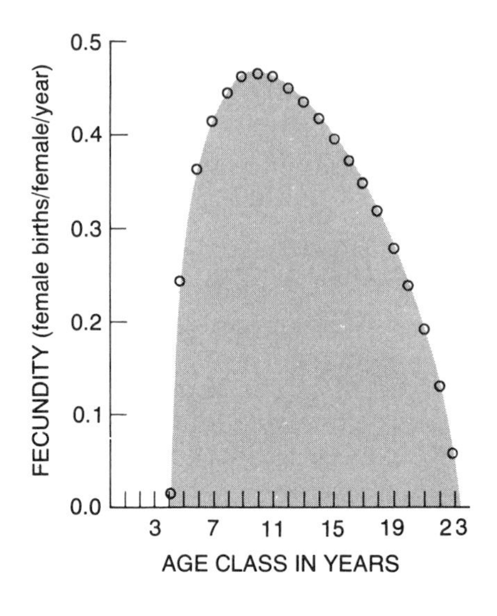

FIGURE 1. Fecundity rates, expressed in female births per female per year in each age class, for the fur seal population during the past few decades of apparent equilibrium.

function satisfied the special case of equation 7 for $\lambda = 1$,

$$\sum_{x=1}^{w} \phi_x = 1 \tag{25}$$

where the net maternity function is calculated from the fecundity and survivorship schedules according to equation 3, and the survivorship schedule is obtained from the survival rates of Figure 2 according to equation 2.

The net maternity function that results is shown in Figure 3. There

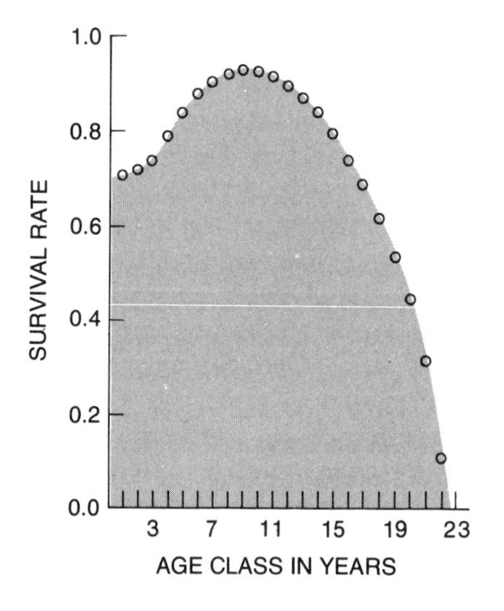

FIGURE 2. Survival rates for the fur seal population at equilibrium, expressed as the fraction of each age class surviving to enter the next age class.

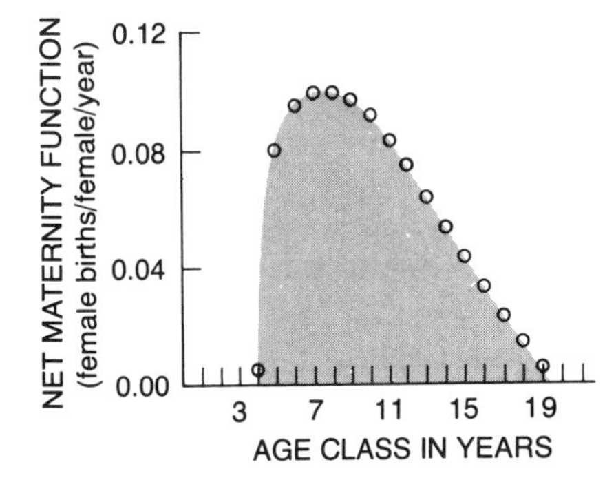

FIGURE 3. Net maternity function for the fur seal population at equilibrium, expressed as expectation of female births per year at each age per newborn female.

we see that the preponderance of reproductive activity attributable to a cohort accrues between ages five and 12, even though fecundity rates (see Figure 1) remain substantial for older individuals. The explanation is that fewer individuals survive to reach these greater ages, as may be seen from the survivorship schedule in Figure 4.

The sensitivity of the factor of increase to changes in age specific fecundity or survival rates, calculated according to equations 9 and 11 is shown in Figure 5. There we see that the growth rate is far more sensitive to modifications in the vital rates applying to younger age classes, as predicted by our abstract evaluation of these sensitivities and by our analy-

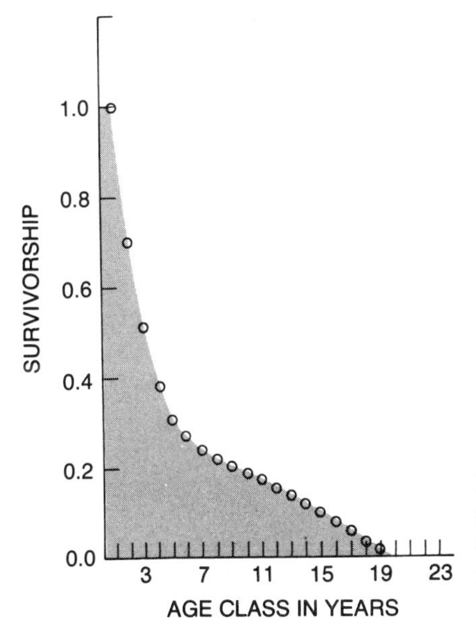

FIGURE 4. Survivorship for the fur seal population at equilibrium, expressed as the fraction of female newborn that will survive to enter each age class.

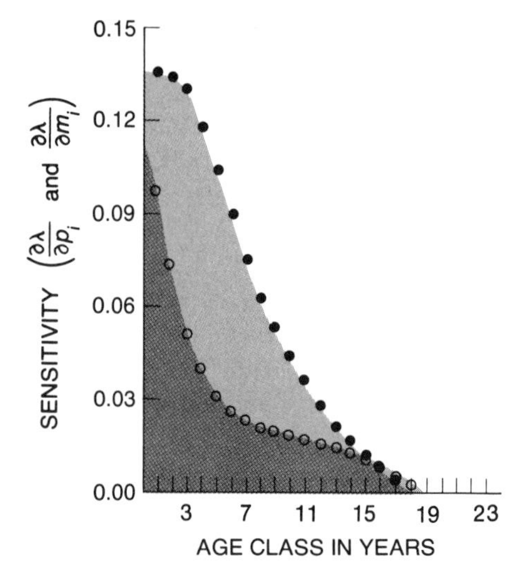

FIGURE 5. Sensitivity of the population growth rate (annual factor of increase) to small changes in the survival rates (●) and fecundity rates (○) calculated for the fur seal population at equilibrium.

sis of the approximate model (equations 16a–d). In all except the very oldest age classes, the growth rate is more sensitive to changes in survival rates than in fecundity rates, often by a wide margin. For example, according to the graphs, an addition of 0.01 female births per year to females in age class 10 would add 2×10^{-3} to the factor of population increase, whereas an addition of 0.01 to the survival rate in age class 10 would add 4.5×10^{-3} to the factor of population increase. The same addition to the survival rate of age class three would add 13.0×10^{-3} to the factor of population increase. Thus, a population manager can use Figure 5 as a means of calculating the yield, in terms of increase in population growth rate, for any particular investment in enhancing age specific survival rates or fecundity rates.

We have little in the way of direct evidence concerning the actual life table that applied to the fur seals during the period when the population was experiencing sustained growth of eight percent annually. On a number of grounds it is believed that the difference between that life table and the life table that applies at present (equilibrium) was confined almost exclusively to higher survival rates for the youngest individuals. A schedule of survival rates, modified in this manner, that, in conjunction with the unmodified fecundity schedule, confers the observed growth rate is shown in Figure 6.

Let us imagine that it had been decided that the population was becoming too dense during this period of sustained growth. Any number of culling programs could be devised to result in a life table conferring zero growth ($\lambda = 1$). It would simply be a matter of multiplying the original survival rates by one minus the imposed artificial mortality, age class by age class, until the realized survivorship schedule yielded a net maternity function conferring a net replacement rate of 1.0. One form of this solution is given in equation 17. In this hypothetical case, the population might be thinned to the desired population level, and then an imposed harvest rate, given by equation 17, would ultimately result in a zero growth rate.

It might be convenient to harvest the females in the prereproductive age classes, say ages three through five, since males in this age range are already being harvested for the fur trade. A solution for h in equation 17, where a is three and b is five, yields a value 0.225. In other words, this fraction of individuals in age classes three, four and five should be removed each year. If, instead, it was decided to harvest only at age four, 0.534 of these animals would have to be removed annually. We will use this latter schedule to illustrate the problem of stabilizing the harvested population.

Associated with the life table realized under this harvest schedule, there will be a new stable age distribution, shown in Figure 7, alongside the stable age distribution of the unharvested population calculated from equation 6. As we can easily surmise from the discrepancy between the two age distributions, a population growing according to the unharvested life table will have an age distribution quite different from the age distribution which would yield a realized zero growth rate under the harvest. There will be some transition period during which the population converges to the new stable age distribution; during this time the actual growth rate may depart considerably from the ultimate rate of increase associated with the harvested life table. In fact, from the time the harvest is imposed on the growing population in our example, the population

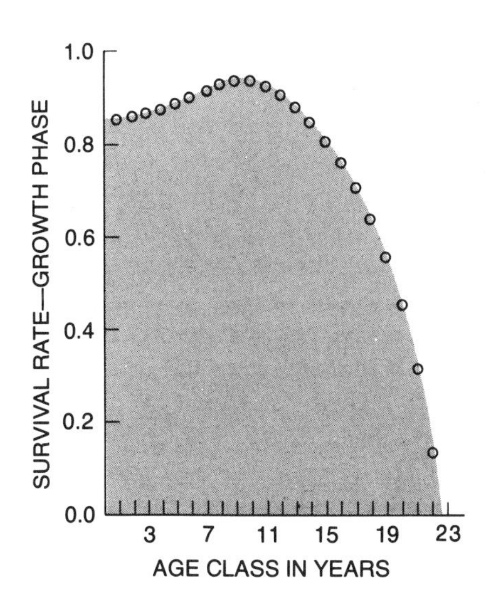

FIGURE 6. Presumed schedule of survival rates applying to the fur seal population during the period of population growth earlier in this century.

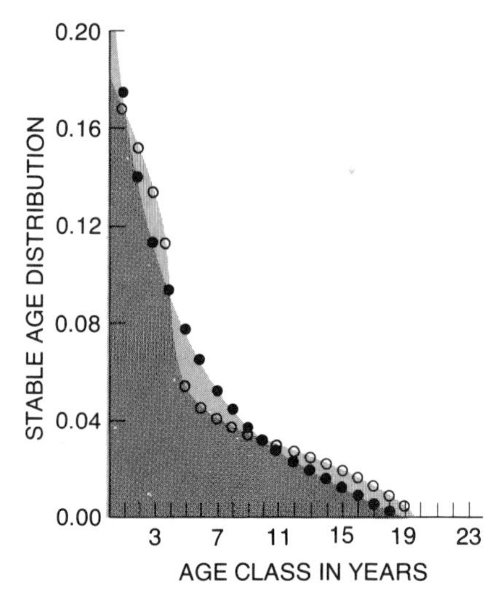

FIGURE 7. Stable age distribution, expressed as the fraction of the female population in each age class, calculated for the fur seal population during the growth phase (●) and under a hypothetical removal program adjusted ultimately to achieve zero population growth through a constant fractional harvest of age class four (○).

will actually increase about 12 percent, reaching a maximum value in eight years. It will then decline about two percent and finally stabilize at a density 10 percent greater than when the harvest was first initiated (Figure 8). If the population enters the period of the "stabilizing" harvest with some arbitrary age distribution (as will occur if no thought is given to tailoring the age distribution during the initial removal episode which culls the population to the desired density), these transient damped oscillations can be much more severe.

If, for example, age classes one through 20 were, in the interests of demographic democracy, equally represented in the founding population after the initial culling, the subsequent population history under the constant harvest would appear as in Figure 9. Here the population overshoots the desired level by almost 25 percent before stabilizing near the target density. That magnitude of overshoot could have very serious consequences.

In this last situation we might be tempted to "correct" the population trajectory by revising the harvest rate each year in response to the observed population growth that year. For example, we could apply equation 11 to the life table realized by harvesting and arrive at some number giving the sensitivity of the growth rate to the realized survival rate of age class four (the harvested age class). This would, in fact, yield the sensitivity of the growth rate to the harvest rate. We might then observe the deviation of the value of the actual factor of increase from one, calculate the change in the harvest rate necessary to change the factor of increase to exactly one, and revise the harvest rate accordingly. The theoretical error in this approach, of course, is that the growth rate referred to in equation 11 is the ultimate rate of increase, not necessarily the realized rate of increase. For the two to be the same, the population must be in a

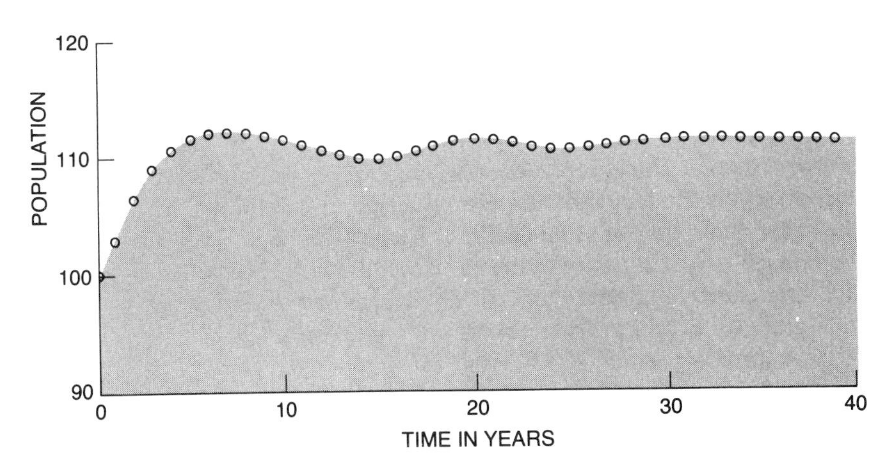

FIGURE 8. Record of the total population size (scaled as the percent of the starting population size) of a hypothetical fur seal population, initially in the growth phase and in the stable age distribution associated with that life table. The population is subjected to a stabilizing harvest beginning at time 0; it continues to grow for several years before actually stabilizing.

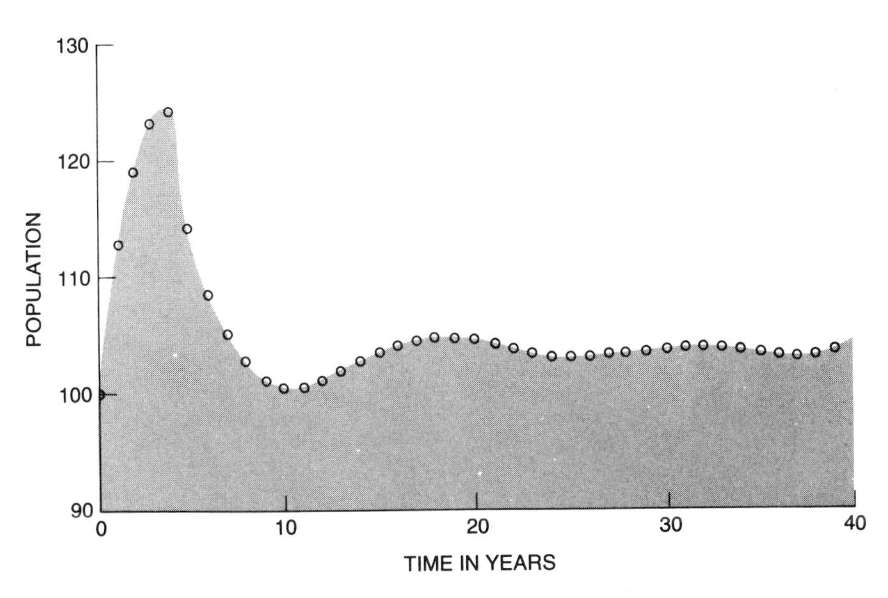

FIGURE 9. Record of total population size (scaled as the percent of the starting population size) of a hypothetical fur seal population initially in the growth phase. The population is subjected to a stablilizing harvest at time 0, where the age distribution at time 0 is uniform for age classes one through 20. This results in a greater initial overshoot than in the example in Figure 8.

stable age distribution, which in fact was the original problem that generated the nonzero growth rate in the harvested population. Thus the proposed "correction" will actually function as a haphazard sort of overcorrection, creating new departures from stable age distribution. The consequences of these departures become apparent only after some time lag—owing to the fact that the reproductive contribution attributable to the individuals that are harvested, or spared the harvest, at age four can be realized only after they reach (or would have reached) their reproductive span, centered around age 10. The trajectory which ensues, as shown in Figure 10, exhibits more persistent oscillations than in the example without "correction" of the harvest rate.

The proper solution, as described above, is to trim the population selectively in the first culling, so that the founding population not only has the desired overall density, but also has as its age distribution the distribution calculated by substituting into the formula for the stable age distribution (equations 18 and 19) the life table realized under harvest. This will result in a truly stable population from the very start.

The recovery potential, V_r, of the harvested population is the sum of the products of the age class frequencies in the harvested population and the age-specific reproductive values calculated for the unharvested popu-

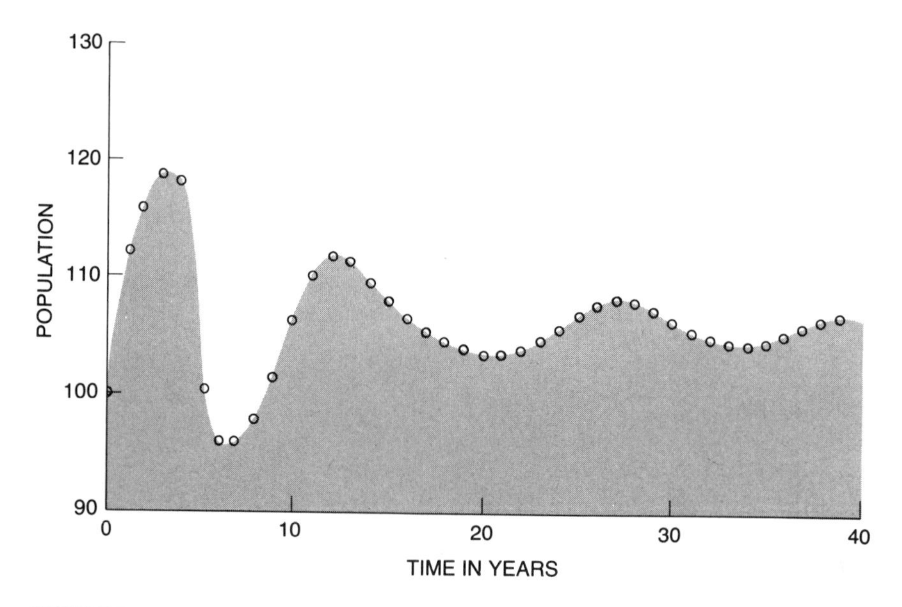

FIGURE 10. Record of the total population size (scaled as the percent of the starting population size) of a hypothetical fur seal population, initially in the growth phase, subjected to a stabilizing harvest beginning at time 0. Here the harvest rate is adjusted each year according to the population growth that year, which results in more persistent oscillations than occurred under a constant harvest rate.

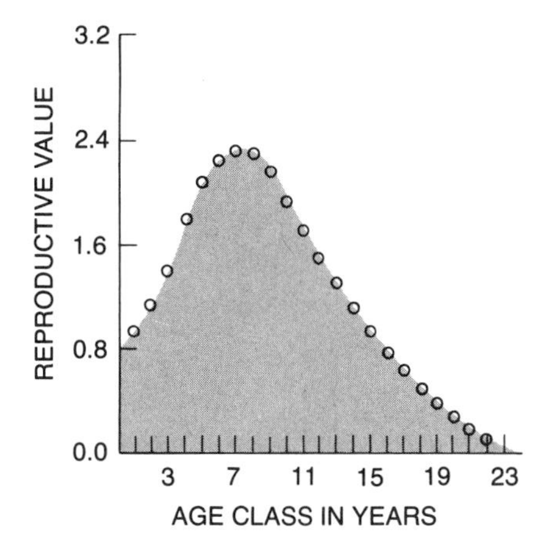

FIGURE 11. Age-specific female reproductive value expressed in female newborn-equivalents, calculated for a fur seal population during the growth phase.

lation, as given in equation 23. This value for the fur seal population with the harvest confined to age class four is 1.55. Spreading the harvest over ages three, four and five improves the recovery potential very slightly. This prompts us to search through other possible harvest schedules in the hopes of finding one which substantially improves upon the recovery potential of our first schedule. But, the number of possible schedules is infinite, so an undirected search is not feasible.

The objective in maximizing V_r is to achieve an age distribution in the harvested population which, to the extent possible, increases the representation of age classes with high reproductive values and reduces those age classes with low reproductive values to a small fraction of the population. In some sense we might want the age distribution to parallel the schedule of reproductive value. The schedule of reproductive value for the unharvested seal population, calculated from equation 8, is shown in Figure 11. Comparison with the age distribution under harvest, shown in Figure 7, makes it obvious that this particular harvest schedule is a rather bad one.

In fact, the optimal harvest, arrived at according to the algorithm described in the previous section of this chapter, yields a recovery potential of 1.78. This is an improvement of 15 percent over the prior harvest schedule, which is to say that the optimally harvested population has the same recovery potential as a 15 percent larger population harvested according to the arbitrary program of removing individuals at age four. By the same token, a captive population released from the optimal removal program and allowed to increase at its potential rate (as in a deliberate repopulation program) would soon come to be 15 percent larger than a population founded by the same number of individuals which had, at the

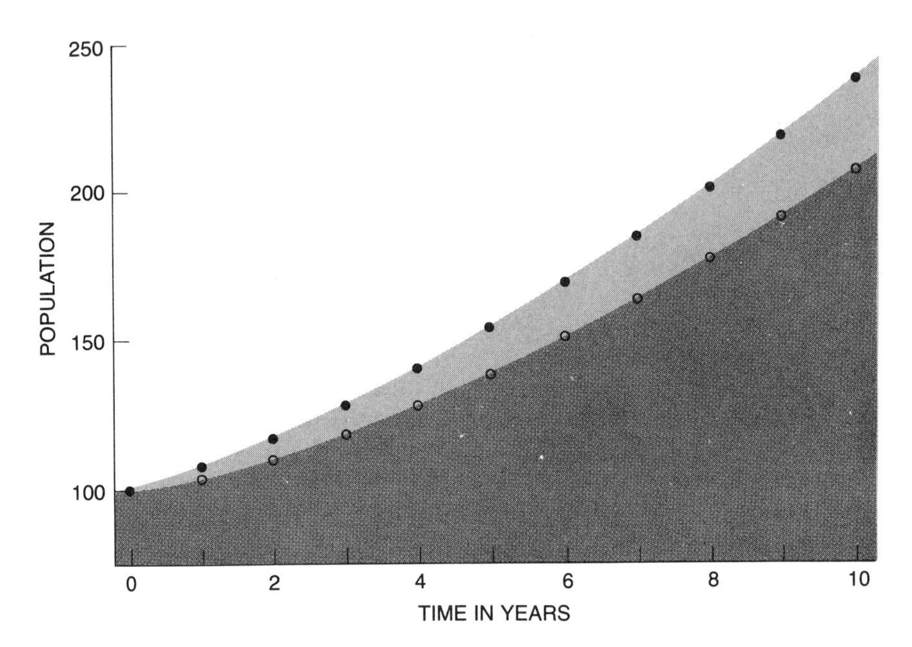

FIGURE 12. Two hypothetical fur seal populations, previously maintained at 100 individuals by removal programs. In one (○) the harvest was confined to age class four, while in the other (●) the removal schedule maximized the recovery potential (see text). At time 0 both populations are released from their respective harvests and they increase according to the same, growth-phase vital rates. The more favorable initial age distribution in the optimally harvested population soon becomes evident in its more rapid growth.

same time, been released from the arbitrary removal program, as shown in Figure 12.

Some other arbitrary removal programs will prove even worse, others not quite so bad. The optimal schedule for this fur seal population consists in removing 8.8 percent of the individuals in age class nine and all in age class 10.

SUMMARY

In this chapter we have considered three types of situations where the techniques of mathematical demography will be helpful in devising programs for the effective management of a population. In all three the precision and generality of the mathematical models permits the formulation of strategies that, compared with management schemes arrived at by more intuitive means, are superior with regard to the rate of population growth, the probability of population survival or the ease of program implementation. The data demands for successful application of these analyses are high, but not unrealistic in the context of closely managed endangered populations.

The first situation considered was that of a population, the growth rate of which we wished to increase. This problem was investigated by inquiring into the response of the population's growth rate to very specific modifications of birth rates and survival rates, age class by age class. The formulae permitted identification of particular stages of the life history which are especially important in determining the population growth rate. With these formulae, a decision maker working with a constrained budget of time and material resources may calculate how best to invest a limited management effort so as to achieve the greatest possible increase in population growth rate.

The second situation considered was that of a wild population which required regular culling to prevent habitat deterioration or other deleterious concommitants of overcrowding. The problem was to devise a program of removals which was both simple to institute and which, to the extent possible, resulted in stable population numbers. Here the solution was developed in terms of the stable age distribution associated with a convenient harvesting schedule.

The third situation considered was that of a captive population which required culling to keep it within the limits of available facilities. Here the key problem was minimizing the probability of extinction of a small population, rather than concern with ease of carrying out the removal program. The solution was developed by showing how to find an age specific pattern of culling that would maximize the recovery potential of the population that remained.

All three classes of calculation were illustrated in the example of the life table of the Pribiloff fur seal population.

SUGGESTED READINGS

Goodman, D., 1979, *Management Implications of the Mathematical Demography of Long Lived Animals,* National Technical Information Service (NTIS No. PB-289 673), U.S. Dept. of Commerce, Springfield, Virginia. A review of evolutionary and management aspects of the literature on mathematical demography.

Keyfitz, N., 1968, *Introduction to the Mathematics of Populations,* Addison Wesley, Reading, Massachusetts. A very thorough presentation of the mathematical methods used to model age structured populations.

PART THREE

CAPTIVE PROPAGATION AND CONSERVATION

CHAPTER 11

AN OVERVIEW OF CAPTIVE PROPAGATION

William G. Conway

Recently a committee appointed by the American Ornithologists' Union and the National Audubon Society reviewed the status of the endangered California condor *(Gymnogyps californianus)* and recommended a program of captive propagation for its preservation (Ricklefs et al., 1978). The committee proposed that "A large portion of the present population should be trapped and bred in captivity . . . [and that] . . . propagation should be continued long enough to produce at least 100 second-generation or third-generation offspring for the release to the wild." In effect, it proposed the creation of a living gene bank.

Captive propagation has had a simple-minded directness; an immediacy that offers an attractive reinforcement to the complex geo-political tasks of habitat preservation. It has been a kind of "ark." Such programs pretend to offer no overall cure for the epidemic of extinction but provide topical treatments of the symptoms expressed by the loss of higher animals. Simply expressed, captive propagation offers another way of fighting the continuing reduction of the earth's diversity—an opportunity to preserve options.

Nearly one-twelfth of all the species of birds and one-sixth of the mammals have been bred in zoos in the past two years (Olney, 1976*a* and *b*; 1977*a* and *b*). At the same time, the continued decline and isolation of relatively large, conspicuous animals combined with the availability of new techniques in wild animal husbandry have led to more manipulative approaches to species preservation (Conway, 1978; Chapter 13). For some animals the only immediate hope of survival is in captivity. While the concept of captive propagation and release to the wild is encouraged by recent advances in animal care, it is challenged by inadequate existing facilities and socio-economic mechanisms, and by the disappearance of the "wild."

Captive bred stocks of wild animals could fulfill four functions in biological conservation: (1) as substitutes for wild populations in basic research in population biology and sociobiology (for example, Chapter 14); (2) as substitutes for wild populations in the development of care and management techniques; (3) as demographic and genetic reservoirs from which infusions of "new blood" may be obtained or new populations founded; and (4) as last redoubts for species which have no immediate opportunity for survival in nature.

Already there are several examples of these kinds of programs producing satisfactory results. Between 1907 and 1917 American bison *(Bison bison)* from the Bronx Zoo's breeding herd were used to restock empty ranges in Oklahoma, Montana and South Dakota, while European bison *(Bison bonasus)* bred in zoos in Sweden and elsewhere were used to return that species to nature after World War II (the last wild individual had been killed in 1921). Captive populations of several species—the Père David deer *(Elaphurus davidianus)* and the Mongolian wild horse *(Equus przewalski)*—are fulfilling the final function: sustaining their kind in captivity in the absence of a population in nature.

Programs to save endangered species of birds exemplify the second function of captive propagation, that of using captive animals of one species to help preserve a rare species. Red-tailed hawks *(Buteo jamaicensis),* Harris' hawks *(Parabuteo unicinctus),* prairie falcons *(Falco mexicanus)* and gyrfalcons *(Falco rusticolus)* have acted as substitutes in developing programs or even as surrogate parents for the peregrine *(Falco peregrinus)* in the pioneering propagation work of Cade and his associates at Cornell University. Sandhill cranes *(Grus canadensis)* have been used by researchers at the Federal endangered species breeding program at Patuxent as stand-ins for the whooping crane *(Grus americana).* More recently, Cade's captive peregrines have acted as a reservoir to provide chicks for an attempt to reestablish a peregrine population in the Eastern United States where the species is extinct. Patuxent's captive whoopers are providing eggs for the attempt to establish a new whooping crane population in Idaho.

OBTAINING A GENE BANK

Biologically, operation of captive propagation programs demands that two problems be solved: how to get animals and how to maintain them. Obtaining wild animals for captive herds and flocks has usually been a fortuitous (if not random) process. Zoos and researchers have felt lucky to have even a rough idea of their animals' provenance when they obtained them from dealers. Even when biologists have collected their own founder groups, they have rarely attempted to assess more than the probable demographic make-up of their captures. Few physiological parameters other than those demanded for health reasons have been measured and no attempts have been made to appraise degree of genetic interrelatedness (Chapter 12). Captive propagation programs should be founded upon an in-depth evaluation of genetic, demographic and behavioral factors (Chapter 14). Each individual should be marked and its relationship with others, insofar as it can be observed at the time of capture, should be noted with a view towards constructing sounder breeding groups in captivity (Chapter 12).

MAINTAINING A GENE BANK

Once a species is obtained, maintaining it relatively unchanged over long periods is the heart of the gene bank task. Wilson's suggestion (1975) that man might reach an "ecological steady state" by the end of the 21st Century offers a useful benchmark, albeit a dubious one. Thus a minimal goal for the maintenance of gene banks might be 120 years. In any case, captive stocks will probably begin to show signs of inbreeding depression if kept a long time as small populations (Chapters 9 and 12).

Many captive groups and many introduced animal and plant immigrants have clearly started from small founder groups (Elton, 1958). Only occasionally, however, have animals sustained themselves where their populations were not permitted to expand rapidly. This provides evidence for the current belief (Chapters 8, 9 and 12) that founder (bottleneck) size is relatively unimportant compared to the maintenance population size. Tahr *(Hemitragus jemlahicus)* and red deer *(Cervus elaphus)* populations in New Zealand (now very large) were started with small numbers, as were various introduced pheasants, partridges, honeybees and black swans.

Only a few animals, in contrast, have been maintained for long periods in zoos. Countless captive populations have been lost in a matter of a few decades, mostly from a combination of stochastic events occuring in the demography of very small isolated populations and from inbreeding depression. Of course, serious attempts to maintain wild animals on a long-term basis are of very recent origin, with the exception of the much noted efforts with the Père David deer, European bison and Mongolian wild horse (Pinder and Barkham, 1978). In consequence, little is known of the characteristics of captive animal populations compared to those in nature. In planning, however, it is reasonable to assume that minimum critical population sizes in captivity can be much smaller than in nature (Chapters 8, 9 and 12).

GENETIC CHANGE IS SLOWED

As a result of captive care and protection, the life expectancy of most wild animals in captivity is many times greater than in nature. This discrepancy is widening with improvements in the technology of captive care (Conway, 1974; Comfort, 1962; Chapter 13). Disease, competition, successional changes and predation are minimized in captivity. While this protection makes its subjects vulnerable to random biological events such as genetic drift and pressures for selection for atypical characteristics, captive management should reduce the opportunities for genetic change. An animal that is part of a captive breeding program can produce more young over many more breeding seasons than its wild counterpart, and more of these young can survive to breed themselves. Thus captive parents may have the opportunity of passing on more of their genes in more combinations.

This potentially rapid expansion of a captive population could help to mitigate the "founder effect" in comparison with wild populations of comparable size. Most importantly, with the slowing turnover rate of the generations in captivity, the opportunity for selective processes to exert pressure upon the genotype is reduced compared with short-lived natural populations of the same size. Taking the length of prime reproductive life of females (that period when the greatest number of viable young are produced) as a measure of the length of a generation, a Père David deer generation at the Bronx Zoo is eight years and that of a tiger *(Panthera tigris)* nearly ten years. But if captive groups are not subjected to management by rigorous long-term reproductive strategies based upon suitable genetic (Chapter 12), demographic (Chapter 15) and behavioral-ecological (Chapter 14) models, all advantage of longevity is likely to be lost.

The crisis facing the Przewalski horse is an example. Now believed extinct in nature, the vast majority of the founding Przewalski herd was obtained at the turn of the century. In 1945, there were only three stallions and seven mares, from which most of the present animals have descended (Bouman, 1977). Maintained mostly in European zoos until after World War II, the horse was mixed with a domestic stallion at one zoo and subjected to arbitrary standards of external morphology at others—standards which were obtained by restricting breeding to a few stallions.

Today there is a population of only about 265 horses where there might have been more than 1,000. A substantial portion of the population is beginning to show the effects of inbreeding depression in lowered vitality and various anomalies. The relatively fast-breeding Père David deer, established with only eighteen individuals by the Duke of Bedford at Woburn Abbey, may have passed through two genetic bottlenecks successfully. Its population had increased from the fifty which survived World War II to about 300 by 1948. There are approximately 800 today.

GENETIC AND DEMOGRAPHIC PLANNING

Would-be gene bank operators in zoos have been provided with straight forward guidelines for a genetic management strategy (Flesness, 1977; Chapter 12): mate the least related animals to each other in each succeeding generation; provide for an equal number of offspring from each parent; and maintain an even sex ratio. However, such a scheme is inconsistent with the behavioral requirements of many species (for example, polygynous deer and antelope). It requires intensive and continuing manipulation of specimens within the captive environment with attendant risk of injury and loss, especially where animals are being managed for "wildness." It may strongly conflict with behavioral (Chapter 14) and demographic strategies. Moreover, the response of various species to inbreeding is surely species-specific.

There are about 30,000 mammals in United States zoos (Seal et al., 1977; Perry, 1974) which offer a measure of present capacity. Supposing that half of this capacity could be given over to gene bank purposes, how many species could be successfully propagated? Flesness (1977) points out that captive populations of 50 to 100 animals could preserve half of their genetic diversity over 100 generations under ideal genetic management. Both Franklin and Soulé (Chapters 8 and 9) claim that 50 breeding individuals constitute the minimum group size for short-term programs, but that more animals will be needed in practice. Given the present capacity of zoos, perhaps 100 species of mammals could be maintained at 150 specimens per species, hardly an impressive response considering the needs foreseen in this book.

Just as important as genetic management is demographic planning. Population size, age and sex-specific survivorship must be carefully controlled if a steady gene bank population is to be sustained in equilibrium with space and other resources (Foose, 1977; 1978). The implied changes in zoo management include not only further reductions in the number of species exhibited, but costly revisions of physical facilities. They may prove less attractive to the visitors who support zoos.

Under any management, diversifying selection will occur between different groups of the same species in different institutions. Outcrossings between such groups and selection against individuals or strains with deleterious traits are among the many strategies which will require national and international coordination if serious programs are to become a reality.

Coordination is the special objective of the International Species Inventory System (ISIS) which was developed as a census and vital statistics inventory for the American Association of Zoological Parks and

Aquariums (Seal et al., 1977). The chances of a single institution maintaining a species for a long period of time is remote. Programs must be multileveled, specialized and interrelated.

While existing zoos will have an important role to play, coordination with rural breeding farms and ranches will also be needed. City zoos may prove to be the best places to accomplish initial acquisition, acclimatization and the development of primary care and breeding techniques. They have comparatively large and broadly experienced staffs. Moreover, their locations make them accessible to participating experts; their facilities make many of them suitable for the early stages of intensive study and intensive propagation. However, they are land poor—even with coordination. All the zoo animal enclosures in the world could fit within the Borough of Brooklyn, comfortably. Nevertheless the advent of ISIS in the zoo field lays the groundwork for the implementation of species by species, national and international reproductive strategies and the disposal of surplus—in terms of both optimum genetic and demographic models.

BEHAVIORAL-ECOLOGICAL REQUIREMENTS

Animals cannot be managed in captivity as though they were interchangeable ciphers. Problems of compatibility bedevil attempts to mate individuals of some species. It required several months of trial and error, seven birds and the cooperation of three other zoos for the Bronx Zoo to develop two compatible pairs of Andean condors *(Vultur gryphus),* for example. Every zoo biologist is familiar with countless instances of animals, from geese to antelope, which have proven individually incompatible, no matter how strongly computers and curators have blessed their union. Worse yet is the almost complete failure of second-generation breeding in some species, such as the orangutan *(Pongo pygmaeus),* where genetic and demographic problems can probably be ruled out. There is the worry that not only behavioral but also unidentified physiological and developmental insufficiencies are affecting success. For these reasons and others, conservation biologists working with captive populations are using an increasing number of manipulative techniques to enhance reproductive success in nature as well as in captivity. They are entering the reproductive process at earlier and earlier stages.

Egg transfers from successful populations to unsuccessful populations, artificial incubation and stimulation of replacement egg clutches are being used with birds, and cross-fostering between related species and individuals, artificial insemination and even imprinting are being used with birds and mammals to increase breeding success. Artificial incubation, for example, is essential to the multiplication of the production that can be obtained from many birds through "double clutching." A bird's first clutch or even first several clutches are removed and artificially incubated. Removal stimulates the laying of replacement eggs and the trick seems to work with a majority of species. Individual cranes have laid as many as six clutches at the Bronx Zoo in place of the normal one. In 1978, an Andean condor was stimulated to lay three one-egg clutches—a sextupling of the normal one-egg-every-two-years pattern of this species. The three chicks produced by this one bird probably exceeded the number produced by the entire remaining wild population of the California condor. This demonstrates one reason for increased interest in captive gene banks for faltering species.

More subtle, if not more complex, must be the design of captive environments to respond to the species-specific, behavioral-ecological requirements of wild animals in captivity (Eisenberg and Kleiman, 1977). While intensive manipulation of individuals may prove practical with some species, it would be difficult to support the management of significant numbers of species in such fashion. However, zoo biologists have too often sought manipulatory approaches after complaining that a wild animal did not adapt to captivity when in fact the first step in rational acclimatization programs is to provide captive conditions adapted to the animal. A bird that has evolved to nest on a marshy substrate, a sandy beach, in a seaside burrow or in a tree crotch is far more likely to nest if the right kind of site is provided. In 1964, when the Bronx Zoo provided a series of simulated habitats for waterbirds, many species began breeding. These species had been displayed in old-fashioned zoo bird cages for more than half a century without laying a single egg.

The provision of key ecological "furniture" can bring animals into breeding condition overnight; the maintenance of animals in appropriate social groupings is no less significant. Flamingos in zoos often breed when kept in flocks but almost never when kept in very small groups. Cheetahs *(Acinonyx jubatus),* it has recently been shown, reproduce more successfully when the females are isolated except during heat periods (Manton, 1975), at which time they may be more likely to breed if provided access to a group of males. Thus, the 1965 International Zoo Yearbook (IZY) census (Jarvis, 1967) recorded no captive bred cheetahs while the 1975 census lists 102 (Olney, 1976*b).*

Although zoos do better than nature in increasing recruitment rates and lowering death rates, ignorance still prevents them from doing as well as they should. According to a recent analysis by Foose (1978), the population of zoo okapi *(Okapia johnstoni)* in January 1977 was 62. It has been derived from an imported population of 31 and has not yet been subjected to significant inbreeding. Nevertheless, of 126 births from 1941 through 1976, only 70 young survived their first year. The population's survival is questionable.

With small populations, specialized areas of ignorance in animal care technology assume great importance. For many mammals no safe pregnancy tests are available. For many birds methods for the determination of sex are only now becoming available (Chapter 13). The incubation of the eggs of some declining avians, cranes for example, is notoriously difficult. Pharmacological methods of inducing receptivity and ovulation in females and spermatogenesis and copulatory behavior in males have not moved beyond the rhino horn and beetle-wing (aphrodisiac) stage for many forms. Knowledge of optimum diet is lacking for many species. New tests, such as those for heterozygosity now being employed by experimental geneticists, have just begun to come into use in wild animal propagation (Chapter 13) and collecting procedures. Concerted efforts have not yet been made to quantify and compare successful care programs with unsuccessful ones; only now are behavioral profiles for a few species from nature becoming available which will permit comparative evaluations of zoo animal behavior patterns. Where establishment of captive groups is indicated, researchers in conservation biology have much to do.

RETURNING ANIMALS TO NATURE

Once obtained and securely maintained, gene banks offer the opportunity for returning animals to some sort of natural state (see Chapter 15 for a more extensive discussion). The replacement of lost wild populations with captive-bred animals within their original homelands has already occurred with species as diverse as bison, blackbuck *(Antilope cervicapra),* peregrines and giant tortoises *(Geochelone elephantopus).* However, the opportunities for such replacements cannot help but become more rare.

Behavioral and ecological obstacles must be overcome. The success of certain introduced animals, even in historically unfamiliar environments, includes many now rare in their own homelands—blackbuck, Himalayan tahr *(Hemitragus jemlahicus),* aoudad *(Ammotragus lervia),* axis deer, *(Axis axis),* Bali myna *(Leucopsar rothschildi)*—and demonstrates that some forms do well with little help. Where behavioral difficulties occur, imagination and a new technology of animal care may hold the key. Genetic variability or demographic balance could even be introduced into an isolated reserve for tigers, for example, by release of captive-reared males in temporary breeding cages on the site of the reserve.

Unfortunately, it is conceivable that the introduction of captive bred animals might pose a threat to some wild populations of the same species. The comparatively rapid turnover of animals in reserves (each diverging in the character of its ecology from others—where species of the same kind survive—and from those in gene banks) may give rise to ecotypes whose fitness would be reduced by interbreeding with other animals.

The creation of new habitats by man offers other possibilities for release. Deserts can be made to support life, dams create new habitats even as they destroy old ones, lands stripped of their vitality can be reclaimed (Chapter 5) and other habitats may be found to contain elements sufficient for species from distant places, if no longer for native forms. In view of the scope of change now being imposed upon the earth's ecosystems and the impossibility of natural evolutionary response within its time frame, man has a responsibility to do more than stand back and wring his hands. He can act to preserve biological diversity, even to the extent of transplanting species, constructing new ecologies and preserving diversity by whatever means. In a thought-provoking discussion of applied biogeography, Wilson and Willis (1975) discussed the future of ecosystem manipulation through "planned biotic enrichment," "species packing" and the "creation of new communities." Captive propagation is a necessary adjunct to this kind of thinking.

AN ECONOMIC BASIS FOR PRESERVATION

Unfortunately, however, the preservation of the majority of animals that man finds attractive probably cannot be sustained on any provable economic basis; plants offer more compelling arguments. Eloquent noneconomic pleas for wildlife preservation are legion (Curry-Lindahl, 1972; Shepard, 1978), but they are more likely to be effective with people relatively free from the daily struggle to survive. Moreover, some of the traditional functions of wild animals in man's economy, such as hunting, are obsolescent. Whereas rarity commonly confers value on many objects of human interest, the decline in utility of an animal, along with increasing rarity and ecological change, may reduce its worth. Recent ventures in game ranching (Chapter 16) may help a few species, particularly because they could produce jobs in Third World Countries, but esthetic use by zoos, aquariums or tourism now has become the greatest economic utility for many species.

The San Diego Zoological Society, the Chicago Zoological Society, the National Zoological Park and the New York Zoological Society have a combined annual budget of approximately $60 million and provide direct employment for more than 2,000 people. The overall economic impact of the Bronx Zoo in New York City has just been calculated at $37.5 million. Thresher and Henry (unpublished), studying tourism in Kenya's Amboseli National Park, recently estimated the present national value of a single maned lion there at $515,000 for viewing purposes, only $8,500 for hunting purposes, and a mere $1,150 for commercial purposes (as a skin). In Kenya, a maned lion over a period of fifteen years has a present value equivalent to a mixed herd of 30,000 cattle. Arguments like these seem more likely to arouse sympathetic interest in Third World Countries for the preservation of wild animals than noneconomic justifications alone. Eventually, these nations might support reintroduction programs. For the present, they must cooperate in the establishment of such banks in developed nations, for an assumption inherent in gene bank precepts is their location in areas thought to be relatively stable and financially supportive.

INADEQUATE SUPPORT

Although many kinds of wildlife conservation efforts profit from a commonality of objective and even international orchestration, zoos have been viewed more as potential despoilers than as conservation educators and potential wildlife preservers, sometimes with good reason. Today, this attitude is changing much too slowly. Significant grants or encouragement from the conservation community have been unavailable for propagation projects in zoos. Nor have conservationists outside zoos attempted to establish realistic propagation objectives until, as with the California condor, it was nearly too late. Moreover, almost all zoos are municipally supported, whether by private or public funds. There is, as yet, no national commitment to the captive propagation of exotic species as there is for works of art through the National Endowment for the Arts. And there are few precedents for the exercise of international altruism by municipally oriented institutions. Even the implications of common ownership required by national and international breeding program coordination are unclear.

It is clear that the development of viable captive groups for some of the world's vanishing wildlife will require difficult choices and the abandonment of most species in favor of a few—because of limited resources. It is obvious that sufficient technologies exist for zoos to embark upon such efforts, but the socio-economic mechanisms required to keep them on course have not yet been developed. The reluctant skippers of these genetic barques and the ultimate arbiters of future evolutionary options cannot help but have many misgivings, not the least of which is that by far the most attractive and most economical gene bank is the proper management of animals in their native habitats.

SUGGESTED READING

The literature on breeding animals in captivity is scattered and often unavailable except in specialized libraries. The most complete compendiums are various volumes of the *International Zoo Yearbook,* published by the Zoological Society of London. Volume 17 concentrates, in part, on endangered species. Another source is *Breeding Endangered Species in Captivity,* 1975, Academic Press, London.

CHAPTER 12

INBREEDING DEPRESSION AND THE SURVIVAL OF ZOO POPULATIONS

John W. Senner

This chapter describes a theoretical study of the genetic factors affecting the survival of zoo populations. The study has two goals: to anticipate the probable impact of various management strategies on the survival of a captive population and to provide an estimate of the risks involved in undertaking a project to maintain a particular species in captivity. To achieve these goals I constructed a mathematical model of a zoo population with nine parameters representing characteristics of a species or of a management decision. I will summarize the essential features of this model and show how variations of each of the parameters affect success of a captive breeding program. (Mathematical derivation and treatment may be found in the appendix.) The results will suggest that the survival of small, captive populations is relatively sensitive to changes in some variables (fecundity, viability, their respective genetic loads and maintenance size) and relatively insensitive to changes in others (sex ratio, its genetic load, and founder size).

The eventual fate of a small closed population of animals is nearly always extinction. Figure 1 illustrates some of the details of the growth and the inevitable decline of a zoo or reserve population founded with four breeding animals and maintained at a limit of 10 animals. Heterozygosity drops as inbreeding increases (shown as line H, where 100 percent represents the heterozygosity level of a large population of unrelated animals). As heterozygosity declines, the average survival of offspring and the fecundity of parents (initially five viable offspring) declines until

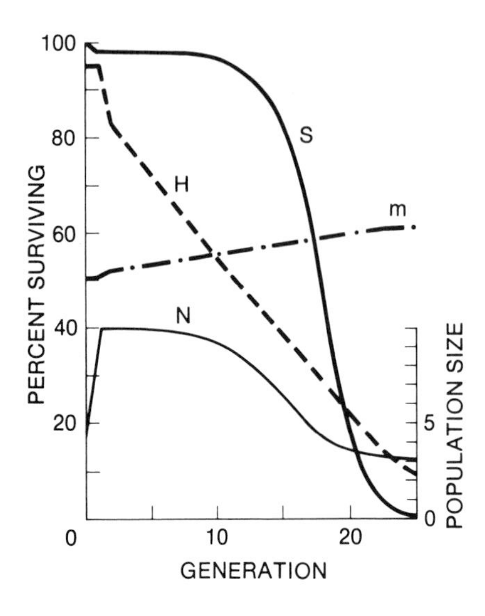

FIGURE 1. Essential features of a model of captivity-bred populations.

the population size (line N) can no longer be maintained at its limit. The proportion of male offspring (line m) increases (in mammals) because females are more susceptible to inbreeding depression. The probability of survival (line S) is high for about 15 generations but drops sharply and approaches zero by the 25th generation. Extinction occurs when there is less than one surviving offspring of each sex in a generation. The probability of extinction is therefore a function of the number of offspring in a generation (a function of the number of breeding females and their fecundity), the survival of the offspring (a function of inbreeding) and the sex ratio of the surviving offspring (also a function of inbreeding).

Any model is a mix of simplifying assumptions and overly detailed analysis of parameters which reflect the particular biases and interests of the model's builder. Important simplifying assumptions made here are:

1. Nonoverlapping generations. Building a model with overlapping generations is possible, but would require additional, though not enlightening, complications.

2. Simplified scheme of population size. Each population is founded with I animals which then multiply to become M animals in the next generation. The population then stays at this maximum limit, M, until inbreeding depression is so strong that replacement rates are not reached and the population size drops.

3. Fecundity follows a random distribution. The number of offspring produced by each female follows the Poisson distribution, as expected in the ideal case that all females are equally fecund (Cavalli-Sforza and Bodmer, 1971—page 311), with the mean of the distribution equal to the expected fecundity of a female. Inbreeding depression is translated into a lowered value for expected fecundity, equivalent to specifying that all

deaths from genetic load occur before the animals are put into the pool of potential breeders. The next generation of parents are selected without regard to ancestry and the sex ratio is made as even as possible. The sex ratio of newborns, however, is represented by a binomial distribution dependent on the expected proportion of males, *m,* in a sample of newborns.

4. Numbers of offspring refers to offspring surviving to adulthood which are fit for breeding. Animals dying before reproductive age—for any reason—are not counted in this model. Animals which show any detectable genetic abnormality are not counted as breeding adults although they may be of value for other uses. (In actual practice there may be some problem in deciding whether a phenotype is a rare but normal variant of the species or is a harmful defect.)

5. Both initial and maintenance population size are meant to be effective population sizes (Kimura and Ohta, 1971), particularly the inbreeding effective population size.

RESULTS

The following cases illustrate the sensitivity of the model to each one of the nine variables.

Fecundity

When reproductive rates are high there is an excess of offspring, thus permitting selection for vigor or other traits. When reproductive rates are low every animal is needed for the next generation of breeders and no such selection is possible. Figure 2 shows the effect of varying the number of viable offspring per female. The lines show probabilities of population survival where female fecundity is three, five and seven offspring. The model is quite sensitive to fecundity, suggesting that good husbandry and an environment conducive to breeding can have a dramatic impact on survival of zoo populations. It is perhaps no coincidence that the most successful examples of captive propagation have been two species closely related to common domestic animals: Przewalski's horse (*Equus przewalskii*) and the wisent (*Bison bonasus*).

Viability Depression

The number of viable offspring declines as the inbreeding coefficient increases. This decline, known as inbreeding depression, can be divided into three separate effects: viability depression, fecundity depression and sex ratio depression. The first of these, viability depression, is the failure

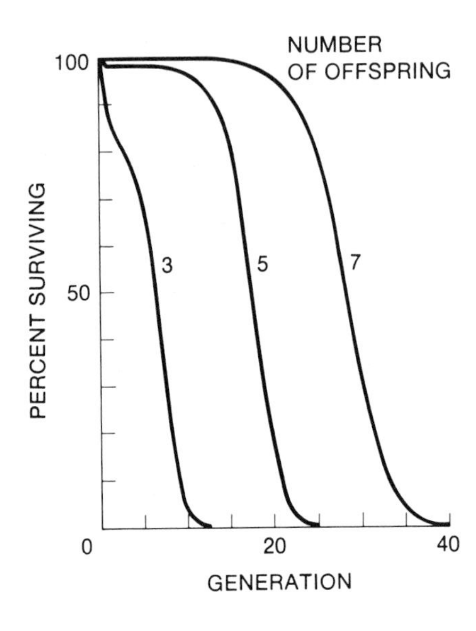

FIGURE 2. The sensitivity of the model to changes in fecundity (mean number of viable offspring per female).

of offspring to survive to maturity as a function of the offspring inbreeding coefficient. If a straight line is fitted to a plot of the inbreeding coefficient versus the negative logarithm of the proportion of the offspring surviving to reproductive age, the slope of this line (B_1) is a measure of the intensity of inbreeding depression. Hence, it is a measure of the genetic load of the source population. I assume a linear relationship because it is expected theoretically (Morton, Crow and Muller, 1956), although Kosuda (1972) has found a concave upward curvature to the relationship between the negative log of survival and inbreeding in *Drosophila* (a discouraging sign for the future of captive populations if it is generally true). Torroja (1964) has less extensive evidence of a concave downward curvature to the inbreeding-survival relationship in *Drosophila* (an encouraging sign consistent with Slatis' (1975) view that if a population survives an initial episode of inbreeding, major problems with further inbreeding depression will probably not arise). Schull and Neel (1965) found no significant departures from linearity in a large human sample, but the low levels of inbreeding in their sample also limited their ability to detect departures from linearity.

Figure 3 shows the effects of variation in viability depression on survival of captive populations. The line labeled 1.0 shows the viability depression level used as a standard for all of the calculations whose results are shown in other figures. Extinction is almost certain within 25 generations. The remaining lines show the effect of variation in B_1 over a range of likely values. This range can be compared to a number of published estimates of B_1 (recalculated to give comparable measures). A half dozen species of *Drosophila* have B_1 values ranging from 0.23 to 1.8 (Mettler and Gregg, 1969). Viability depression in four kinds of fowl range from 1.06 in the domesticated chicken to 3.74 in the Hungarian chukar partridge (Abplanalp, 1974). The mean of 10 studies on Japanese humans is 0.67 (Schull and Neel, 1972). Slatis (1960) found the low value of 0.06, not significantly different from zero, in the wisent.

Fecundity Depression

Inbred animals are more likely to be sterile than outbred animals and inbred mothers are poorer mothers than outbred mothers. Both factors reduce the ultimate fecundity of a pair of animals as a function of the level of parental inbreeding. Figure 4 shows the probability of population survival for three values of fecundity depression. The value of 0.5 for B_2 (Obtained in a similar fashion to B_1) was used as a standard on all other figures. In comparison, Abplanalp (1974) found that the coefficient relating depression of egg hatchability to maternal inbreeding ranged from 0.12 for chickens to 0.55 for the chukar partridge. Studies on Japanese humans give values of 0.5 and 0.6 for B_2 (Schull and Neel, 1972; Tanaka, 1977).

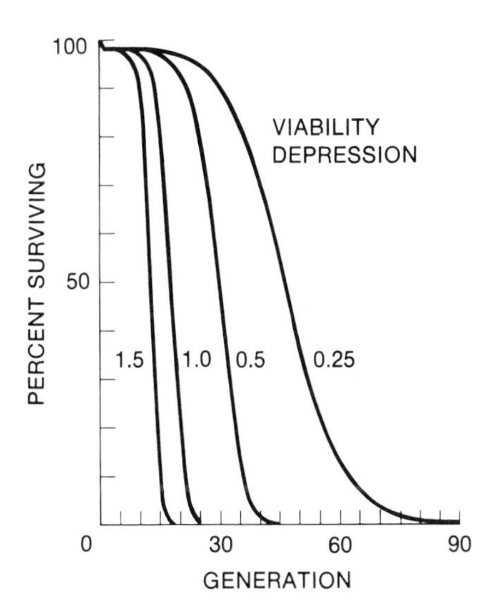

FIGURE 3. The sensitivity of the model to changes in the level of viability inbreeding depression.

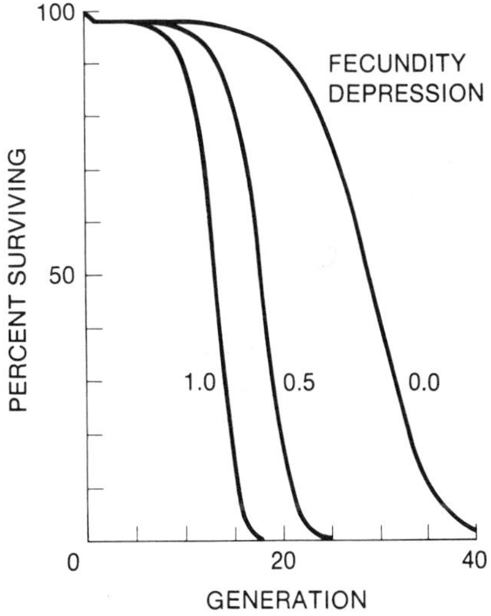

FIGURE 4. The sensitivity of the model to changes in the level of fecundity inbreeding depression.

Sex Ratio Depression

Sex ratio also changes during inbreeding, apparently because the male X chromosome (in mammals) is always hemizygous, independent of inbreeding. A female X chromosome pair can become increasingly homozygous with an increase in inbreeding. Therefore, males become increasingly common among the survivors at higher levels of inbreeding. This shift in sex ratio can be seen in Table 2 in Chapter 9. A more precise study of the sex ratio change resulting from sex chromosome inbreeding is given by Hook and Schull (1973). The data in this paper and that in Table 2 of Chapter 9 result in estimates for the value of B_3 (obtained in a similar fashion to B_1) of about 1.5, although 0.5 is used as the standard value in all other figures. An uneven sex ratio affects population survival in two ways: directly, by lowering the probability of at least one individual of each sex surviving to reproduce and indirectly, by increasing the rate of inbreeding. Nevertheless, Figure 5 shows that the intensity of sex ratio depression is not sufficient to greatly affect the probability of population survival.

If inbreeding depression results from the mutation of "good" genes to "bad" genes, which are lethal or deforming when in the homozygous condition (mutational load), it should be possible to "cleanse" the animal of its deleterious mutations by inbreeding, then culling defects and mating survivors to restore the animal's natural variability and evolutionary potential. On the other hand, some loci confer greater fitness when heterozygous (segregational load). At these loci inbreeding increases the proportion of the genome which is homozygous and inevitably decreases mean fitness. Sickle-cell anemia, which occurs in some human populations exposed to malaria, is an example. Inbreeding could "cleanse" a population of the sickle-cell defect, but would also result in lower resistance to malaria. Fitness of the population as a whole is higher with the sickle-cell gene than without it when malaria is present.

While there is considerable controversy on the relative contribution of the two types of genetic load to inbreeding depression, there is little doubt that at least some load is due to the second (balance) cause (Lewontin, 1974). For that reason, it would be imprudent to change the level of inbreeding depression in a captive population by attempting to purge the population of its load (Chapter 9).

Genetics of the Founders

What can be done to enhance the survival of a species that has been picked for a captive breeding program? If the relationship between genetic variability and productivity found in most studies of inbreeding depression is generally true, two rules immediately follow:

1. Avoid starting a population with animals which are already inbred.

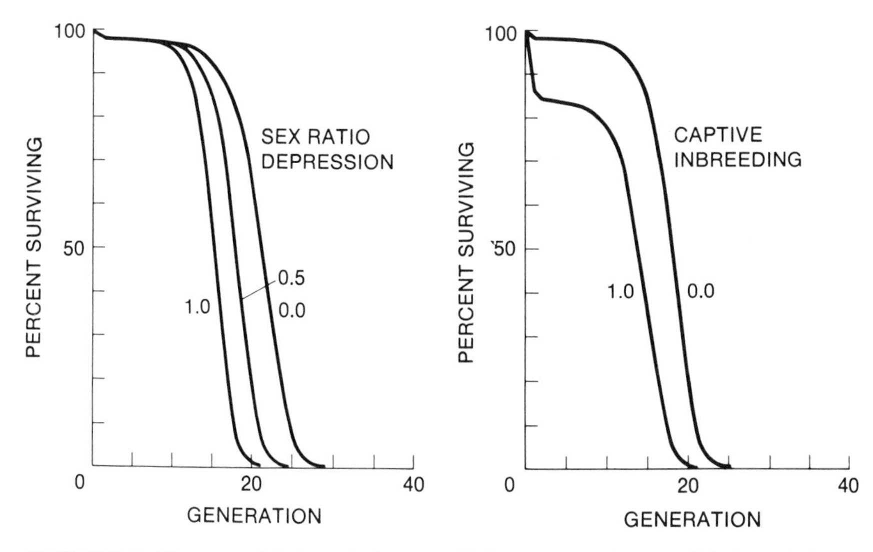

FIGURE 5. The sensitivity of the model to changes in the level of sex ratio inbreeding depression.

FIGURE 6. The sensitivity of the model to the inbreeding coefficient of founder animals. The lower curve reflects the effect of starting with founders which are completely inbred but unrelated.

A group of unrelated but completely inbred animals has half as much variability as unrelated animals that are not inbred. The difference is illustrated in Figure 6. The standard case assumes that the initial population is slightly inbred so that average f is 0.05 as might be found in rare and declining species.

2. Do not start a population with related animals. Figure 7 illustrates the effect population relatedness has on survival. The extreme value of 0.25 represents the case where the founder group is composed of sibs. The standard case assumes that relatedness of the initial population is 0.05, as might be found in a rare species. One need not look far to find cases in which these rules are violated, usually because of the realities of the acquisition process. A serious captive breeding program should invest enough care in the acquisition phase to assure that as many animals as possible are as unrelated as possible.

Initial Population Size

Two population size parameters have an important effect on survival: the initial effective population size *(I)* and the maintenance effective pop-

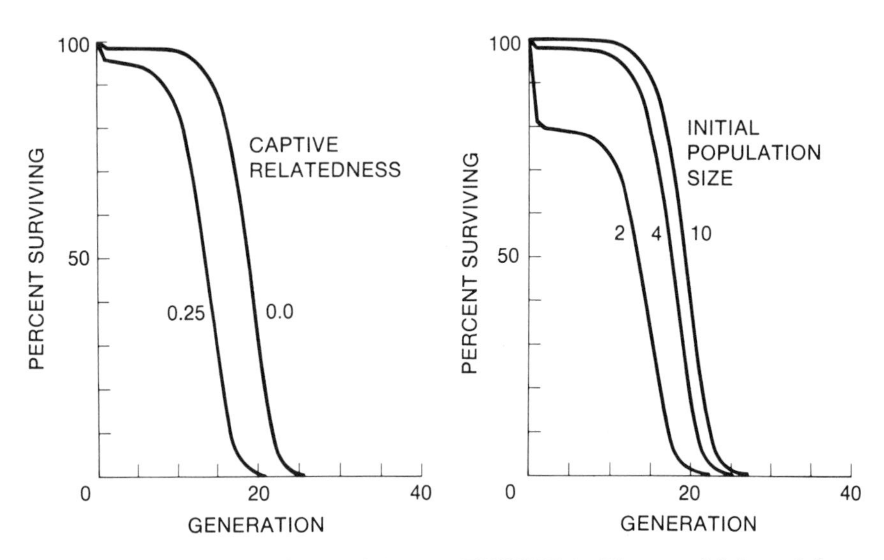

FIGURE 7. The sensitivity of the model to the relatedness of founder animals.

FIGURE 8. The sensitivity of the model to variation in the number of founders.

ulation size *(M)*. I assume that the population has effective size I when captured, then increases to size M in one generation and remains at that population size until reproduction falls below the replacement level. Figure 8 demonstrates the effect of varying the initial population size around the standard value of four used on all other figures. The probability of survival for an initial effective population size of four (the standard value) is much closer to the survival rate of an effective size of 10 than to that of an effective size of two. This implies a nonlinear relationship between initial size and survival. In Figure 9 this relationship is viewed in another way by calculating the increase in the inbreeding coefficient per generation for various population management scenarios, and then determining the number of generations required to reach the arbitrary value of $f = 0.60$. Each line represents a fixed maintenance population size (shown

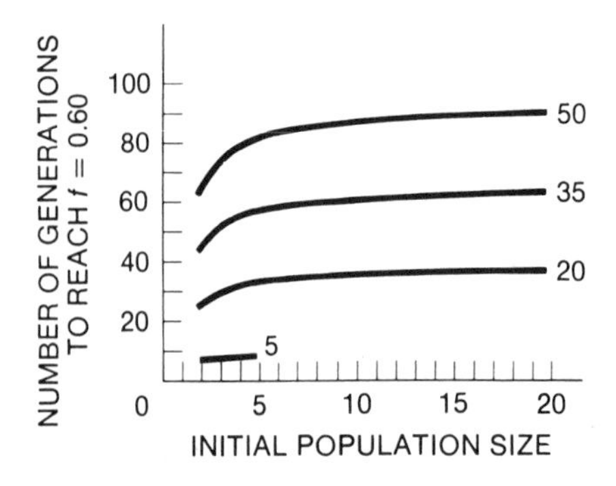

FIGURE 9. The effect of increasing the initial population size on the numbers of generations it takes to reach $f = 0.60$. The number at the end of each curve is the maintenance size.

at the right) initiated at the population size given on the abscissa. The number of generations taken to reach f = 0.60 is given on the ordinate. For all maintenance population sizes, increasing the initial population size from two to four or five greatly lengthens the time for the inbreeding coefficient to increase to 0.60, but increasing the initial population much above five has little affect on survival.

Maintenance Population Size

In contrast, the corresponding Figures 10 and 11 demonstrate that increases in the maintenance population size result in proportional increases in survival. Figure 10 shows how varying the maintenance population size affects the probability of survival to the generation given on the abscissa. The standard population size used on all other figures was 10. In Figure 11 each line represents sets of populations started at the given initial population size (2, 5, 20, 50). Each population was then maintained at the size given on the abscissa, with the number of generations required to reach an f of 0.60 given on the ordinate. This shows that effort put into

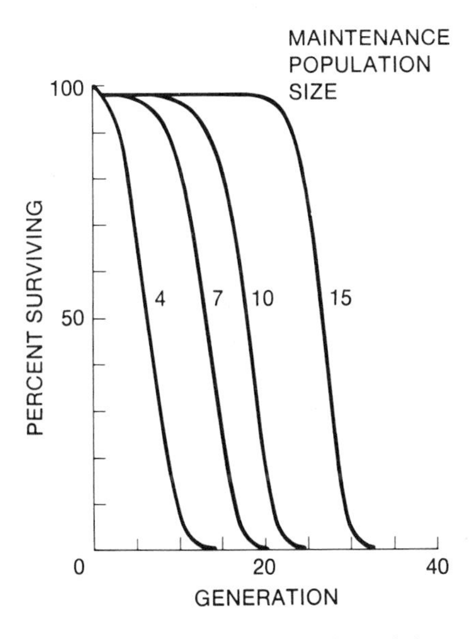

FIGURE 10. The sensitivity of the model to different maintenance sizes.

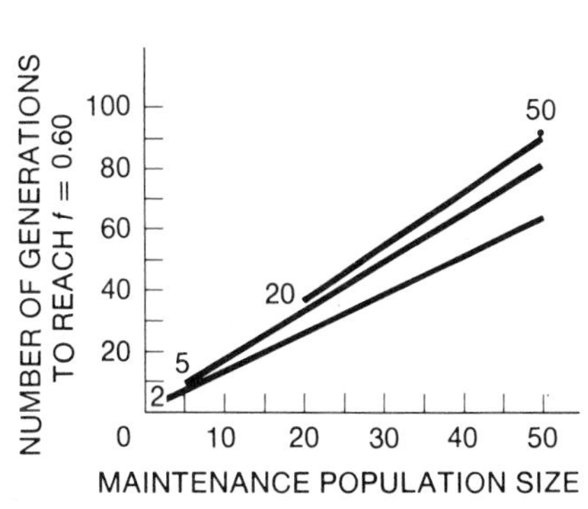

FIGURE 11. The effect of maintenance size on the number of generations it takes to reach f = 0.60. The number at the origin of each line is the founder size.

increasing the maintenance population size results in proportional increases in survival. As seen in Figures 8 and 9, however, effort put into increasing initial size leads to diminishing returns.

Manipulation That Increases N_e

Can anything be done to improve the chances of survival once the animals have been collected and have reached captivity? Given an upper limit for the actual number of animals that can be maintained, one variable with a great impact on effective population size (N_e) is the distribution of offspring among individuals (Chapters 8 and 9). The simulations discussed above all used random-mating sexual populations with Poisson distributed numbers of gametes per animal. This means that neutral heterozygosity is lost approximately at the rate of $1/2N_e$. The more general formula is

$$N_e = \frac{(N_{i-2}\bar{g}-2)}{(\bar{g}-1+\frac{s_g^2}{\bar{g}})}$$

where N_{i-2} is the number of grandparents of the ith generation, g is the number of gametes per individual, and where $\bar{g}$ and s_g^2 are the mean and variance of gamete production, respectively.

Because a Poisson distribution of offspring has been assumed so far, for which $s_g^2 = \bar{g}$, it follows that when the population is of constant size, $N_e = N - 1$ where N is the population (census) size. But if it is possible to manipulate the offspring distribution so that all parents have equal numbers of offspring, then $s_g^2 = 0$, and $N_e = 2(N - 1)$. That is, by the expedient of managing reproduction so that all parents contribute equally to the next generation, one can double the effective population size.

Another approach to increasing N_e (thereby decreasing the rate of inbreeding) is to exercise control over mate selection. Such control requires the calculation of inbreeding coefficients. Exact coefficients of inbreeding can be calculated from pedigree information (Wright, 1977), but these are not particularly useful in this general model. Exact coefficients of inbreeding in each generation can also be calculated when matings follow a regular scheme (for example, circular half-sib mating, double first cousin mating). These schemes require complete control over the pattern of mating and fixed population sizes. They will not be very useful to the zoo breeder unless breeding pairs can be closely managed. One such system is maximum avoidance of inbreeding as shown in scheme A of Figure 12 (for a population size of four). In this scheme as in the elimination of differential reproduction, there is a doubling of the effective size over random mating and $N_e = 2N$.

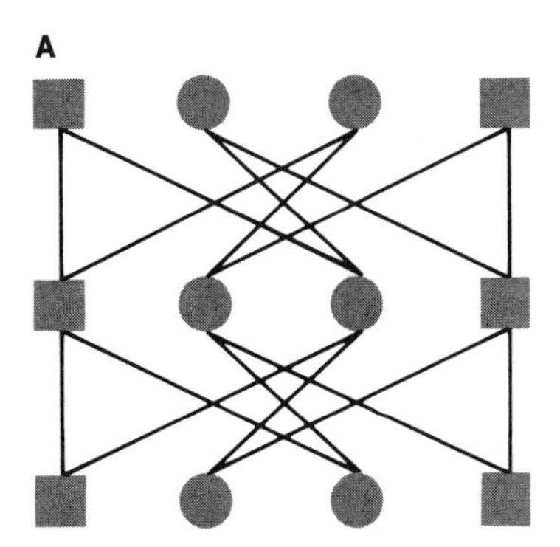

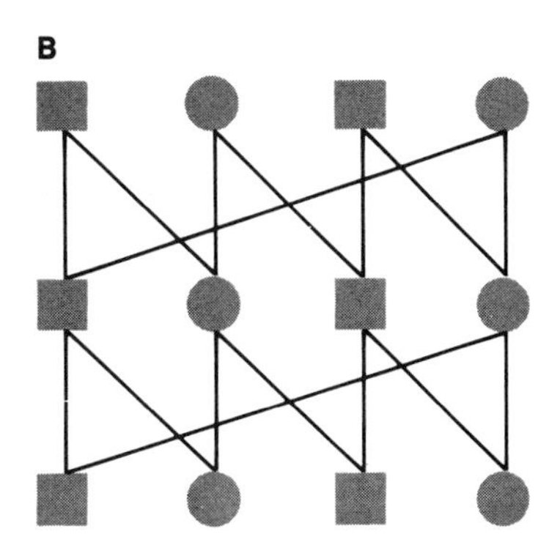

FIGURE 12. Mating schemes used to decrease the rate of inbreeding in captive populations.

Scheme B in Figure 12 is an example of circular half-sib mating. Although it starts out deliberately breeding half-sibs, this scheme eventually and paradoxically results in less loss of heterozygosity. Robertson (1964) shows that for circular half-sib mating

$$N_e = \frac{2(N + 2)^2}{\pi^2}$$

Circular half-sib mating does not improve on maximum avoidance of inbreeding until after generation 16 in this example, and then only slowly. By this time many groups will be extinct. For this reason alone, circular half-sib mating will not be of practical value to the captive breeder. In addition, both this breeding scheme and the former require specific matings which could be highly disruptive in social species, particularly where there are pair bonds and dominance hierarchies.

Finally, it should be noted that much of the increase in effective size produced by such schemes can be accounted for by the schemes' inherent enforcement of equal genetic contribution of parents to the next generation. Therefore, there is little to be gained by the use of such protocols so long as the variance of the progeny distribution is kept low.

SUMMARY

1. Extinction of small populations is inevitable; its probability depends on fecundity, viability and sex ratio. These in turn depend on population size which roughly determines the rate of inbreeding.

2. The extinction model is quite sensitive to female fecundity, which in turn means that the quality and sophistication of husbandry has a dramatic impact on survival of captive populations.

3. Inbreeding experiments on domestic and semi-domestic species show that viability and fecundity decline upon inbreeding, and that the rate of such inbreeding depression can vary widely among species.

4. Unbalanced sex ratio increases the rate of inbreeding and increases the probability of stochastic extinction in small populations.

5. In starting a captive breeding program, it is wise to avoid founders that are already inbred (even if unrelated). Captive breeding is expensive, so it is false economy to use any but the best available founders.

6. The sensitivity of the model to the number of founders (above a founding group size of four or five) is much less than to the maintenance size of the colony. That is, survival is proportional to maintenance size, but increasing founder size leads to diminishing returns.

7. Breeding systems can be manipulated to increase effective size (which decreases inbreeding), but the costs of such manipulations in social species will usually be greater than the genetic gains. The most important manipulation of captive stocks is to equalize the genetic contributions of parents to the next generation.

APPENDIX: MATHEMATICS OF THE EXTINCTION MODEL

If the probability of extinction in any given generation is $Pr(E_i)$, the probability of survival to the end of generation t is

$$Pr(S_t) = \prod_{i=1}^{t}[1 - Pr(E_i)]. \tag{1}$$

Extinction at generation i occurs when either no male offspring or no female offspring live to the age of reproduction. The probability of this event can be calculated if the distribution of the number of males and females is known. Usually, population geneticists have assumed that the total number of offspring per female follows a Poisson distribution with a mean equal to the expected number of offspring, R_i (Cavalli-Sforza and Bodmer, 1971; Kimura and Ohta, 1971). For any given cohort size, the number of males and females is assumed to follow a binomial distribution dependent on the expected proportion of males, m_i. Then the probability of a particular combination of male and female offspring is the Poisson probability of a cohort of the given size multiplied by the binomial probability of the particular male-female combination in that cohort.

An equivalent approach is to calculate two independent Poisson probabilities, one from the distribution of male offspring and the other from the distribution of female offspring. The probability of the combination of *j* males and *k* females is then the product of the two probabilities:

$$Pr(\textit{ j male offspring}) = \frac{e^{-mR}(mR)^j}{j!},$$

$$Pr(\textit{k female offspring }) = \frac{e^{-R(1-m)}R^k(1-m)^k}{k!}. \qquad \textbf{(2)}$$

The probability of extinction in generation i is the probability of producing no male or no female offspring:

$$Pr(E) = Pr(j = 0) + Pr(k = 0) \ - Pr(j = 0)Pr(k = 0). \qquad \textbf{(3)}$$

Factors affecting R_i and m_i result in changes in the extinction probability over time. The total number of offspring reaching reproductive age is determined by the number of breeding females in the parent population, k_{i-1}, their fecundity, Z_i and the viability of the offspring, V_i:

$$R_i = k_{i-1}\, Z_i V_i. \qquad \textbf{(4)}$$

The number of breeding females is affected by both the number of offspring and the proportion of males in the previous generation. When reproductive potential exceeds the limits to the population size and many females are available, the number is set to the smallest integer greater than or equal to $(1-m_{i-1})M$, where M is the maximum population size. Inbreeding reduces fecundity to some fraction of its outbred rate so that

$$Z_i = Z_0 e^{-B_2 f_{i-1}} \qquad \textbf{(5)}$$

where Z_0 is the fecundity of outbred animals, $f_{i-1}$is the inbreeding coefficient of the parents and B_2 is the coefficient relating loss of fecundity to inbreeding level (see text)—the coefficient of fecundity depression. Viability of offspring is dependent on their own inbreeding level so that

$$V_i = V_0\, e^{-B_1 f_i} \qquad \textbf{(6)}$$

where f_i is the inbreeding coefficient of the offspring and B_1 is the coefficient of viability depression. Since Z_0, the fecundity of outbred animals, is defined to the number of offspring living to the age of reproduction, V_0, the viability of noninbred offspring, can be set to 1 and equations 4, 5 and 6 combined to give R_i as a function of the inbreeding level of parents and their offspring:

$$R_i = k_{i-1}Z_0e^{-(B_1f_i+B_2f_{i-1})}. \tag{7}$$

The proportion of males in generation i, m_i, (when males are the heterogametic sex) increases when a population is inbred (Hook and Schull, 1973). Males are always hemizygous for their X chromosome but females can be heterozygous or homozygous for X chromosomes. Homozygosity for an X chromosome is harmful, just as homozygosity for an autosomal chromosome is harmful. If the proportion of living males to male zygotes is

$$V_{ji} = V_{j0}e^{-B_1f_i} \tag{8}$$

and the proportion of living females to female zygotes is

$$V_{ki} = V_{k0}e^{-(B_1f_i+B_3f_{si})} \tag{9}$$

with V_{ji} and V_{ki} the viability of males and females, respectively, f_{si}, the inbreeding coefficient of the sex chromosome (see Wright, 1978, for details on its calculation) and B_3 the coefficient relating loss of viability to level of inbreeding, then the natural logarithm of the ratio of living males to living females follows the linear relationship:

$$ln(m_i/(1 - m_i)) = A + B_3f_{si}. \tag{10}$$

The logarithm of the primary sex ratio, $ln(m_0/(1 - m_0))$, found in outbred populations, is A. Equation 10 is more conveniently expressed as

$$m_i = m_0/(m_0 + (1 - m_0)e^{-B_3f_{si}}) \tag{11}$$

giving the proportion of males in any generation as a function of the outbred proportion and the level of sex chromosome inbreeding. Values of R_i and m_i from equations 7 and 11 can then be used with equations 2 and 3 to find a probability of population extinction as a function of offspring autosomal and sex chromosome inbreeding and parental inbreeding.

In a population with separate sexes, the average inbreeding coefficient of each generation's offspring depends on their grandparent's consanguinity, their grandparent's average inbreeding and the effective population size. The expression is simplified if given in terms of heterozygosity *(H)* rather than inbreeding. If $H_i = 1 - f_i$ and the coefficient of consanguinity of parents equals the inbreeding coefficient of offspring, then

$$H_i = ((N_e-1)/N_e)\, H_{i-1} + (1/2N_e)\, H_{i-2}. \tag{12}$$

The effective population size, N_e, depends on the number of reproducing individuals in the grandparental generation as well as the mating system and the distribution of reproductive success over individuals. Nonrandom mating is considered briefly in the text; here random mating is assumed. Under this assumption effective population size is

$$N_e = (N_{i-2}\bar{g} - 2)/(\bar{g} - 1 + s_g^2/\bar{g}) \qquad \textbf{(13)}$$

where N_{i-2} is the number of breeding individuals two generations back (grandparents), g is the number of gametes contributed by them, $\bar{g}$ is the mean and s_g^2 is the variance of that contribution. For a population which deviates from the ideal only in sex ratio, expression 13 simplifies to the more familiar

$$N_e = 4j_{i-2}k_{i-2}/(j_{i-2} + k_{i-2}) \qquad \textbf{(14)}$$

where j_{i-2} and k_{i-2} are the number of male and female grandparents (Crow and Kimura, 1970; Kimura and Ohta, 1971; Chapter 3). The inbreeding coefficient of the offspring generation is the weighted average value of $1 - H_i$, with H_i calculated from equations 12 and 14 for each possible value of j males and k females and weighted by the probability of that combination, as given by equation 2. Program limitations required that f_{si} be approximated by f_i instead of using Wright's (1978) more exact formulae analogous to equations 12 and 14. When the sum of the males and females exceeded the management capacity, j and k were adjusted to not exceed the capacity population size and to equalize the sex ratio as much as possible before effective population size was calculated.

SUGGESTED READINGS

Kimura, M. and T. Ohta, 1971, *Theoretical Aspects of Population Genetics,* Princeton University Press, Princeton, New Jersey. How mating systems and offspring distributions change effective population size.

Schull, W. J. and J. V. Neel, 1972, The effects of parental consanguinity and inbreeding in Hirado, Japan, V., Summary and interpretation, *Am. J. Hum. Genet.,* 24, 425–453. A summary and interpretation of an important series of studies in Japan.

Slatis, H. M., 1960, An analysis of inbreeding in the European bison, *Genetics,* 45, 275–287. One of the most complete studies of inbreeding in a captive wild animal.

Wright, S., 1969, *Evolution and Genetics of Populations, II., The Theory*

of Gene Frequencies, The University of Chicago Press, Chicago. The theory of inbreeding, including the calculation of inbreeding coefficients from pedigrees and the analysis of mating systems.
Wright, S., 1977, *Evolution and the Genetics of Populations, III., Experimental Results and Evolutionary Deductions,* The University of Chicago Press, Chicago. Summarizes effects of inbreeding on plants and laboratory and domestic animals.

CHAPTER 13

THE TECHNOLOGY OF CAPTIVE PROPAGATION

Kurt Benirschke, Bill Lasley and Oliver Ryder

There is little disagreement that our present concern over conservation issues is principally the result of the rapidly expanding human population. Our need for ever-increasing territory and natural resources has diminished our wildlife and necessitated safeguarding some animals in reserves or zoos, lest they vanish in the next decades (Ziswiler, 1967). Although the Earth's human population has been increasing for a long time, the explosive growth in this century must be blamed largely on two factors: the remarkable progress in technological skills and the rapid increase in medical knowledge.

We will argue here that conservation and the captive reproduction of endangered species would be much enhanced if the same technology and medicine were applied to these vanishing animals. If the reproductive physiology of zoo animals were as well understood as is that of domestic species and man, and if freedom from infection and parasites could be assured for these animals as we are capable of doing for ourselves, we could look ahead with greater assurance that perilously endangered species could be saved from extinction. We are not now in this position. In part, this is due to the low priorities given to these enormous tasks and also because of the very small number of scientists working in this field, compared with those working in research on domestic species and human medicine.

If it is true that a last refuge for much of our wildlife will require skillful management by man—be it in reserves, wild animal parks or zoos —then we must rapidly acquire a great deal of knowledge. Moreover, this knowledge must be in many different areas of scientific endeavor.

ENDOCRINE STUDIES

When captive reproduction and exchange of breeding stock is discussed, the suggestion is often made to use artificial insemination rather than ship animals across the globe for the exchange of genetic stock. After all, this procedure is successful in humans, most cattle are now reproduced by this means and semen banks assure the availability of genetically qualified semen of a variety of domestic species. Although this method of artificial reproduction seems to be an eminently reasonable approach to the task at hand, it has not been practiced successfully, as yet, with endangered species. First, it must be realized that an enormous amount of research has gone into bull semen technology. After initial successes with electroejaculation, it is now feasible to use props with artificial vaginae for the collection of many domestic species' semen. The need for chemical immobilization and electroejaculation has been overcome. Many different media for the dilution of semen and cryoprotectants were experimented with before uniform success was achieved with the freezing of intact cells that could be unfrozen at liberty.

In wild animals it is now also feasible to collect semen, albeit with electroejaculation and usually under sedation (Seager et al., 1975). It is even feasible to freeze semen of some species and reawaken its motility later. In other species, particularly the apes (with which such a practice would be most desirable), the rapid coagulation of ejaculated semen has foiled attempts at freezing and, thus, widespread use of artificial insemination. But even if freezing were widely feasible, when would one inject the freshly thawed, viable sperm into the female? For instance, viable semen from elephants has been collected in Africa and is maintained in liquid nitrogen at the London Zoo (ZSL, 1978). But, it is impossible at present to ascertain the time of ovulation in elephant cows and, thus, make the project feasible. Indeed, it is not yet known whether many species are, like rabbits, reflex ovulators (where the egg is shed only after copulation) or whether the process of ovulation is spontaneous as it is in primates.

The Role of Endocrine Studies

Having recognized these difficulties and the enormous variation of the reproductive cycles of mammals, it has been our belief that precise delineation of these cycles is needed before substantial progress can be made. We have developed an endocrine research unit at the San Diego Zoo. Its principal objective is to identify the hormones and cyclicity of a wide variety of species and to correlate these with behavior and, ultimately, with the success of pregnancy. In contrast to studies of domestic animals and man, work with captive wild animals cannot be readily undertaken from serum samples since immobilization for bleeding is not only traumatic but also may interfere with the general hormone balance.

Consequently, most of our studies are now being performed from regular urine collections. Three specific examples will illustrate the scope of what can now be done to monitor pregnancies, establish reproductive cycles and gain a deeper insight into comparative reproductive physiology.

Leaf-eating monkeys (langurs) are prized possessions of zoos. They are, however, difficult to maintain. Repeated abortions and deaths among Douc langurs have plagued our colony. Placental gonadotropin is, as yet, poorly defined in langurs and early pregnancy cannot be reliably diagnosed by "pregnancy tests." On the other hand, when total estrogens are measured in the urine of langurs, it is seen that substantial amounts are excreted until the animals are near term (Figure 1). After having ex-

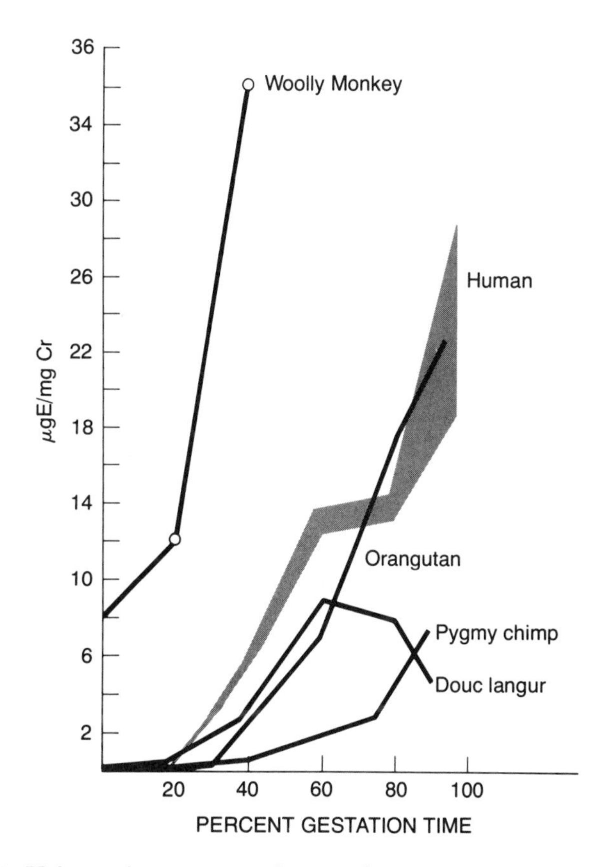

FIGURE 1. Urinary immunoreactive total estrogen content expressed as μg estrogen (E) per mg creatinine (Cr) during normal pregnancies of several primate species. Douc langur estrogen levels remain above 2μgE/mgCr in the last two thirds of pregnancy.

amined a number of pregnant and nonpregnant animals, it is possible to diagnose pregnancy with assurity. Moreover, when the urinary estrogen excretion falls markedly during pregnancy (Figure 2) it is possible to forecast fetal demise and to alert the veterinary staff to the possible need for extraction of a dead fetus (Resnik et al., 1978). Before these studies were undertaken, it was not possible to make this diagnosis unequivocally because closely related cercopithecids (rhesus monkey, baboon) produce only small quantities of urinary estrogen in pregnancy. In this respect then, the "feto-placental unit" of langurs is more like that of humans than of rhesus monkeys, even though the rhesus and the langur belong to a closely related group of old world monkeys. These findings have led to further endocrine studies of the placentas and adrenals of these animals and should give us the background for better management in the future.

Although not endangered, the nine-banded armadillo has been selected as a study model because of its notorious captive sterility. Perhaps

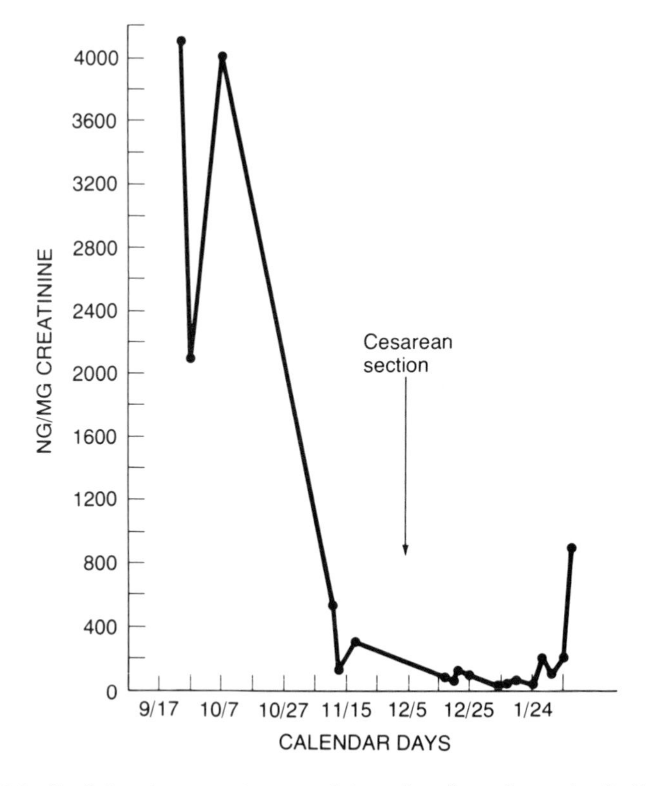

FIGURE 2. Serial urinary estrogen determinations in a single Douc langur pregnancy. Approximately one month prior to expected delivery the abrupt fall of estrogens indicated fetal death and prompted Cesarean section of a macerated fetus. Note the difference in scale (ng/mg = 1/1000 μg/mg).

by better understanding its reproductive problems, insight might be gained which will help other species (for example, the giant panda). In addition, the nine-banded armadillo exhibits some human-like features. It has, for instance, a uterus simplex, usually found only in primates; monozygotic quadruplets that possess a huge fetal adrenal gland with an involution at birth that is similar to man; hemochorial placentation; and (unlike man) a body temperature of 33°C! These are among many other unusual and challenging attributes. Intensive study in our laboratory has shown that neither the corpus luteum nor the placenta produce the progesterone needed to maintain pregnancy. This hormone emanates primarily from the fetal adrenal gland. This unusual capacity, a very new and exciting find, shows that the nine-banded armadillo contrasts with most other mammals whose fetal-placental-maternal relationships have been explored. It may be a very "primitive" trait but it illustrates the diversity of the reproductive cycles in mammals. Of greater pertinence to captive propagation is the recognition that female captive sterility (not shared by males) is associated with a high progesterone output (Lasley et al., 1977). Various studies point to the adrenal gland of captive adult females as the site of progesterone production; one may speculate that this is a carry-over of the fetal capacity to produce progesterone. Current studies are designed to disrupt this inappropriate behavior and manipulate the captive group towards intentional reproduction.

A third group of species whose endocrine system is of interest are the lemurs bred in our park. These species have well-defined winter breeding seasons with fully displayed sexual cycles. First blood, and now urine, from the monkeys were examined for estrogen, testosterone and progesterone. With these measurements it was possible to pinpoint the time of ovulation and to demonstrate differences in the cycle lengths and types of the three species studied (Bogart et al., 1977). Although our initial attempts at artificial insemination have failed, we will try again, not only to prove this technique feasible, but also to overcome the unwillingness of one specific female ruffed lemur to mate.

Techniques for Hormone Assay

Our endocrine studies are performed in a laboratory equipped for specialized radioimmunoassay procedures (RIA). These are standard in most hospitals. RIA directs specific antibodies against the substances to be measured.

Assume that testosterone is to be quantitated in a biological sample. An antibody directed against testosterone plus a known quantity of radioactive testosterone are added to a prepared sample. The antibody absorbs competitively equivalent amounts of the unknown (unlabeled) and radioactive (labeled) testosterone. After an appropriate equilibration period, the antibody-bound fraction is separated and its radioactivity is quantitated with a scintillation counter. From established displacement curves, the unknown quantity is then assessed.

RIA work with specific antibodies is feasible with widely differing wild animals because of the great similarity between most reproductive steroid hormones in mammals. When marked differences exist in the nature of steroid hormone metabolism, as is the case for perissodactyla and cetacea, other techniques–such as high pressure liquid chromatography–are employed for the definitive identification of such steroids. The point is that urinary work with steroids can achieve meaningful results in a wide variety of species.

The problem is altogether different with protein hormones since their structure differs more widely in the animal kingdom. Thus, it is unlikely that the antibodies for human or monkey pituitary luteinizing hormone (LH) could reliably measure the concentrations of that hormone in armadillos, gazelles or elephants, to name a few species of concern. Nevertheless, the normal human reproductive cycle and its disorders have been defined by protein hormone measurements. They have been studied with (10 minute) serum samples, though not with urine samples–a need we perceive and are now pursuing. Fortunately, it was recently realized that the biologic activity of LH can be determined by its capacity to stimulate rat testicular interstitial cells to produce testosterone; testosterone is readily measured subsequently (Dufau et al., 1974). The rat interstitial cell assay is widely employed in this laboratory and has been successful in interspecific studies. Fortunately, enough LH appears in the urine to be measured. Parallel dose response curves have been achieved from pituitary extracts, providing credence to plasma and urinary LH profiles. Most excitingly, it has been shown that the fetal armadillo possesses much higher LH levels than its placenta or mother, providing a basic clue to the mechanism of fetal sexual development.

These initial studies of the reproductive cycles of wild animals give us confidence that the modern techniques of endocrinology are not only applicable, but will be enormously helpful in the future management of captive endangered animals.

Sex Determination and Endocrine Studies

We recognized that many birds in our collection could not be sexed. We conjectured that, since mammals excrete testosterone (T) and estrogen (E) in their urine, these steroids might be found in the feces of birds, making sexing feasible. Systematic studies of fecal T and E have indeed shown that the sex of mature, healthy birds can be determined by such measurements. This technique is particularly successful when the steroid levels are expressed as the ratio of E/T (Czekala and Lasley, 1977). In most specimens with documented age and gonadal status, the ratio is greater than two in females; much lower in males. However, the absolute value varies between species. Figure 3 shows the results of studies with Hispañolan parrots. The samples were sent as unknowns, though the animals had previously been sexed by other means. The findings clearly indicate the utility of this technique since all specimens were correctly identified. (It should be emphasized that the feces were preserved in formalin—we now use a one percent solution—to inactivate any Newcastle virus contamination.)

Since some 30 percent of birds are sexually monomorphic, this tech-

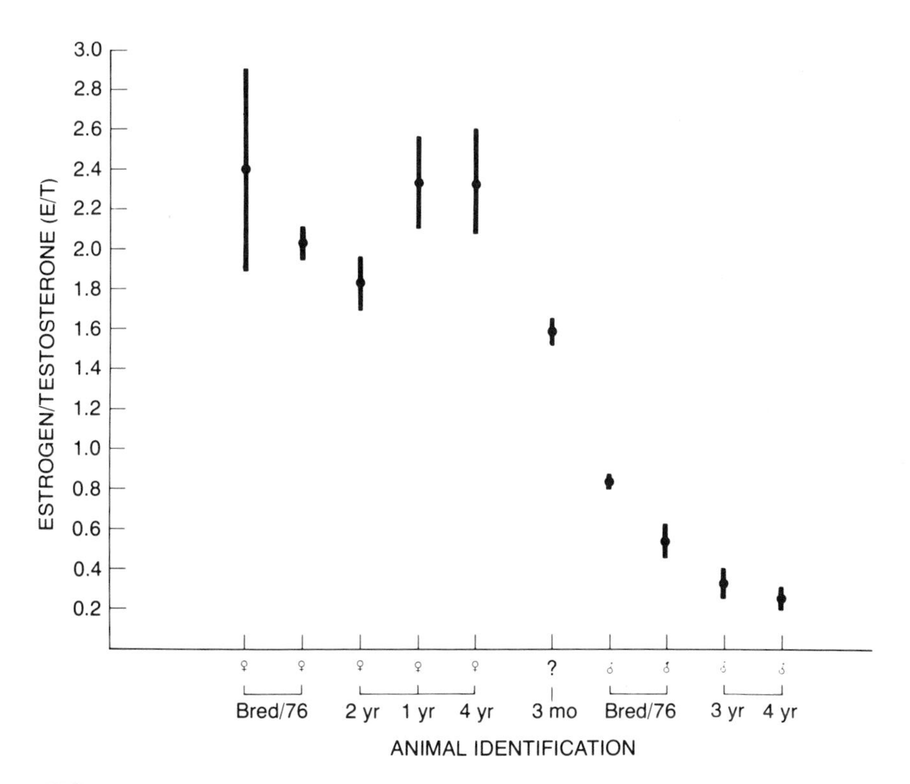

FIGURE 3. Estrogen (E)/testosterone (T) values of three fecal samples from ten Hispañolan parrots *(Amazona ventralis)* with standard errors indicated by bars. The five animals at left are clearly females (E/T ± 2); the four animals at right are males (E/T<1); the sixth animal with a ratio of 1.6 is immature and cannot yet be sexed by this method. (From Bercovitz, 1978)

nique is of great importance for future bird reproduction. To be sure, ornithologists can sex birds by laparotomy—veterinarians have recently observed avian gonads with great ease through the laparoscope. Still, the ease with which feces can be collected and shipped to endocrine laboratories makes fecal sexing a more widely applicable method. Through serum T and E measurements, Komodo dragons have also been sexed; it is likely that other species will benefit from these techniques as well. Furthermore, if endocrine derangement is suspected, these methods will aid in diagnosis and therapy (as has been amply demonstrated with human subjects.)

GENETIC STUDIES

As with endocrinology, enormous strides have been made in cytogenetics by medical researchers during the past two decades. Just as a wide variety of anomaly syndromes have been tied to chromosomal errors, frequent abortions can be linked to chromosomal rearrangements due to malsegregation at meiosis. When investigators have examined comparable events in domestic species (sheep, cattle, horses) they have found similar syndromes. We can therefore assume that these syndromes will be found among species we wish to conserve.

To begin with, it was necessary to define the normal chromosome number of each species, an overwhelming task in itself. Of the approximately 3,000 species of mammals, perhaps one third to one half have now been superficially so characterized and documented (Hsu and Benirschke, 1967–1977). This is usually done on one or two specimens of each species. Techniques for examining chromosome bands, which provide much more data, have not been applied in most instances. When more specimens became available and when banding procedures were applied to animals from different regions, it was found that considerable chromosomal polymorphism exists in any given species (which is consistent with recent findings among humans). At times, numerical differences in chromosome number have been found in taxa that have nearly identical phenotypes, so one can assume that crossbreeding may be deleterious as a conservation measure. Since few investigators engage in such crossbreeding studies however, proof for the disadvantages of mating chromosomally inappropriate specimens has been rare. A few examples may point out the possible future implications.

Chromosomal Polymorphism Within Species

The chromosomal polymorphism of South American primates is particularly striking. One would assume from the findings to date that many are involved in a process of active speciation. Thus, squirrel monkeys from different regions possess different karyotypes which correlate with slight differences in facial pelage (Jones et al., 1973). Similarly, spider monkeys, a large group with many color variants and a distribution that includes wide regions in northern South America, differ in the structure of several chromosomes and fall into at least four separate genotypes. While these two species have the same number of chromosomes in all members, the structure of individual chromosome pairs, as ascertained by banding analysis, differs and is explainable only by assuming that segments have inverted or translocated (Figure 4).

The phenomenon in the owl monkey is more complex. To date eight different karyotypes have been identified that correlate with regions of origin. In this species the chromosome number varies from $2n = 46$ to $2n = 56$. More important, it has been shown that pairing the appropriately karyotyped individuals leads to significantly improved reproduction (Cicmanec and Campbell, 1977). Hybrids of animals with different karyotypes are known to exist. If experience with human translocation carriers having 45 chromosomes is indicative, one can assume that mal-

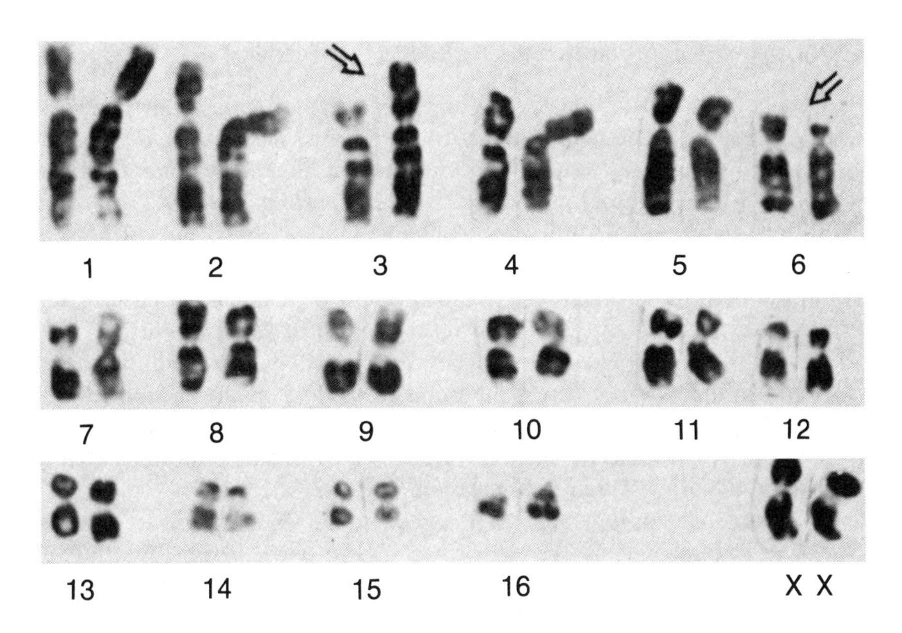

FIGURE 4. Giemsa-banded karyotype of Bolivian female spider monkey *(Ateles paniscus belzebuth)* collected in the Bolivar province by Dr. R. A. Cooper, 1973. Most of the 44 chromosomes are paired. The arrows indicate presumed exchange between short arms of chromosome 3 and 6. This translocation is heterozygous in this specimen, homozygous in others.

segregation of unpaired chromosomes at meiosis would lead to an increased frequency of abortions. Direct proof of such aneuploid abortuses is not available; such data will be important for future conservation strategies.

It can be argued that such chromosomal rearrangements are a natural evolutionary phenomenon and that mating only "matched" pairs would be setting the clock back. Far too little information is available about this to make sound recommendations. The discovery of cattle with only 59 chromosomes (the normal number is $2n = 60$) and possessing a 1/29 translocation provides a demonstration of this. This genotype appears to have arisen in a Swedish bull whose semen was widely dispersed. It is now heavily represented in some inbred lines of cattle. To date, the potentially significant deleterious effect of this genotype has not been clearly identified in the reproductive history of the cattle although its existence is suspected (Gustavsson, 1969). At the same time, "new" cattle possessing a balanced set of 58 chromosomes have been produced.

If similar events occur in the inbreeding of endangered species, would they have an undesirable or a neutral impact on the future of such species? The answer to this important question is not known, but the findings from domestic species argue strongly for continued study.

Chromosomal Evolution and Phylogeny

We have been studying the chromosomes of antelopes and horses to better understand the evolutionary relationships between these species. In these investigations, a number of unusual findings were made that are pertinent in this context. For instance, the barking deer of India and China, although reasonably similar in phenotypes, have quite different chromosome numbers. The Indian muntjac *(Muntiacus muntjac)* has a chromosome number of $2n = 7$ (males), $2n = 6$ (females), while its Chinese counterpart *(Muntiacus reevesi)* has a number of $2n = 46$.

In similar though usually less dramatic situations, such findings allow us to determine whether certain animals in collections are of hybrid origin. This is a diagnostic test we are asked to perform with increasing frequency. Often, the parent species differ only by minor banding pattern changes (felidae) or C-band polymorphism (cetacea).

A surprising finding was made recently in Soemmering's gazelles *(Gazella soemmeringi)*. We received two such specimens from Busch Gardens and found that their chromosomes had remarkable differences, much more than is usual in artiodactyla (Figures 5 and 6). The finding became potentially important when it was learned that the ancestral group suffered from an excessive frequency of abortions and stillbirths. If such reproductive mishaps can be shown to result from aneuploid karyotypes, separation of the stock into appropriately matched genotypes would be mandatory. However, catching and karyotyping such fickle ani-

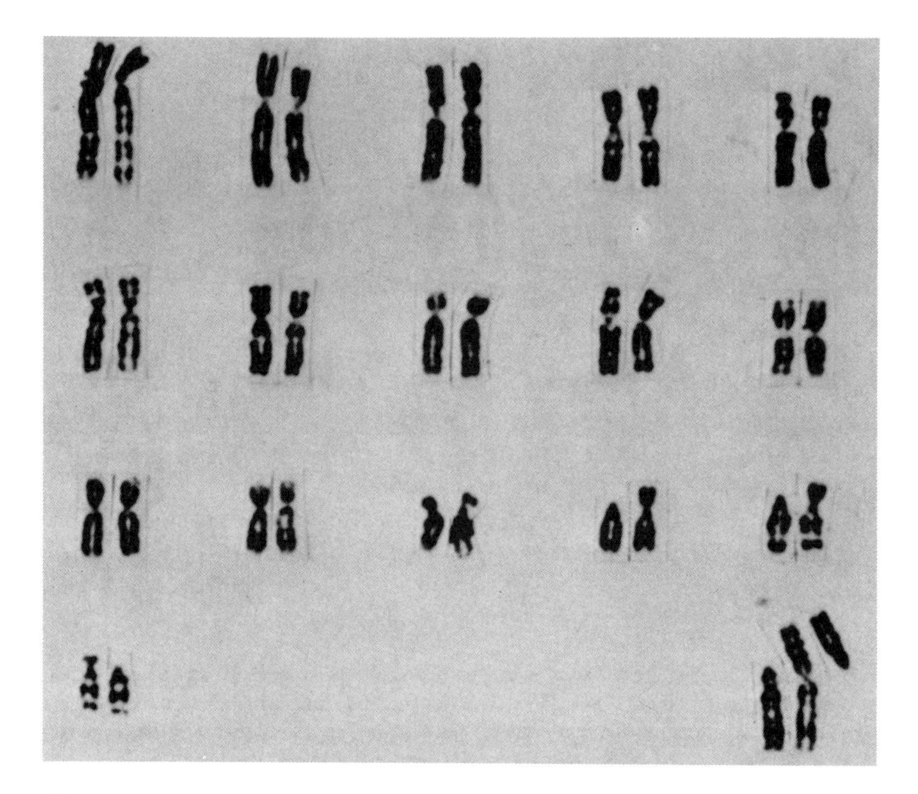

FIGURE 5. Karyotype of female Soemmering's gazelle *(Gazella soemmeringi)* with 35 chromosomes (compare with Figure 6). This animal has only five truly acrocentric elements. In this Giemsa-banded karyotype the presumed X-chromosomes are placed last. In many gazelles X/autosome fusions exist, a process which may be underway in this species.

mals is a difficult task. Nevertheless, if proven to be of great significance, karyotyping may become part of all initial quarantine studies.

There are some 60 okapis now in captivity and it is unlikely that new stock will be acquired. Of the few animals karyotyped, some have a chromosome number of $2n = 46$, others $2n = 45$—the latter possess one "Robertsonian fusion." At present, we are uncertain whether this fusion has a deleterious impact on reproduction but we are concerned about it.

A somewhat similar polymorphism was found in the kulan, a close relative of the onager (wild ass). The few specimens examined so far either had 54 or 55 chromosomes, differing from the onager ($2n = 56$) by one Robertsonian fusion step. It is conceivable that this is chromosomal evolution caught in the process. Alternatively, one might wonder whether

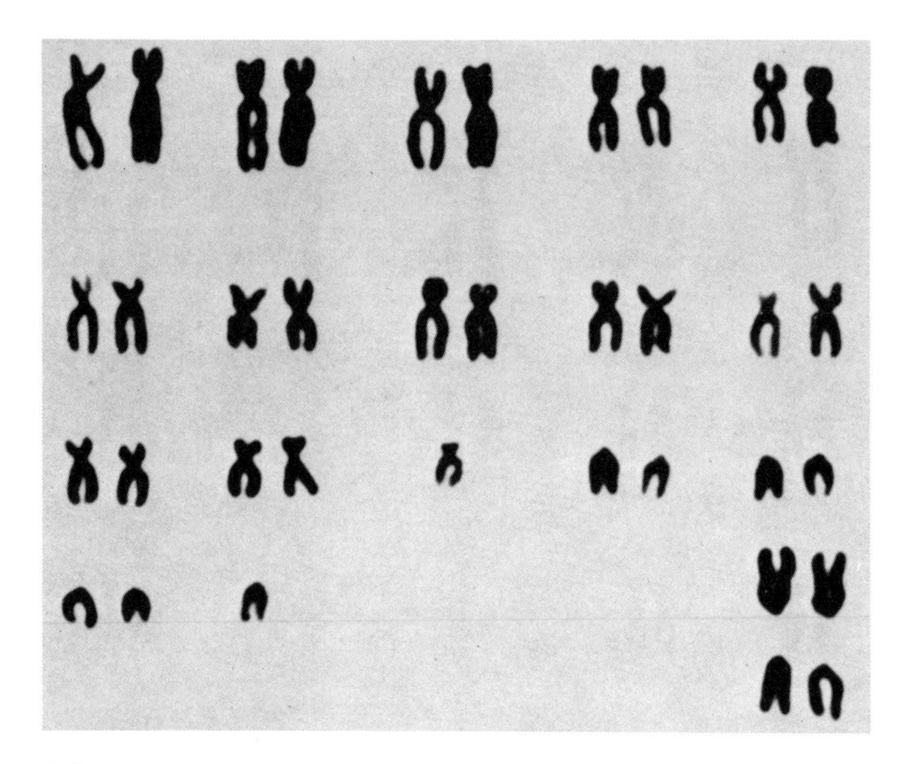

FIGURE 6. Karyotype of female Soemmering's gazelle from the same herd as the animal in Figure 5, but with 36 chromosomes and 11 acrocentrics. The last four elements presumably make up the compound X shown in Figure 5. These are complex chromosome problems that are currently under detailed investigation.

these animals are hybrids of a species with 56 chromosomes and one with 54, but it is too early to tell since only six specimens are available.

All domestic horses examined to date, and these include a large variety of breeds, have 64 chromosomes; the presumed ancestral Przewalski's horse always has 66. The fertile hybrids are easily identified since they have 65 chromosomes. Here we can insist that such hybrids should not enter the gene pool of the few animals remaining in zoological gardens (Ryder et al., 1978).

Cytogenetic studies have also been useful in determining some taxonomic relationships. Because it can be expected that unusual fusions of chromosomes occur only on very rare occasions, it is possible to deduce the descendency of some species. For instance, among the antilopinae the extraordinary X/autosome fusion links at least eight species, but is not found in four others. It is probably valid, then, to postulate the relationship shown in Figure 7 (Effron et al., 1976). Similarly, Y/autosomal fusion links the tragelaphinae together, but the additional X/autosome translocation is common only to a few of these animals and, therefore, is presumably of later origin.

The point of these studies is that the cytogenetic assessment of mammals has not only theoretic scientific interest, but can also be of great value in conservation efforts. Unbelievably, for instance, trisomies (the equivalent of human mongolism or Down's syndrome) have already been detected in a chimpanzee, a gorilla, and an orangutan. The potential breeding of these animals must be carefully weighed. Amniocentesis and cell culture, which guide human pregnancies, are possible in captive wild species. Indeed, the L/S ratio of amniotic fluid, a determination of lecithin and sphingomyelin that allows the judgement of fetal maturity, has been employed repeatedly in gorillas prior to optional Cesarian section (O'Grady et al., 1978). Likewise, antenatal chromosome analysis of amniotic fluid is feasible in pregnant cows (Bongso and Basrur, 1977) and could be used in other species.

An important question for the future is: Should we determine the fetal sex of some species and abort the male pregnancies? For example, too many male Arabian oryxes now exist and are maintained at great expense. This is also true of a variety of other ungulate species and other animals with a male harem group breeding behavior.

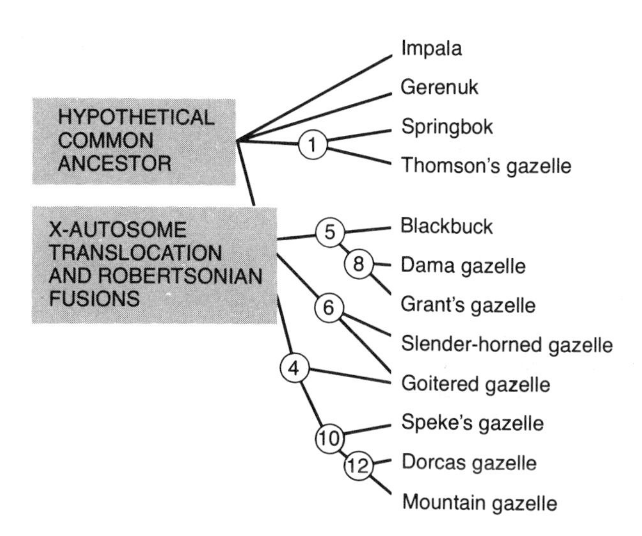

FIGURE 7. Diagram of the presumed evolutionary relationship among Antilopinae as suggested by banded karyotypes (Effron et al., 1976). A translocation between X and autosome is common to all eight lower species. The numbers in circles indicate common metacentric elements formed by Robertsonian fusion. (From ZooNooz, August, 1976)

Cytogenetic factors as causes of repeated failures of pregnancies can be illustrated by our oldest male Douc langur which has an abnormal karyotype—a reciprocal translocation among two autosomes. Repeated abortions and stillbirths (identified by a rapid fall of urinary estrogens during pregnancy) have been witnessed in our colony. We hope to identify an aneuploid abortus at some future date to verify the translocation as the cause of the reproductive failures.

A Frozen Zoo

Progress is slow in so complex an area of investigation, and the most desirable specimens are not always accessible. Who would want to immobilize a giraffe or a rhinoceros, for instance, merely to study its chromosomes? At the same time, though, new techniques are being developed continuously that invite a more detailed examination of previously studied karyotypes. The solution to this problem is the liquid nitrogen preservation of cell cultures. With this technique, which is very much like the preservation of semen, cells can be "resuscitated" at any later time.

It is for this reason that we have endeavored to store in our "frozen zoo" (the "ark of the twentieth century," as it has been called) strains of cells from biopsies collected from mammals that die in our zoo. This collection is of value to investigators who study the process of cell aging and those who hybridize cells or DNA to establish homologies and localize genes to individual chromosomes. While similar studies with reptiles and birds are now going on in a few laboratories, much more effort is needed to realize our dream: characterizing the genomes of most wild species. In the meantime, the conservation by freezing of strains of cells with known origin at least helps to conserve these genetic sets for later study.

Inbreeding and the Assessment of Genetic Diversity

Another matter of great concern for those involved with the long-term propagation of species in zoos is the potential development of animals with defects due to excessive inbreeding (Chapters 8, 9 and 12). This question is beginning to receive attention but much additional work is required. In the past, breeding efforts involving highly inbred animals have been suspected of playing a part in a number of disease processes and reproductive failures.

One consequence of inbreeding is a reduction of variability in the descendants of inbred individuals. Genetic variability can be assessed in captive and natural populations by blood grouping and electrophoretic techniques. It is clear that species vary greatly in their genetic variability. Only very sparse data exist about genetic diversity in the noncaptive populations of most species kept in zoos and aquaria. If such data were available, the genetic management of exotic species could proceed more rationally. In the meantime, genetic data between captive populations in different institutions can be compared. As a practical measure, we can assume that the maintenance of polymorphisms within a species will better its chances for prolonged captive breeding success.

Our examination of genetic variability in Przewalski's horse, *Equus przewalskii* is an example of the practical application of this type of genetic monitoring of captive endangered species. A substantial herd of this endangered species has been propagated at the San Diego Zoo and Wild Animal Park in recent years. While probably extinct in the wild, some 270 of these animals survive in zoos and animal parks. Our herd traces its ancestry to nine individuals captured in Mongolia and brought to Europe in 1901 and 1902. Genetic variability in blood groups, hemoglobins and serum proteins of Przewalski's horses (Trommershausen-Smith et al., in press; Ryder, 1977) have been monitored in collaboration with scientists at the University of California at Davis and Los Angeles.

Figure 8 displays the results of the separation by polyacrylamide gel electrophoresis of serum proteins in some Przewalski's horses. Polymor-

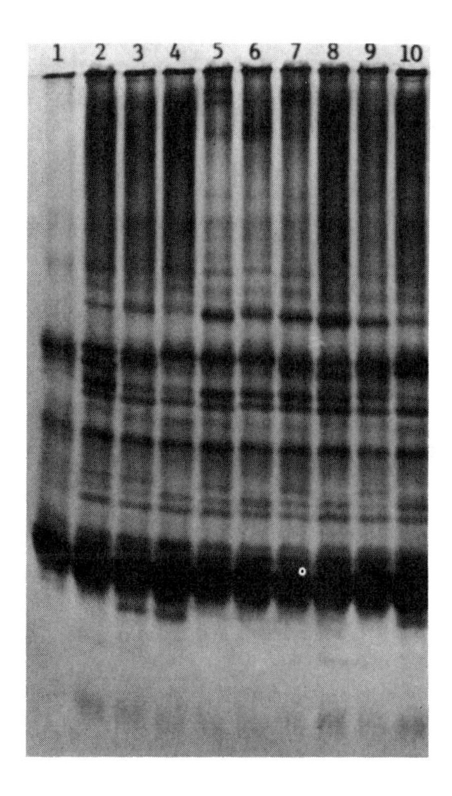

FIGURE 8. Polyacrylamide gel electropherogram of serum proteins in 10 *Equus przewalskii* individuals. Przewalski's horse serum samples were electrophoresed in 7.5 percent polyacrylamide at pH 8.4; proteins were made visible by staining with Coomassie blue. Slot 1: foal; slot 2: Henrietta; slot 3: Bogatka; slot 4: Jeanhold; slot 5: Bosaga; slot 6: Belkina; slot 7: Belzar; slot 8: Bolinda; slot 9: Bogdo; slot 10: Belina. Serum albumins are the darkest staining bands; serum transferrins are the darkly staining bands midway between albumin and the top of the gel. Prealbumins run ahead of (below) albumins. Note heterogeneity of protein bands. (From Ryder, 1977)

phism in several proteins (notably transferrins and prealbumins) can be observed. Taken together, these studies have revealed rather remarkable degrees of polymorphism in the characters assessed. In the offspring born at the San Diego Zoo and Wild Animal Park, polymorphism was present in 12 of the 17 examined loci in spite of rather large inbreeding coefficients (f = 0.218 to 0.334). The parental stock forming the basis of these herds displays polymorphism at 14 of the 17 loci. The decrease in number of polymorphic loci can be attributed to the fact that a single mare, Roxina, the sole source in this herd for polymorphism in two systems, has failed to reproduce.

In order to maintain genetic diversity at the examined loci (and, presumably, other loci as well) it would appear that concerted efforts should be made to see that Roxina reproduces. These studies directly illustrate the value—in terms of preservation of genetic variability—of having each individual participate in herd reproduction. We feel that this is particularly important with species such as Przewalski's horse for which the number of individuals in the total population is limited and the chances of obtaining additional animals is remote. In such cases, the future existence of the species will depend on successful captive breeding efforts. In light of these considerations, it behooves us to apply the most modern technologies available for assessing parameters of genetic diversity and heterozygosity, and to apply pertinent findings in the management of endangered species.

VETERINARY MEDICAL CONTRIBUTIONS

Because of the large variety of species under their care and the bewildering diversity of as yet poorly understood animal diseases, the zoo veterinarian faces formidable difficulties in his daily task. It is gratifying to see, though, that within the last decade more emphasis has been placed on exotic species in Veterinary School curricula and that many more training opportunities are now being provided to expand knowledge in this important field.

Along with this expanding knowledge, considerable strides are being made in the care of exotic species that aid substantially in conservation efforts. Infectious and parasitic diseases are being treated more efficaciously and—a particular concern of zoos—the spread of agents from one group of animals to others having no previous contact with such agents is being prevented. Laparoscopy and roentgenography are widely employed in the care of exotic species and, when needed, very modern surgery (for instance, bone plating) has been successfully executed with a variety of endangered wildlife.

New programs currently under active development will make further contributions, the beneficial effects of which are difficult to anticipate at this time. We are thinking particularly about sophisticated dental care. With the frequently altered diet and social conditions common in captive species, dental problems exist in many specimens. Their recognition and therapy are important. Besides tooth extraction, root canal therapy, capping and other therapeutic measures can be undertaken to restore animals to optimal conditions (Robinson, 1978).

As this specialty evolves, it is gratifying to see that physicians (with their elaborate equipment) are not only willing, but anxious to contribute their specialized skills to the efforts of conservation. Thus, the gap between the sophistication of human medicine and care for exotic species is narrowing. What hinders progress most is the relatively small number of individuals working in this complex field and the general paucity of funds available for a more rapid expansion of instrumentation, research activities and personnel.

An example is the employment of ultrasonography as a diagnostic tool in zoological parks. The sophistication of this methodology has expanded rapidly in the past decade. It is now possible not only to measure the size of a fetus, determine the location of the placenta and follow growth, but also to measure such things as bladder size and cardiac chambers. Real-time ultrasonography is used to detect tumors, cysts, foreign bodies and a host of other medical problems without the dangers of irradiation. In addition, it allows the physician to monitor physiological events without having to rely on stationary images as with x-rays. It is expensive, but so helpful that modern reproductive medicine can no longer exist without it. Its application to conservation should not be delayed because of possible monetary restrictions; if we take this challenge seriously, it is our responsibility to educate the public appropriately.

CONCLUSION

The world's wildlife is rapidly diminishing. It has become increasingly likely that a substantial number of species may become extinct in the near future. Zoological gardens can serve as the last reservations for a small number of these animals, but are not yet uniformly prepared to assume this enormous task. By and large, not enough is known of the biologic properties of most species to assure that they can be reproduced in perpetuity.

It is our belief that abundant biomedical technology exists to acquire the necessary information for captive reproduction of endangered species. This must be brought to the zoos quickly, public support must be solicited. A wide variety of endocrine, genetic and infectious disease work needs to be undertaken. The results of such studies combined with behavioral information can then be applied to save our dwindling wildlife.

SUGGESTED READINGS

Antikatzides, Th., S. Ericksen and A. Spiegel, 1976, *The Laboratory Animal in the Study of Reproduction,* Gustav Fischer Verlag, Stuttgart. This book presents a variety of comparative endocrine work in such animals as primates and rodents, and gives helpful insight into the topic. No work on comparative endocrinology of exotic species has yet been compiled.

Benirschke, K. (ed.), 1969, *Comparative Mammalian Cytogenetics,* Springer-Verlag, New York. A conference monograph, this is the first attempt to address the evolutionary significance of changing mammalian chromosome number.

Chiarelli, A. B. and E. Capanna, 1973, *Cytotaxonomy and Vertebrate Evolution,* Academic Press, New York. Authorities review in different chapters the chromosomes of various vertebrate classes. Particularly helpful is the tabular listing of chromosome numbers with references to practically all animals karyotyped before 1973.

Fowler, M. (ed.), 1978, *Zoo and Wild Animal Medicine,* W. B. Saunders Co., Philadelphia. This is the most up-to-date text on exotic species veterinary medicine and includes recent advances made in this complex field.

White, M. J. D., 1973, *Animal Cytology and Evolution,* Cambridge University Press, London. The most comprehensive treatise on the significance of chromosome rearrangements.

Yen, S. S. C. and R. B. Jaffe, 1978, *Reproductive Endocrinology,* W. B. Saunders Co., Philadelphia. This text discusses all principal endocrine mechanisms and assays in a very up-to-date position, but as applied only to human reproductive endocrinology.

CHAPTER 14

THE SOCIOBIOLOGY OF CAPTIVE PROPAGATION

Devra G. Kleiman

For the captive propagation of an animal to be successful, one must take into account the species' spatial requirements, dietary specializations, environmental needs (temperature, light cycle, humidity), specific housing requirements and medical problems. This knowledge must be obtained from detailed studies in the natural habitat. Without it, a captive breeding program will proceed on a trial and error basis. Serendipity is at the root of some successful propagation programs, but many species which have particular but unknown requirements have been impossible to maintain or breed in captivity. Historically, they were referred to as "delicate." Improved field techniques (for example, radio telemetry) and a remarkable increase in the amount and quality of field research have provided clues about life history requirements which, in some cases, have made previously "delicate" species prolific breeders.

Knowledge of a species' social requirements has always been considered important for successful captive propagation. A superficial analysis of a species' social system, however, can be misleading and can result in serious errors when establishing captive colonies. For example, two species in nature might live in groups which have an identical age and sex composition, yet have entirely different mating and parental care systems as well as social organizations. Attempts to apply the same captive management techniques to both would probably result in the failure to maintain and propagate one of the species.

Moreover, just as species may differ in their ability to adapt to the absence of specific nutritional or housing requirements, there are unknown species-specific differences in the social lability of animals (for example, the degree to which an abnormal social environment can be accommodated without pathological consequences). Differences in the degree of social lability may have greatly influenced which species have historically thrived in captive conditions, where there has been ignorance about their social requirements.

Sociobiology has been defined by Wilson (1975) as "the systematic study of the biological basis of all social behavior." As such, it has its roots in modern evolutionary theory, ethology, physiological psychology, population genetics and ecology. The study of social behavior must, therefore, encompass knowledge of a species' phylogenetic history, ecology, life history parameters, population genetics and behavioral adaptations.

Studies of the behavior of captive animals have a long history, but have mainly contributed to the disciplines of ethology and physiological psychology. Because of small sample sizes and restricted conditions, behavioral research in zoological gardens has concentrated on the dynamics of social interactions, animal communication, the evolution of behavior and the analysis of the motivation and function of behavior. Thus, some of the major branches of study within sociobiology have not been explored in zoos. One of the major purposes of this chapter is to suggest not only how studies from other branches of sociobiology may benefit captive propagation programs, but to suggest how studies of captive animals—especially in zoos—may contribute more broadly to the field of sociobiology.

SOCIAL ORGANIZATION

Social organization is an umbrella term which encompasses a number of other characteristics, some—though not all—of which are independent of each other. Subsumed under the term social organization is the mating system of a species, be it monogamous, polygynous, polyandrous or promiscuous. Mating strategies must be differentiated according to whether a particular male-female relationship is constant over several reproductive efforts, or changes with each reproductive effort. For example, numerous bird species are serially monogamous, with partners changing each season, while others form long-term or even lifetime bonds.

Rearing strategy will also affect the ultimate social organization of a species. Most important are: (1) the degree of contact between the mother and young during rearing; (2) the degree to which females will rear young together; and (3) the degree of participation by the male or older nonreproducing offspring in parental care.

Group size is partly affected by mating and rearing strategies, but is also influenced by feeding habits, food distribution and the method of acquisition, shelter requirements, antipredator needs and phylogenetic constraints such as a species' size and mobility.

The successful establishment and maintenance of a species in captivity depends upon detailed knowledge of social organization. For example, the average group size, sex and age composition of a wolf pack *(Canis lupus)* and a lion pride *(Panthera leo)* might be similar, yet the mating systems, parental care techniques and social organizations of these carnivores differ dramatically (Kleiman and Eisenberg, 1973). These differences will have a major effect on the development of captive breeding programs for these two species. Wolves are essentially monogamous; only the single founding pair will usually reproduce. The adult male and older offspring aid in rearing younger siblings, but the latter are themselves reproductively suppressed. Female wolves are as fiercely competitive as males (perhaps more so) and may have serious conflicts after puberty. Females as well as males are likely to disperse from the natal group (see Kleiman and Brady, 1978).

Several males and several females in a group of lions may reproduce. The males of a pride are typically brothers and must compete with other, unrelated males for control of a pride whose core is a group of related females and their offspring. There is, therefore, considerable turnover in the males from year to year. Young males disperse from the natal pride while females normally remain to reproduce within their natal group. Adult males do not participate in parental care (Schaller, 1972; Bertram, 1976).

In captivity, such differences may be expressed at several stages in the development of a breeding group. A wolf pack is best founded upon a single unrelated pair. A lion pride can probably be established immediately with a group of related, tolerant males and a group of related females. Maturing males and females within a wolf pack may develop conflicts with the father and mother, respectively, while such conflicts will develop mainly among males in lions. Among wolves, the onset of reproduction may result in fighting within each sex; this is rarer among lions. Several female lions can usually raise their litters cooperatively, while such an event is less likely in wolves. Each of these factors has relevance to captive management decisions.

Mating Strategy

Definitions of mating strategies are, to some degree, simplistic. As Jenni (1974) has pointed out, males and females may have different strategies within a single breeding system which may depend on the temporal organization of reproduction. Among mammals, if a female's period of estrous is short, and only one male at a time is likely to mate within a social group (for example Père David's deer, *Elaphurus davidianus* or Hamadryas baboon, *Papio hamadryas*), the female can be considered to be monogamous in her relationship with the male while the male is considered to be polygynous. If the major male breeder changes annually or every two years, the female could be considered to be serially monogamous. Similar effects of temporal changes in the structure of a species breeding population make it clear that time must be included as part of any definition of a mating system.

The temporal organization of reproduction is important for captive colonies because of the potential genetic consequences. For example, Père David deer males compete for the position of harem master during the breeding season. During the rut, one male is likely to do a majority of the breeding, but other males may also reach the status of harem master when the dominant bull is fatigued. The top position is achieved by fighting among the males which can—and has—led to the death of captive individuals (Wemmer, 1977; Wemmer and Collins, in preparation). The same bull may achieve control of the harem in successive years, but eventually it cannot compete effectively with younger bulls reaching their prime. A management policy which isolates the majority of bulls from the harem during the rut and which allows only one or two males to copulate with the females, prevents intrasexual competition and thus does not permit the natural elimination of potentially unfit males. Although permitting mate selection to develop without excessive interference should be the goal of any long-term propagation program, the potential for mortality from fighting is great. Thus, curators and keepers usually decide which animals will breed.

The process of mate selection may not only result in fighting among males, but may be extremely disruptive to entire social groups. For example, in several langur species (*Presbytis* spp.) all-male groups may periodically take over established breeding units (Rudran, 1973; Sugiyama, 1965; Blaffer-Hrdy, 1977). The phenomenon has also been described for other colobine and cercopithecine monkeys (for example, Struhsaker, 1977), and it is likely to be found in other primate species that tend to live in age-graded male troops or one-male troops (Eisenberg et al., 1972). Takeovers are usually accompanied by great social upheaval and tension —including infant mortality—in part from infanticide by the invading male(s) (Rudran, 1973; Blaffer-Hrdy, 1977).

Rudran (1973) indicated that such takeovers may occur, on the average, every three years in the purple-faced langur, *Presbytis senex,* while Blaffer-Hrdy (1977) estimates that takeovers occurred as often as every 28 months in one population of *Presbytis entellus.* One of the effects of such takeovers is the prevention of inbreeding within a deme.

Clearly, it would be unacceptable to orchestrate such changes in captive colonies of colobines—that is, to permit groups of subadult and adult males to invade established breeding units. More appropriately, the adult male could be replaced with a different male every two to three years. Even such planned replacements, however, might result in juvenile and infant deaths due to aggression from the introduced male; zoological parks should perhaps expect an increased juvenile mortality at such periods or decide to delay the introduction of a new male and thus forgo reproduction until most infants and juveniles in a group are weaned. In either case, a management plan for species with this reproductive strategy must include the expectation that there may be no infant cohort every third or fourth year when males are exchanged.

Interestingly, there are reported (via the zoo grapevine) cases among some captive primates of males and other group members killing infants. Documentation of the events surrounding such deaths, including the names of the species in which infanticide has occurred, might aid captive management and contribute to our understanding of the function and evolution of infanticide in wild primate populations.

Although it may be possible to artificially limit intrasexual competition and still achieve the outcrossing benefits of this process by manipulating individuals or groups, some species or individuals may not reproduce adequately without competition. Among the males of some mammals there is a complex feedback system regulating androgen levels through social behavior—for example, the rhesus monkey, *Macaca mulatta* (Rose et al., 1975). This system may affect the strength of the male libido. The performance of socially dominant behaviors (for example, threats and fighting) and the experience of winning aggressive encounters may be essential to achieve hormonal levels adequate for the performance of sexual behavior. The absence of competition may depress male libido sufficiently to inhibit reproduction.

Females may be similarly affected; the absence of male-male competition, for example, could result in somewhat lower androgen levels in a male which in turn could depress female reproductive function. Such a phenomenon could simply be an extension of the "Whitten" effect in which female mice housed in groups without a male exhibit irregular or no reproductive cycle until exposed to a mature male or his odor (Whitten and Bronson, 1970). Thus, a female's estrous cycle may be irregular or depressed in the presence of a male whose urine or other glandular secretions indicate low androgen levels.

Most zoos have experienced situations where a single pair has not reproduced despite every effort to provide adequate housing, diet and other environmental needs. Currently, at the U. S. National Zoological Park there is a single pair of lowland gorilla *(Gorilla gorilla)* and a pair of Indian rhinocerus *(Rhinocerus unicornis)* whose lack of reproduction may be attributable to the absence of male-male competition. In both species there was a single successful reproductive effort followed by a depression of the female estrous cycle and relative disinterest on the part of the male.

The successful propagation of monogamous mammals has posed a problem for most zoological institutions. First, this mating strategy has been difficult to recognize in the field since monogamous species can exhibit a group size varying from one to 15 individuals, as shown in Figure 1 (Kleiman, 1977; in press, *a*). For example, elephant shrews *(Elephantulus rufescens)* tend to be solitary (G. B. Rathbun, in press) while hunting dog packs *(Lycaon pictus)* are extremely large due to the presence of subadult and adult "helpers" and a normal litter size averaging six to seven (occasionally reaching 15 pups) (Kleiman and Eisenberg, 1973). Both species were originally assumed to be polygamous until detailed field studies revealed the true mating system (*Lycaon:* Kühme, 1965; van Lawick, 1973; Frame et al., in press; *Elephantulus:* G. B. Rathbun, in press).

To propagate monogamous species one must accept the fact that no more than one female will usually breed in a group. Thus, several pairs or family groups should be maintained to ensure that reproduction is not halted with the death of a breeding male or female. This requires extra cage space and keeper efforts as well as attention to appropriate group size, age and sex structure.

Reproduction in monogamous mammals can be negatively affected by problems of mate selection and intrasexual competition. Although one can usually expect a successful mating when a pair is newly established, occasional pairs are incompatible. For example, in rufous elephant shrews, females are typically dominant over males (G. B. Rathbun, in press). In captivity, pairs with reversed dominance status will rarely reproduce successfully (G. B. Rathbun, personal communication). In the absence of detailed observations of pair interactions, failure of reproduction can only be ascertained five to six months after the initial pairing. Even with a successful breeding pair a time lag may occur before the first pregnancy.

With some species, it may be possible to encourage mate selection by permitting one individual to choose between two potential mates. How-

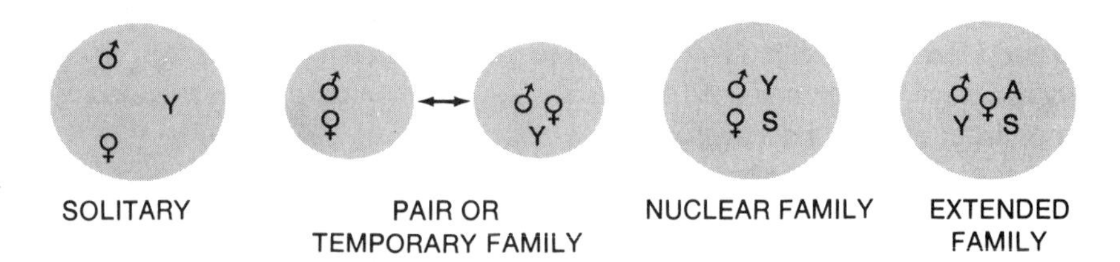

FIGURE 1. Group size and structure in some monogamous mammals. The circles represent joint territory. The placement of the symbols for the breeding male (♂), breeding female (♀), immature offspring (Y), subadult offspring (S) and related adults (A) within a circle indicate the distances between individuals. In the "pair" condition, young may only be with the parents temporarily, until dispersal. There may be more than one Y, S, and A, thus leading to group sizes of 10 to 15 individuals.

FIGURE 2. The monogamous golden lion tamarin *(Leontopithecus rosalia)* is an endangered primate from the southeastern coastal rain forests of Brazil. Males and other relatives aid in parental care by helping to carry young. (Photo by National Zoological Park)

ever, this can only be accomplished where adults of the same sex can be housed together without serious fighting. Among lion tamarins *(Leontopithecus rosalia),* an endangered primate species from the southeastern coastal rain forests of Brazil, this is more easily accomplished with males (Figure 2). In a study designed to examine the process of mate selection, I housed adult or young adult females with two males. Of ten such trios, only one had to be dissolved due to overt aggression between the males (Figure 3, Trio G). Yet, in the majority of trios only one of the males exhibited sexual behavior (Figure 3). Sexual behavior by the sexu-

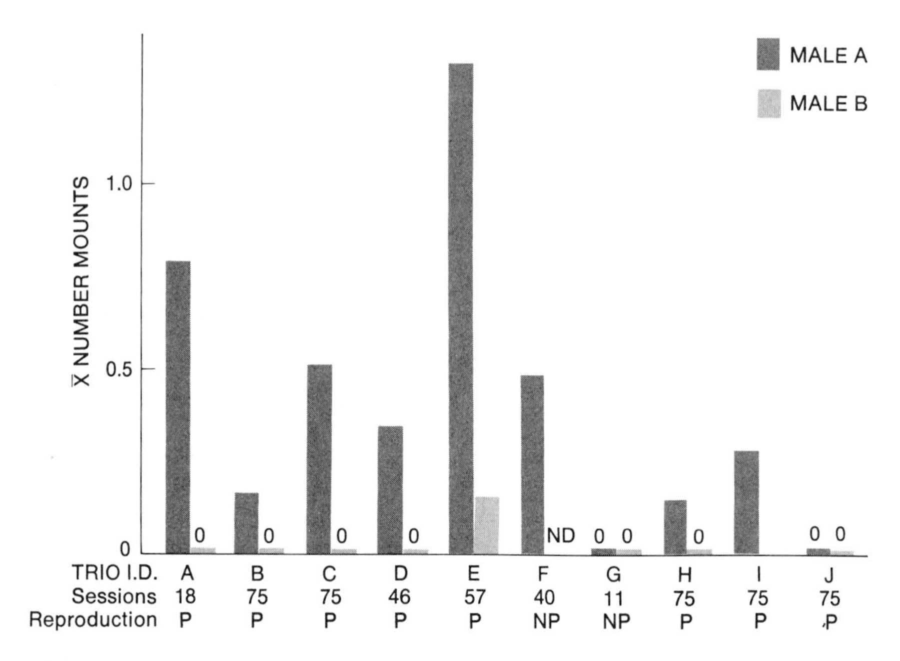

FIGURE 3. The mounting frequency of each male in ten trios of lion tamarins composed of two males and one female. NP : no pregnancy; P : pregnancy occurred during the observations; Sessions refers to the number of half-hour observation periods; ND : no data available.

ally inactive males in Trios E and I often consisted of mounting without pelvic thrusting; thus copulation was certainly not successful. In Trio G, which was dissolved, the female became pregnant by male A soon after observations terminated. In Trio J, male A was known to have impregnated the female soon after the trio was established even though no mounting was observed during formal observations.

There were several other interesting results of this study. First, overt signs of dominance were rarely seen between the males; the sexually inactive male was not isolated from the pair. Indeed, the trio rested and slept as a group. Second, there was only minimal evidence that the female deliberately chose one of the males. Figure 4 details the grooming interactions of the female and males in the ten trios. In the majority of cases, the sexually active male groomed the female significantly more than did the sexually inactive male. In three of the four exceptions, the female was related to the sexually inactive male (sister, Trios H and F; mother, Trio J) and had been living with him and other family members prior to the establishment of the trio. In only four trios did the females preferentially groom the male with whom they mated; in two of these four trios, the female was the sister of the sexually inactive male. Thus, female preference may be more strongly exhibited when a female is choosing between a relative and nonrelative.

The lack of overt male competition and of clear-cut female choice makes some sense in the context of the tamarin social organization. Family groups may include subadult and, perhaps, adult offspring. These and perhaps other relatives aid in the rearing of offspring by carrying infants and feeding weanlings. It is to the advantage of the breeding pair to be tolerant of relatives as long as they do not challenge reproductive dominance since helpers may increase the survivorship of the reproductive pair's offspring. Thus, nonbreeding subadults and adults can be integrated into family activities without aggression from the dominant reproductive pair.

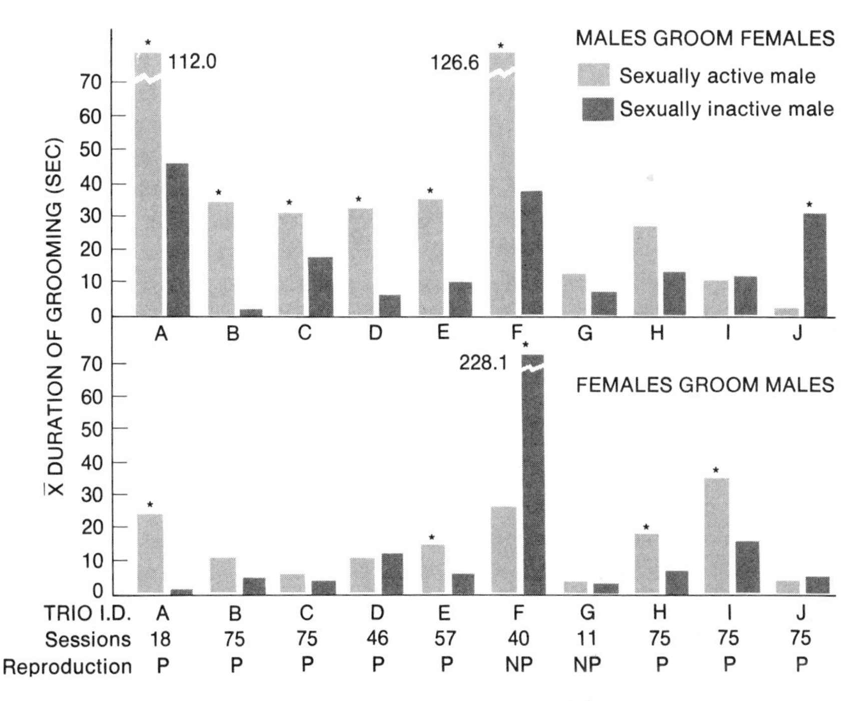

FIGURE 4. Grooming interactions in ten trios of lion tamarins. NP: no pregnancy; P: pregnancy occurred during the observations; Sessions refers to the number of half-hour observation periods. An asterisk indicates a significant difference between the two males of a trio, both in those grooming females and in those being groomed (Wilcoxon Matched Pairs Signed Ranks Test).

In some monogamous mammals intrasexual competition may be more intense among females than males (Kleiman, in press,*a*). The competition may be expressed in a conventional manner (by fighting, for example) but differential reproductive success can occur without overt agonistic behavior, thus maintaining the reproductive superiority of a single female. A review of the methods a female uses to maintain reproductive dominance may indicate how subtle some of the mechanisms are (Figure 5).

Among common marmosets *(Callithrix jacchus),* only one adult female in a group will exhibit a normal reproductive cycle. In families or artificial groups, the estrous cycles of all females except the dominant one will be suppressed (Hearn, 1977; Lunn, 1978). Among timber wolves *(Canis lupus)* the dominant female may prevent a subdominant female from copulating through overt threats and attacks, which disrupts mating attempts (Rabb et al., 1967). Should a subdominant female of a monogamous species become pregnant, the stress of living with a dominant female can cause abortion and stillbirths. I have noted this form of reproductive inhibition several times in the green acouchi *(Myoprocta pratti).*

Lastly, although subdominants may mate successfully, become pregnant and give birth, the survival of their offspring may be jeopardized by the dominant female. For both wolves and hunting dogs, there are reported cases where dominant females either killed the offspring of a second female or were sufficiently disruptive of the mother-young interaction that the infants died of neglect, starvation or harassment (Altmann, 1974; van Lawick, 1973).

The effect of the dominant female's presence may be subtle, and differential reproductive success may only be noticeable after several reproductive efforts. Dik-diks *(Madoqua kirki)* can be maintained in trios (or even groups) containing several reproductive-age females, all of whom may breed, even though they are monogamous in nature (Hendrichs and Hendrichs, 1972).

At the U.S. National Zoological Park, the death of pre-pubertal juveniles was often attributed to many causes, including severe weather. No one considered that offspring of subdominant females may be under greater stress and therefore more prone to a variety of illnesses. Because of difficulties in identifying the offspring of different females, we cannot say with certainty that mortality was greater in the young of subdominant females. Yet, this is certainly a factor which must be examined in the future.

To summarize, in monogamous species (as well as with other mating strategies) adult females housed with dominant females may be reproductively suppressed by a variety of mechanisms, including estrous cycle disruption or inhibition, mating prevention, pregnancy interference or disruption and mortality of offspring at several postnatal stages. Re-

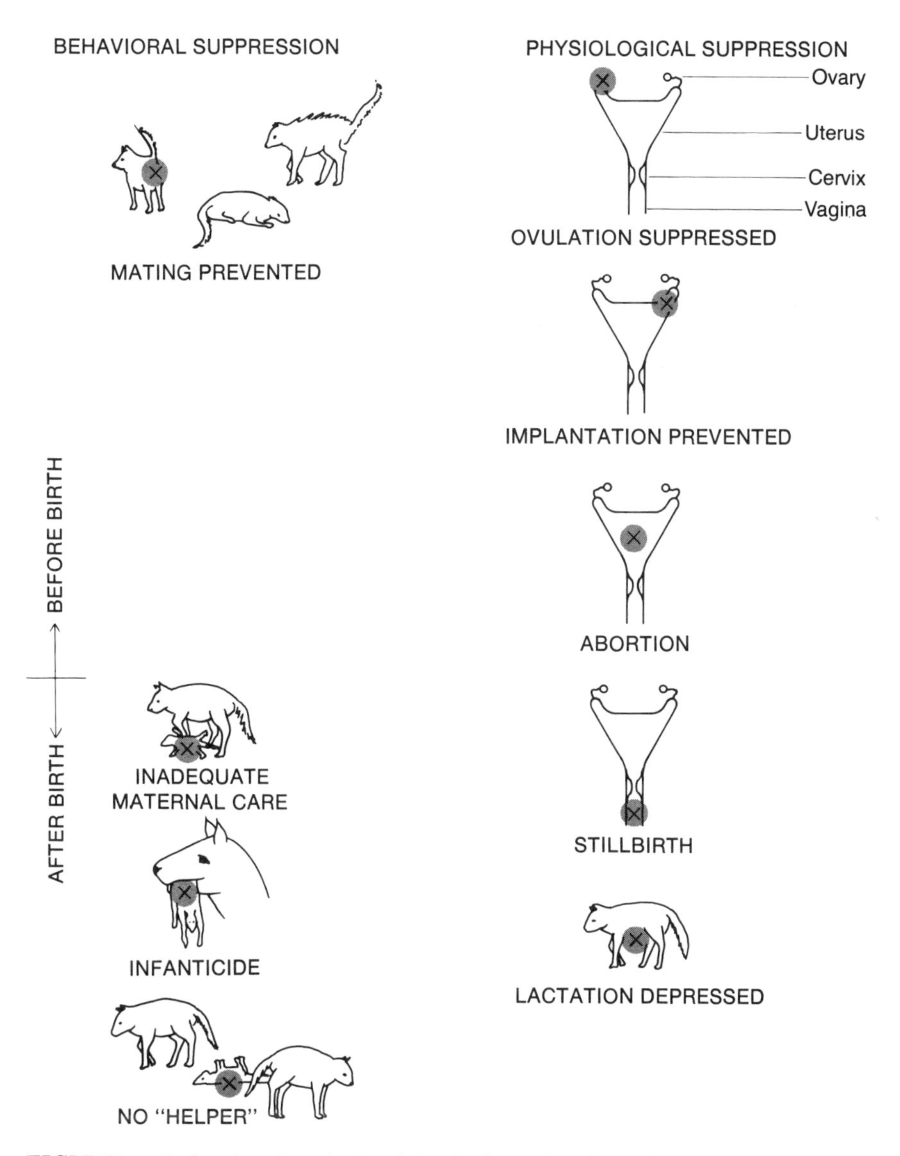

FIGURE 5. Behavioral and physiological mechanisms involved in the reproductive suppression of subordinate females in mammalian species exhibiting monogamy. Species differ in terms of which mechanisms are most common. Reproductive suppression is also seen in mammals exhibiting other mating systems.

productive suppression may occur not only as a result of direct action by the dominant female, but also through the subtle effects of her presence. These effects can cause stress in the subdominant female, negatively affecting pregnancy and lactation, or cause stress in her offspring, thereby increasing their susceptibility to disease and decreasing their likelihood of survivorship.

There appear to be species-specific tendencies in the use of different methods of reproductive inhibition among the females of both polygynous and monogamous mammals (Eisenberg, 1967). The study of these differences could be of enormous benefit in defining species limitations, and, thus, in improving the management of mammals in captivity. Such studies must carefully consider the degree to which reproductive suppression may be mediated by the olfactory sense alone. For example, there may be some species in which olfactory contact with reproductively dominant neighbors may be enough to inhibit reproduction in an otherwise adequate pair. This has been suggested for some species of marmosets and tamarins, although it has never been adequately proven. Olfactory control of reproductive processes has recently been reviewed by Doty (1976).

A consideration of mating strategies must include the degree to which inbreeding occurs in each species in nature. In many mammals, father-daughter and mother-son crosses may be prevented by: (1) dispersal of young at puberty or (2) the disappearance, death or reproductive senility of parents before the sexual maturity of offspring. In zoos, such natural phenomena are often prevented by management constraints, yet there are occasional examples. This indicates that there is a suppression of mating in closely related individuals. For example, in a wolf pack derived from a single litter—established in 1963 or 1964 at the London Zoo—there were no pregnancies until 1973, and no young survived from the first two litters (Olney, 1975; 1976). The females would come into heat annually, but no successful copulations occurred, despite the pack being organized into the typical male and female hierarchies (Kleiman, unpublished observations, 1964–1969).

In three trios of lion tamarins *(Leontopithecus rosalia)* in which the female was related (either sister or mother) to one of two males, mating occurred only with the unrelated male, even though the related males were all sexually mature (Figure 3; Kleiman, 1978). Mainardi (1963 *a; b; c*) has shown that young female mice prefer to mate with unfamiliar individuals rather than conspecifics with whom they were raised. This suggests the existence of a behavioral mechanism for suppressing inbreeding. Clearly, the strength of such inhibitions will vary across species and may even be overcome in the absence of an appropriate partner. But the effects of such inhibitions may be felt in captive propagation programs when unrelated individuals are not available for breeding. In any case, inbreeding should be avoided if at all possible (Chapters 8, 9 and 12).

Group Size and Structure

Knowledge of a species' mating strategy permits sensible decisions about the optimum sex ratio to be maintained in captivity, but does not always provide enough information on the spatial or social needs of a species. For example, although talapoin monkeys *(Miopithecus talapoin)* are polygamous, males and females tend to segregate into sex-specific subgroups; reproduction is poor when the sexes are forced into close contact with each other. Females become extremely aggressive and males may die from the frequent harassment (Rowell, 1973). Talapoins may, therefore, have to be separated by sex except during mating if enclosure sizes are inadequate.

Among monogamous mammals, as already indicated (Figure 1), group size differs from species to species. Management plans must, therefore, take into account such factors as whether a family group can remain socially stable. For example, the captive reproduction of elephant shrews is most successful when weaned offspring are removed from the parents prior to subsequent births; this would be unnecessary and even undesirable in wolves or African wild dogs. In some cases, the need for isolation of the breeding male, female or the pair is not obvious, except for the absence of successful reproduction. Both the cheetah *(Acinonyx jubatus)* and the red panda *(Ailurus fulgens)* can be maintained in groups in captivity without apparent aggression. However, reproduction in cheetahs is most successful when the female is isolated except during the mating period; red panda females reproduce best when either isolated or housed with a single male.

Although there is some information concerning which species need to be isolated—as individuals or pairs—for successful reproduction, less is known about which species need a minimum group size in order to breed. For example, many bat species normally reproduce in large colonies. In two small groups of the long-tongued bat *(Glossophaga soricina)* where only a single female was present, reproduction was poor. In the related short-tailed leaf-nosed bat *(Carollia perspicillata)* two different colonies with 10 to 20 females reproduced successfully under the same environmental conditions. In a small group of *Carollia* (with only three to four females), however, reproduction was erratic (Kleiman, unpublished).

The need for a minimum number of females may have several bases. In some bat species, colonial roosting may act as a heat conservation mechanism. This is especially true for the temperate zone verspertilionids, but may also apply to tropical bats. The duration of pregnancy is known to be affected by temperature in *Pipistrellus pipistrellus* (Racey, 1973), but lactation and the growth of young may also be affected. By living in a nursery colony, females and their young may more easily maintain an optimum temperature at a low energy cost.

Anti-predator strategies may be responsible for the females of some species reproducing in large groups. Often this is accompanied by reproductive synchrony. For example, wildebeest *(Connochaetes taurinus)* give birth synchronously in East Africa; it has been shown that this is an adaptation that floods predators with potential prey (wildebeest infants), thus reducing overall infant mortality (Estes, 1976). Although there are no data to indicate that the reproduction of captive wildebeest is reduced without a minimum number of females, there may be certain species in which the evolution of reproductive synchrony as an anti-predator strategy has proceeded to the point where a large female group is essential for adequate reproduction.

Rearing Strategies

The survivorship of offspring may be affected when the rearing strategy of a species is not carefully considered. The females of many mammals rear young cooperatively or communally. The degree to which the isolation of a lactating female influences the development of offspring has not been carefully documented. For example, the young of many species of colobine monkeys are transferred to "aunts" soon after birth (Blaffer-Hrdy, 1976; Horwich and Manski, 1975) and therefore spend limited time with the mother. It is not known whether the mother is simply tolerant of other animals' involvement with her offspring or whether this rearing strategy is so highly specialized that young born to a single isolated female in captivity will have a reduced chance of survival. Data are simply not available to indicate whether there is higher infant mortality in captivity in species like elephants, dolphins and colobine monkeys where "aunts" are an integral part of the rearing system.

A similar problem exists for species where males and juvenile "helpers" aid in rearing the young. Jantschke (1973) documented the importance of the presence of the father for the successful rearing of young by a female bush dog *(Speothos venaticus)*. Undoubtedly, in other monogamous exotic species similar effects will be seen. Dudley (1974) elegantly showed that male *Peromyscus californicus parasiticus* contribute to pup survivorship by keeping the litters warm while the female is absent. However, males may provide other aid, such as feeding weanlings, providing protection from predation or transporting young (Kleiman, 1977).

In lion tamarins, the adult male carries and shares food with the young and aids in their socialization (Hoage, 1977; 1978; Figure 2). Lion tamarins may have one to three young. An analysis of the number of days after birth when the father begins to carry the most recent offspring suggests that fathers exhibit paternal care earlier when the litters are large (Table I). Thus, the timing and degree of parental care by fathers and

TABLE I. The average number of days after birth when the father begins to carry offspring from different size litters in lion tamarins *(Leontopithecus rosalia)*.

	Litter size		
	1	2	3
Mean	14.2 days	8.6 days	3 days
Range	12–17 days	5–11 days	–
Number of litters	6	9	1

Data are from four females.

other relatives may be dependent on the needs of the mother at any given time.

In many mammals, females rear their young isolated from conspecifics. Forced contact or close confinemment have been shown to reduce infant survivorship in certain species. Martin (1968; 1975) first discovered that a female tree shrew *(Tupaia belangeri)* housed with a single male needed a separate nest box for the young. Tree shrews exhibit an "absentee" parental care system (Eisenberg, 1977); young are cannibalized if the female is forced to nest with them and the male. Similarly, reproduction in red pandas *(Ailurus fulgens)* is nearly always unsuccessful when two or more females are housed together. Although the females become pregnant, the young are inevitably ignored or destroyed. The mating strategy of red pandas is not known, but such findings suggest that red pandas may be monogamous in nature.

THE PROBLEM OF DISPERSAL

One of the least considered problems in the successful maintenance of captive populations is how and when to dispose of adults and offspring while still maintaining an optimal age and sex structure and not disrupting the social dynamics of a group. Little is known about the life history of groups in the wild (their formation, maintenance and dissolution) although such factors may be extremely important in understanding population regulation in a species (Eisenberg et al., 1972).

Field studies on a variety of mammals have recently been concentrating more closely on the long-term changes in groups of known composition, and information on group life histories is beginning to appear. Already such studies have shown that methods of dispersal differ from species to species and may not be entirely predictable from social organization or mating strategy. Moreover, Bekoff (1977) has recently suggested that individuals of the same species may be behaviorally polymorphic with respect to dispersal strategies. He cites observations of individual differences in coyote litters *(Canis latrans)* which result in both the most dominant and most subordinate littermates being least interactive with siblings. Bekoff suggests that these noninteractive individuals may be most likely to disperse from the natal group even though the reasons for their isolation from littermates are entirely different.

Some recent observations of lion tamarins indicate how difficult it is to predict which animals are likely to disperse (and which, therefore, should be removed from a family group). Adolescent lion tamarin males begin exhibiting signs of sexual maturity, such as scent marking (Kleiman and Mack, in press) and arch displaying (C. D. Rathbun, in press) while still in the family group. By contrast, females do not exhibit signs of sexual maturity until they are removed from the parents and paired with an unrelated male; thus they are socially and reproductively inhibited. However, in both sexes dominance relations are established and can be identified among siblings of the same sex. Dominance relations among sibling males occasionally include fights which result in scratches and minor wounds. The young dominant male may pursue the mother while in heat as well as scent-mark and arch display more than the father, yet still be tolerated by the parents with only minor squabbles.

By contrast, in two family groups a dominant sister was killed; in both cases the mother was strongly implicated in the death. Interestingly, the attacked females were not even sexually mature (both were about a year of age) and both females either had a twin sister or were in a family group where there were several other female offspring. Among young dominant males, there have never been any deaths. A casual observer of a family group might recommend the removal of pubertal males due to their greater sociosexual activity. However, it appears that females are at greater risk and at an earlier age. Presumably, such females would disperse from the family group before being killed. Thus, in this species strife among males seems to be less damaging in the long run and is resolved overtly through squabbles without serious damage.

THE EFFECTS OF CAPTIVITY ON BEHAVIOR

Studies of captive animals or attempts to extrapolate from the behavior of captive animals to the behavior of wild animals (or the reverse) must take into account the long- and short-term effects of captivity on behavior. Right from the initial choice of which animals to bring into captivity, human selection is applied to maintaining and breeding individuals with certain behavioral phenotypes (for example, tractability and tameness). Individuals which exhibit either extreme fear or extreme aggression towards humans are usually weeded out of a breeding pool because they are difficult to manage. Such human selection may be unconscious.

Regardless of human selection, behavioral types such as easily stressed individuals may not reproduce successfully; thus their contribution to the gene pool will be lost. This may alter the behavioral genotype of the captive population, decreasing the tendency to avoid humans or other predators.

Genotypes which would not survive in nature may be maintained in captivity. Hand-rearing rejected or weak young retains genes in the captive population which might have been eliminated in the wild. A classic case of this is the inbreeding of partially albinistic tigers *(Panthera tigris)* at several zoos. It is not known whether this rare mutant has peculiar behavioral characteristics associated with it, although highly inbred individuals do exhibit abnormalities of the visual system (Guillery and Kaas, 1973).

The behavior of captive animals may be altered permanently by having young reared in inappropriate social or environmental conditions. Hand-rearing often prevents individuals from later forming adequate social attachments. Goldfoot (1977) recently reported that the greater the degree of social deprivation in hand-reared rhesus monkeys *(Macaca mulatta)* during development, the lower the eventual reproductive success. Similar correlations between other complex adult behaviors and deprivation during development have been reported. Predators learn to immobilize and kill prey slowly with the aid of parents and other relatives. Such experience cannot be duplicated once an animal becomes an adult nor can such an "untrained" individual properly "teach" its own offspring. The success of such techniques as artificial insemination and hand-rearing have the potential of increasing the population size of a captive species, but at the expense of the normal behavioral repertoire. With increasing dependence on artificial means of reproduction, it is possible to maintain a captive species in which individuals can neither mate properly, nor rear their young.

A final factor which influences the captive behavior of species is human interference in the normal patterns of mate selection and dispersal of adults or young. Arbitrary decisions based on management considerations may significantly affect the gene pool and result in altered behavioral genotypes.

There are a variety of factors which have long- and short-term effects on the behavioral phenotype of individuals. Some are irreversible. Behavioral alterations may not only be acceptable, but desirable for populations of species which are to be retained in captivity in perpetuity.

However, zoos are breeding certain species with the expressed intent of eventually returning them to the wild. In these cases the unconscious damage to the gene pool and behavior of the captive population may prevent a successful reintroduction. Efforts at the reintroduction of species have been few and of limited success for the most part. Brambell (1977) details some of the considerations which must go into such an effort (see also Chapters 11 and 15).

SUMMARY

Inadequate consideration of mating and rearing systems and of the dispersal of adults and young from breeding groups can negatively affect reproduction in captivity (Eisenberg, 1967). Unfortunately, it is extremely difficult to pinpoint the reasons for a lack of reproduction in a species, especially if the characteristics of the species' life history are poorly understood. Zoos rarely publish an analysis of a propagation program that failed, as a pathologist would publish the results of an autopsy. Zoos do not like to discuss their failures. However, lack of publication is often due to small sample sizes or to the anecdotal nature of the observations. One can never be sure whether a finding from a single individual or group in captivity is representative of a species or idiosyncratic. For example, there may be numerous cases of infanticide in zoo primates which are directly comparable to similar events in the wild. Yet, it would be extremely difficult to locate details in zoo records which might be relevant to our understanding of this phenomenon—such as the age and sex structure of the group and social changes which occurred prior to the infanticide. Moreover, to determine in which primate species infanticide is most likely to occur would require comparative data detailing the prevailing social conditions when infanticide did not occur. Such an analysis could be done, but only with great effort.

Interestingly, a captive propagation program which is a failure may contribute more to sociobiological theory or knowledge than one which is a success. For example, behavioral mechanisms which decrease the likelihood of inbreeding could be analyzed more easily in a zoological park setting. A study of the life history characteristics of species in which mating between close relatives is inhibited could provide comparative information which would be valuable for our understanding of how inbreeding is prevented.

Current knowledge of the mating and rearing strategies of the monogamous marmosets and tamarins is derived from numerous failures to successfully propagate these primates. Zoos and other institutions had attempted to maintain them in artificial groups of unrelated individuals, only to find that no more than a single female would reproduce. Moreover, young which were removed from the parents after weaning, but before a subsequent birth, exhibited improper parental care once they themselves reproduced; the young lacked as juveniles the infant care experience which is essential for adequate parental care in adulthood (Hoage, 1977; 1978).

Captive studies can contribute to sociobiology in other ways. Longitudinal studies of species in zoos can provide important information on the social dynamics of groups over several generations. Such an approach has been followed in captive wolf studies (Rabb et al., 1967), and recently with Père David's deer (Wemmer and Collins, in preparation).

Unfortunately, such studies are rarely conducted in zoos because a commitment of a decade or more is required, especially with longer-lived species. However, some of the most important problems in sociobiology relate to the interactions of kin and the degree to which the genetic relationships of individuals affect social behavior. And such problems can effectively be examined in zoo populations.

SUGGESTED READINGS

Crandall, L. S., 1964, *The Management of Wild Mammals in Captivity,* University of Chicago Press, Chicago.

Eisenberg, J. F., 1966, *The Social Organization of Mammals, in Handbuch der Zoologie,* VIII (10/7), Lieferung 39, De Gruyter, Berlin.

Hediger, H., 1950, *Wild Animals in Captivity,* Butterworth's and Co., Ltd., London.

Hediger, H., 1955, *Psychology and Behaviour of Captive Animals in Zoos and Circuses,* Butterworth's and Co., Ltd., London.

Hediger, H., 1969, *Man and Animal in the Zoo,* Seymour Lawrence/Delacorte Press, New York.

Research in Zoos and Aquariums, 1975, Institute of Laboratory Animal Resources, National Academy of Sciences, Washington, D.C.

International Zoo Yearbook, Volumes 1–17 (1960–1977), Zoological Society of London.

Wilson, E. O., 1975, *Sociobiology: The New Synthesis,* Harvard University Press, Cambridge.

Der Zoologische Garten, Leipzig, East Germany.

CHAPTER 15

IS REINTRODUCTION A REALISTIC GOAL?

Sheldon Campbell

One accepted goal of captive breeding by zoos and other agencies is the eventual introduction of captive-bred animals into wild habitats. Captive-bred animals might be released to restore the species to a former range, to bolster the breeding nucleus of a dwindling population, to introduce a biological control or for some other reason.

SOME EXAMPLES

To date, there have been few planned attempts to introduce captive-bred animals into wild habitats, but some of these have met with success. The wisent (European bison) (Figure 1) was bred in various European zoos and later released into nature reserves in Poland, where the animals have survived through several generations. The Bavarian National Park has been stocked with wolves and other animals bred in German zoos. Ibex and chamois were reintroduced into former ranges from stock bred at the Munich Zoo. The Atlanta Zoo has reintroduced crocodiles into Mexico. The San Diego Wild Animal Park sent four Arabian oryx (Figure 2) for introduction into a reserve in Jordan. And for several years now, various zoos in Germany and Switzerland have cooperated in a program to breed, train and reintroduce the European eagle owl into wild habitats.

A small number of successes, however, should not blind anyone to the numerous problems entailed in the captive-breeding of animals for introduction into wild habitats. Not the least of these problems are economic ones. Conway (Chapter 11) estimates that no more than 100 to 150 mammal species could be captive bred by U.S. zoos simply because of space limitations and the costs of holding and caring for the large groups that

FIGURE 1. The wisent *(Bison bonasus)*. By the end of the Second World War only 15 individuals reportedly remained in the primeval forests of Bialowieza, Poland. These individuals formed the basis of a breeding program and subsequent successful reestablishment of a wild herd. (San Diego Zoological Society Photo)

FIGURE 2. The Arabian oryx *(Oryx leucoryx)*. This spectacular antelope, the victim of game massacres, is now virtually extinct in the wild. Recently, the San Diego Wild Animal Park sent four to Jordan for introduction into a reserve. (San Diego Zoological Society Photo)

would be needed to assure sustained propagation and to minimize the deleterious effects of inbreeding.

The effort by Cade and his associates at Cornell University to propagate and introduce the peregrine falcon (Figure 3) at selected sites in the United States serves as an example of some of the difficulties in captive breeding, including the costs. The Cornell program, begun in 1970, did

FIGURE 3. The peregrine falcon *(Falco peregrinus)*. This cosmopolitan species once occurred throughout North America, but is now scarce in the western United States and absent from the eastern states. Presently it is the object of an intensive captive propagation and reintroduction program. (San Diego Zoological Society Photo)

not achieve success in breeding until 1973, when 20 offspring were hatched. The number of hatchlings has since risen to 300. While propagation has been successful, however, the program of introducing young birds into wild habitats has so far been inconclusive. To date 152 hatchlings, released before they have learned to fly, have been placed in selected nests. Of these only one pair has become established, and the female is still too young to breed. Another female has a territory, but hasn't captured a mate. Six to eight more falcons have been accounted for and there may be 30 or more that survived the winter of 1977–78. Because the program is barely underway, no true evaluation of its success will be possible for another two or three years. Meanwhile, the costs mount. So far the project has absorbed approximately one milllion dollars; it will take another four or five million to continue it for another 10 to 15 years. It was observed, however, that while these costs seem high, they are small compared with the amounts raised to save or restore historical structures and works of art—to say nothing of the amounts spent for military equipment.

Captive breeding programs for insects, fish, amphibians and reptiles should not be overlooked. At present there appears to be a tendency to concentrate on more popular animals—mostly birds and mammals. Captive bred insects, for example, could well prove decisive as biological controls or pollen disseminators in agricultural areas.

PROBLEMS ASSOCIATED WITH REINTRODUCTION

Institutions or individuals that undertake captive breeding programs will, sooner or later, encounter some or all of the following:

1. Introduced animals sometimes leave the release area and migrate to places they are not supposed to be. In July of 1978 the Canadian Wildlife Service released 28 wood bison into Jasper National Park. At the end of six weeks they had traveled 200 miles north of the release area—well beyond the boundaries of the park.

2. Animals to be released may carry diseases that can be transmitted to wild populations of their own or other species. Every care, therefore, should be taken to release only disease-free animals.

3. Follow-up studies will be necessary after release to ascertain whether or not captive-bred animals are adapting to the wild (and if not, why not).

4. Many programs will have to embody the holistic approach pioneered by the U.S. Fish and Wildlife Service in breeding and releasing masked bob whites. Field biologists studied behavior, preferred foods and other matters related to survival of the species in the wild. These studies were incorporated with the work of laboratory specialists in nutrition, physiology, captive behavior, veterinary science and other pertinent dis-

ciplines to develop a plan designed to assure the survival of most of the released birds.

The masked bob white program began in 1968/69 with 57 birds. Now the Fish and Wildlife Service maintains a breeding group of approximately 300 to 350 birds that produce around 2,000 offspring a year. This number cannot be greatly increased because of the difficulties in training the birds for life in the wild. Before release the young masked bob whites are subjected to simulated attacks by coyotes, raptors and men with guns "to scare the dickens out of them" and teach them when to seek cover. Texas bob white, vasectomized to preserve normal sexual habits but prevent interbreeding, are released with the young masked bob whites as "guides" to food, shelter and cover in the event of attack by predators.

5. Laws and bureaucratic procedures often complicate efforts to release captive bred animals. Because this method of conserving threatened and endangered species is new, some nations, prepared administratively to export native species, have no administrative machinery for importing them.

6. Most efforts to release captive-bred animals require more than permits and other legal documents from goverments controlling the release areas. In many instances, public relations are a decisive factor. Those responsible for the release program must consider the social and political attitudes of the people living in or near the area where a release is contemplated. In some instances, national or regional pride may impede or stop release programs. For example, the effort to reintroduce the rosy marsh moth into England was opposed by Welsh nationalists who wanted to retain the moth exclusively in Wales.

7. In some instances, the introduction of captive-bred species may prove difficult if the released animals are threatened with genetic swamping by a related species (the red wolf has been pushed toward extinction as a distinctive form by interbreeding with coyotes).

8. People responsible for captive breeding programs must try to avoid the artificial selection of breeding stock for what are judged, consciously or unconsciously, to be "desirable" characteristics. To the extent possible, conditions akin to those in the wild should be maintained (Chapter 14).

9. Animals in captive breeding groups should be outbred whenever possible (Chapter 12). Experience indicates that inbred insects decline rapidly when they face competition with wild populations of the same species.

10. Animals that result from several generations of captive breeding are likely to be less successful after introduction into wild habitats than wild-caught animals that are translocated, or first generation captive-

bred animals. After several generations the progeny of a species may be altered in form and behavior through: (a) unconscious selection by breeders for characteristics such as tractability, lethargy and early reproduction; (b) failure by breeders to provide conditions allowing the expression of the full behavioral repertoire used by wild members of the same species; (c) hand rearing of sickly infants or juveniles when similar animals born in the wild would probably die without reproducing; (d) being influenced by constant associations with human beings, a situation which, at the very least, might change defensive or flight behavior.

Nevertheless, if enough captive-bred animals are released at one time —even though they are the product of many captive-bred generations— some will probably adapt and propagate so that natural selection may once again become the decisive factor.

The experience at Cornell with peregrine falcons may illustrate many of the difficulties in captive breeding programs. Presently only about half the males hatched in captivity will mate of their own initiative—a fact that could be attributable to unconscious selection by the breeders. Because of this problem, the Cornell program relies on artificial insemination. However, when manipulation is used to induce ejaculation, only small amounts of semen are produced, moreover, the males, once traumatized by being ejaculated, may cease producing semen during that breeding season. To increase the semen production and avoid trauma, personnel in the Cornell program deliberately seek to imprint males on selected handlers. When imprinting is successful, the males try to copulate, voluntarily discharging copious amounts of semen into a fist, shoulder, or other collector (for example, a hat with a specially designed brim).

11. Animals are frequently more adaptable to altered circumstances and environment than we suppose. Members of some species (the whooping crane, for example) have adapted to environmental conditions vastly different from those thought necessary for their survival. Thus, the apparent vulnerability of a species to change in habitat should not keep captive breeding and reintroduction from being attempted.

SUMMARY

If the geneticists are correct—and our limited experience bears them out—captive propagation programs cannot be carried on indefinitely. After a finite number of generations (depending on population size), the stock will begin to show the effects of domestication, random change and inbreeding depression (Chapters 8, 9, 12 and 14). It seems only reasonable, then, to begin planning for reintroduction at the very beginning, even if the reintroduction window won't be open for a century or more.

Some skeptics will point out, with reason, that it is unrealistic to hope for reintroduction opportunities in many nations, especially those now experiencing rapid population growth. No one could argue this point, at least for the next few decades. Nevertheless, conditions change; unexpected opportunities for establishing semi-natural habitats could arise within the lifetime of many captive propagation programs. It would seem prudent, therefore, to propagate as many endangered forms as is economically and technically feasible.

SUGGESTED READINGS

Brambell, M.R., 1977, Reintroduction, *Inter. Zoo Yearb.,* 17, 112–116. A general discussion that cites examples and suggests a protocol for reintroducing species.

Duffey, E., 1977, The re-establishment of the large copper butterfly *Lycaena dispar batava* Obth. on Woodwalton Fen National Nature Reserve, Cambridgeshire, England, 1969–1973, *Biol. Conserv.,* 12, 143–158. Describes the best known and most comprehensive attempt at invertebrate reintroduction.

Kear, J., 1975, Returning the Hawaiian goose to the wild, in *Breeding Endangered Species in Captivity,* R.D. Martin (ed.), Academic Press, London, pp. 115–123. Documents the spectacular recovery of a nearly extinct species.

PART FOUR

EXPLOITATION AND PRESERVATION

CHAPTER 16

AFRICAN WILDLIFE RESOURCES

Malcolm Coe

The African continent covers an area of 28,672,000 km^2 which is occupied by the most diverse mammal fauna in the world. This richness of species can be accounted for in large part by the immense variety of Africa's climates, soil and vegetation. The African continent as a whole is relatively arid (Figure 1). The major African vegetation types are forests (in which we will include rain forest, woodland, thorn forests and temperate brush), occupying 18 percent of the land surface, the great lowland to alpine grasslands, covering 42 percent, and the semi arid scrub and true desert, which make up the remaining 40 percent. True rain forest is limited to the Congo basin and the Cameroons with an area of deciduous rain forest north of the Dahomey gap from Sierra Leone through to the Ivory Coast. These wet areas, whose annual rainfall may be as high as 500 cm, exhibit a very diverse flora. Few of these forests are today truly primary; except for some patches on mountains and other inaccessible spots, the vegetation has been disturbed by either shifting cultivation or forestry operations.

Passing north, south or eastwards from the great forest block we encounter progressively drier environments. To the north and east is the Guinea savanna, an area with a largely fire resistant vegetation and a rainfall in excess of 100 cm per year. Still further north where the annual rainfall has fallen to 50 cm, we encounter the so-called Sudan savanna which forms a great strip from the west to the east coast and is characterized by the *Acacia,* Dom palm *(Hyphaene thebaica)* and the great thick-stemmed Baobab *(Adansonia digitata).* Still further north and eastward to the shores of the Red Sea is the Sahel savanna that comprises a barrier between the true desert and the more extensive Guinea-Sudan savanna.

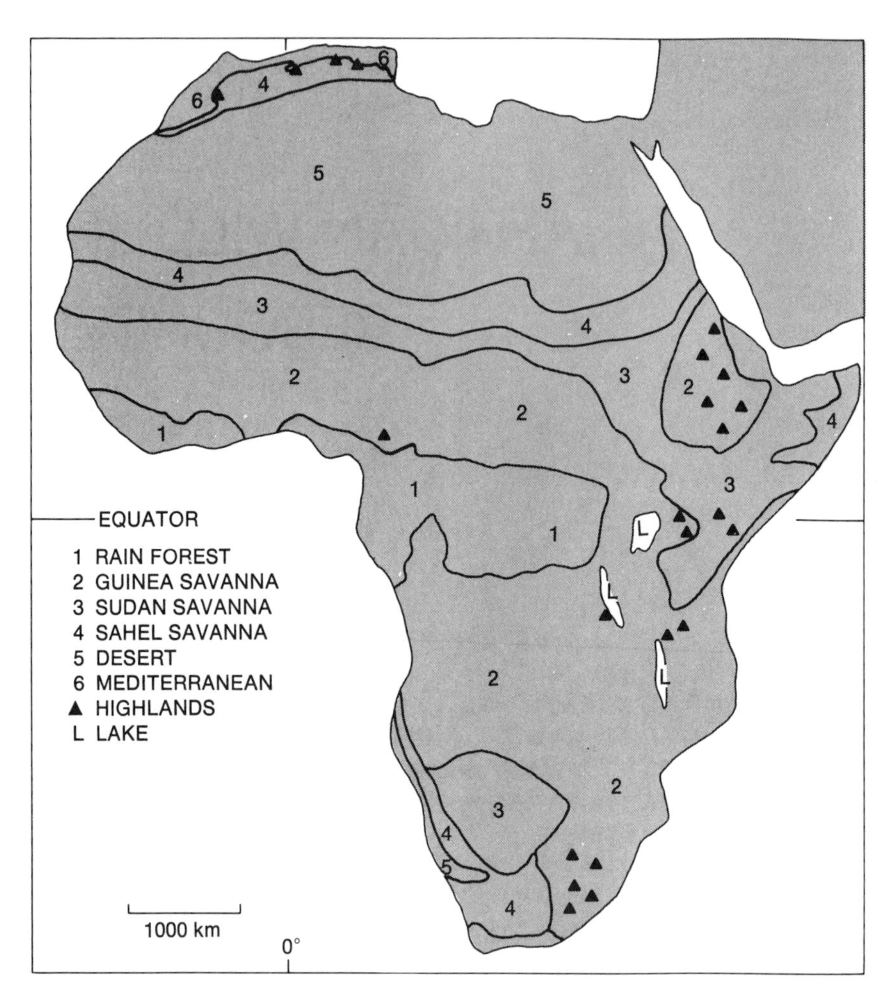

FIGURE 1. The main vegetation zones of the African continent. (Modified from l'Association pour l'Etude Taxonomique de la Flore d'Afrique Tropicale)

The Sahel receives only 25 to 50 cm of rain a year and its vegetation cover consists of scattered (often quite thick) *Acacia* and *Commiphora* spp. and *Salvadora persica* on alkaline ground.

Although this decrease in rainfall appears to be, to a large degree, a latitudinal phenomenon, a similar decrease in rainfall occurs when we travel eastwards—although small outliers of rain forest still occur in the comparatively dry western areas of Kenya. To the south there is also a decrease in rainfall, but here one encounters the very extensive Miombo woodlands *(Brachystegia-Julbernardia),* which are a southern climatic counterpart of the more northerly Guinea savanna. All these features may be related to the amount of rain falling in a particular area, but we must remember that the effect of the rainfall may depend on whether the precipitation has a single peak (as in West Africa) or has a bimodal pattern (as in the east).

The African large mammal fauna has its peak of species diversity in the drier regions of the continent. These areas sustain high standing crop biomasses of moderately long lived herbivores and their predators. These species assemblages of herbivore grade from browsers through mixed feeders to grazers, depending on whether they are feeding in the low elevation savannas (dominated by scrub) or the upland savannas of East Africa (dominated by grass). One notices especially that the desert environments are dominated by water-independent species such as the oryx and the gazelle, while the semiarid environments, such as Tsavo in eastern Kenya, are dominated by mixed feeders such as the elephant, browsing species such as the black rhino *(Diceros bicornis)*, the gerenuk *(Litocranius walleri)*, the lesser kudu *(Tragelaphus imberbis)*, and smaller numbers of grazing species that occupy grass patches in the dense scrub. The open grasslands of the Serengeti are the domain of the grazing species, including the white bearded wildebeeste *(Connochaetes taurinus albojubatus)*, Thomson's gazelle *(Gazella thomsoni)* and the topi *(Damaliscus korrigum)*.

The vegetation and wildlife of Africa are so intertwined with man, and all of these with climate, that any consideration of biological factors alone is precluded. This chapter will be concerned with developing a profile of Africa's past in relation to the man-nature interface upon which a rational perspective for African wildlife in the future will be based.

PAST CLIMATES AND EARLY HUNTERS

The African Sahel is increasing in size and for many years ecologists have attempted to determine whether this trend toward desertification is due to the climate, to human agencies or a combination of both (Aubreville, 1949; Cloudsley-Thompson, 1977; Curry-Lindahl, 1972*a;* Ormerod, 1976). There seems to be good evidence that the Sahara was less arid in the Pleistocene. It has been suggested (Cloudsley-Thompson, 1977) that changes in the area of the antarctic ice cap during glacial and interglacial periods would have resulted in considerable climatic variation in the Sahel zone. In terms of recent changes and the influences of man, the climate of the late Pleistocene periods is more relevant.

We know very little about the causes of geographical aridity, its distribution, or its changes. All we can say is that as long ago as the Cretaceous period (over 130 million years ago) the northern areas of Africa were arid, indicating that those conditions must be related to some fundamental property of atmospheric circulation. The climatic changes of Europe during the last 17,000 years are well documented and the pattern of glacials and interglacials has now been shown to have occurred contemporaneously in East Africa (Coetzee, 1964; Livingstone, 1952) and South America (Hammen and Gonzales, 1960). Cloudsley-Thompson (1977) compared climatic data for Europe with that of East Africa, Sudan and the Sahara, and showed that the cultural events in the Sudan coincide with the conjectured climate events of the period.

Relatively minor changes in climate could have had immense effects on the distribution of plants and animals. It has been suggested that the movement of the caloric equator could account for the immensely rich Cape flora (Aubreville, 1949) which is now compressed into a small area at the southern tip of Africa. Such movements associated with glacial conditions could well have produced a cooling of up to 5°C with associated changes in rainfall. Small changes in rainfall in the arid zones could also have a profound influence on mammal distribution. Even if the temperature changes were somewhat less than those proposed by palynologists (Coetzee, 1967), we must remember that an increase of a few centimeters in rainfall in the desert will have a profound effect on the vegetation, while an order of magnitude increase in rain forest will have very little effect, if any. Cloudsley-Thompson (1977) suggests that there may have been up to a 40 percent variation in rainfall during the drier and wetter periods, the latter probably coinciding with high lake levels in the Sahara. As recently as 7,000 years ago large mammals, such as the elephant (probably *Loxodonta africana cyclotis*—the forest elephant) and hippo *(Hippopotamus amphibius)*, were present in the Sahara up to 500 km further north than today (Moreau, 1963). After this wet period (7,000 to 3,500 years ago) the northern Sudan and southern Egypt became much drier. It is the end of this period of aridity that coincides with the appearance of pastoralists with their cattle who lived in areas east and west of the Nile that are now complete desert (Jackson, 1957).

The sequence of events in the central Sahara during the last 7,000 years seems to follow an interesting cultural-climatic pattern. Hunting cultures existed in the region at about 7,000 years ago, but were replaced by invasions of pastoral people with large herds of cattle between 6,000 and 4,000 years ago. By about 3,000 years ago the Sahara was opened up by various trade routes and military expeditions and by the introduction of horses. Finally, about 2,000 years ago camels appear for the first time, coinciding with the advent of desert conditions that must have been very similar to today (Cloudsley-Thompson, 1977). We can make an educated guess about the origin of the Sahara. On the edge of a desert it takes very little to over-utilize an area and start the slide downwards to irreversible desert. Initially, the human population increases and the stock animals consume the local plants, removing young replacement seedlings and converting perennial grasses to annuals. Cattle and domestic animals tend to be kept in herds as an anti-predator and anti-raider device, resulting in the formation of tracks and localized soil compaction. This reduces water penetration. If we add to these the effects of fire and the need for wood we have a compound effect that in a very short space of time will turn a well-covered area of scrub into an almost barren desert.

Man and animals have interacted in Africa for millions of years. Martin (1966) pointed out that since pollen evidence and radiocarbon dating indicate that large scale climatic changes associated with glacial and interglacial phases appear to have been worldwide, it is surprising that the extinction of the Pleistocene megafauna does not appear to coincide with these more or less synchronous changes. At present the African large mammal fauna only contains 60 percent of the genera that were encountered at the time of the hand axe faunas. In the first one and one half million years of the Pleistocene 19 large mammal genera were lost; during the last 100,000 years or so at least 26 genera have disappeared—a rate of at least 20 times that of the preceding period. Although accurate dating of deposits of this period are difficult, it appears that the remains of extinct mammals have nowhere been found associated with artifacts dated at less than 40,000 years. Thus, there appears to be a strong temporal association between the middle Pleistocene fauna and the Chellean-Acheulean cultures which are associated with a wide variety of tools and the remains of contemporary mammals apparently butchered with them.

In relation to the potential influence of climatic changes on faunal extinction, the case of Madagascar (Martin, 1966) is of special interest. Since the time of the Melanesian invasion of this island about 1,000 years ago 16 genera of vertebrates, all of which were common in fossil deposits, have become extinct. During this period there is no evidence of major local climatic change nor is there evidence of any previous occupation of Madagascar by hominids. If we accept hunting as a prime cause of such extinction, it is of even greater interest to note that 14 species of lemurs, megaladapids and indrids disappeared. These were large, terrestrial and probably diurnal species. The survivors are smaller and largely nocturnal mammals.

Perhaps the best evidence for an association between tool manufacturing man and the extinction of the Pleistocene megafauna is that similar events are much later in North America than in Africa. Thus, in America 35 genera of large mammals disappeared, most of them within the last 12,000 years, while in the preceeding 2 million years only 73 genera became extinct. Recent work in Australia suggests that a large-scale extinction of large herbivorous marsupials occurred in the interior of that continent about 25,000 years ago. This event coincided with a period of intense aridity and the presence of hunting man (Williams, personal communication).

MODERN MAN'S IMPACT ON AFRICAN HABITATS

The Use of Bush Meat

Although *Homo erectus* and *H. sapiens* may be implicated in the extinction of large elements of the Pleistocene megafauna, it was not until man entered the fields of animal husbandry and agriculture that his more drastic alteration and manipulation of the environment must have occurred. At the primary hunter-gatherer phase man still seems to have been very much an opportunist omnivore taking a wide variety of animal and plant foods as they varied with the seasons. Similarly, in rain forest there has probably been a slow transition from the hunter-gathering phase to a culture associated with opportunistic agriculture, out of which man probably developed more regular shifting or slash and burn forms of husbandry.

Today in West Africa a patch of forest is selected and the lianes are cut and allowed to dry before the area is fired. Later, the remains are felled and burned further and the ash scattered before the area is crudely hoed and planted with forest rice, cassava, sweet potatoes, yams and a variety of other crops. The reduction in yield after one or two crops are harvested has led to the development of systems of movement within the forest whereby each area farmed is not revisited for up to 20 years (possibly a lot longer in the past). Today, however, in many forest areas these periods have been reduced to as little as four or five years.

It must be remembered that groups using this type of agriculture are still heavily dependent on hunting and still rely on supplies of bush meat for their protein. In the interior of Liberia, close to Mount Nimba on the Guinea-Ivory Coast border, the Mano and Gio tribes supplement their diet with a large variety of "small beef" which ranges in size from termites and locusts through frogs and lizards to bats and small antelopes. Asibey (1974) lists the animals in Table I as being utilized as bush meat by people in Ghana. To this list should be added the larger amphibians (for example, *Dicroglossus occipitalis).* Note the wide spectrum of sizes of the animals and the very different vegetational strata they occupy, from the highest canopy to almost entirely terrestrial species (Coe, 1976).

The native peoples of Africa have probably utilized the wildlife of their tribal territories for as long as man has been a hunter. Animals have been killed for their meat, skins (utilized for shelter, clothing, containers, household items, weapons and shields), ligaments (stretched and dried as bow strings) and other parts of the body which are used for medical and magical purposes as well as for personal adornment and decoration. Such

TABLE I. Common terrestrial animals eaten in Ghana.

Animal	Remarks
Rodents and lagomorphs	Hares *(Lepus* spp.), large rodents *(Cricetomys* and *Thryonomys)* and porcupines *(Hystrix* and *Atherurus)* commonly utilized. Smaller rodents often caught and eaten by small children. All squirrels *(Protoxerus, Heliosciurus, Funisciurus,* etc.) including scaly tailed flying squirrels *(Anomalurus* and *Idiurus)* eaten.
Bats	All larger genera of fruit bats eaten (including *Eidolon, Epomophorus, Hypsignathus).* Often smoked for long distance transport.
Large insectivorous mammals	All species of pangolin *(Manis* and *Phataginus)* and the aardvark *(Orycteropus)* are high priced sources of bush meat.
Primates	All apes and monkeys are regularly eaten, either fresh or smoked.
Birds	Most of the larger birds or abundant smaller communal species are eaten, as are almost all their eggs.
Reptiles	All varanids, chelonians and the larger snakes *(Bitis* and *Python)* are eaten. Children often hunt and eat small lizards (agamids and skinks).
Invertebrates	Termites and beetle larvae *(Phyncophorus phoenicalis)* are eaten. The giant African land snails *(Archachatina* and *Achatina)* form an important item of diet in some areas.

Note: This list, with minor modifications, would apply to most of West African forest or forest fringe environments.
(Modified from Asibey, 1974)

uses are, of course, not in any way inconsistent with conservation. When human population densities and recruitment were low and death rates high, all these natural resources were utilized on a sustained yield basis.

Curry-Lindahl (1969; 1972) estimated that between 60 and 70 percent of the annual protein consumed by the people of Liberia is derived from bush meat; in Ghana this figure is also about 60 percent, in the Congo (Brazzaville) 50 percent and in Botswana 60 percent (Richter, 1969). Charter (1971) in a study in southern Nigeria reported that fish made up the largest part of the local protein intake (60 percent) followed by domestic animals (21 percent) and bush meat (19 percent). Charter considered the value of the bush meat eaten at the time of his study (1965–1966) to be at least $15 million, though the price varies a great deal from place to place and from time to time. These figures give some indication of the great importance of wildlife as a source of protein for the generally meat hungry people of Africa.

Asibey (1974) estimated the importance of bush meat at a number of centers in Ghana and concluded that in Accra, the capital, 8,747 kg per month are offered for sale, while smaller quantities (473 kg/month at Damango and 1,310 kg/month in Techiman) were offered for sale in smaller centers. Of the material on sale in Accra 89 percent was made up of meat derived from the cane rat (grass cutter or cutting grass, *Thryonomys swynderianus),* a rodent that weighs up to 6 kg. In areas where patches of forest still support a variety of duiker *(Cephalophus* spp.) and other medium-sized antelopes such as the bushbuck *(Tragelaphus scriptus),* these animals also comprise an important proportion of the bush meat offered for sale.

The survival of larger species is clearly dependent on the local demand, which is a function of human density or the proximity of urban areas. At Mount Nimba, prior to the discovery of iron ore in 1956 (Coe and Curry-Lindahl, 1965), the human population was relatively low and forest buffalo *(Syncerus nanus),* bongo *(Boocerus euryceros),* bushbuck, duiker and giant forest hog *(Hylochoerus meinertzhageni)* were still relatively common. Since this time, however, a large mining town has developed and all the larger herbivores have disappeared from an area of more than 5000 square kilometers.

The variation between different countries and the consumption of bush meat has as much to do with local laws and regulations as it has to do with dietary preferences. In many territories in West Africa virtually every man over puberty owns a shot gun or similar weapon and in Liberia cartridges are sold on the sidewalk like matches. In East Africa, however, the control of firearms is much stricter, and even if the game legislation cannot be strictly enforced in country areas, a spear or a bow and arrow just are not as effective as a gun against the larger game species.

CREATION OF SAVANNA AND DESERTS

Man affects more in Africa than animal distribution and abundance; he is disrupting and destroying entire habitats. Cloudsley-Thompson (1977) has suggested that the shifting practices of forest fringe peoples—involving both clearing and fire—are almost certainly an important factor which erodes the forest edge and encourages the establishment of the more open Guinea type savanna. The Guinea border at Mt. Nimba in Liberia shows this sharp transition from forest to savanna which is certainly manmade and which is at the present time maintained by the intrusion of domestic animals and fire (Coe, 1975). Such vegetation changes clearly favor the introduction of domestic animals and the use of fire as a tool for managing and maintaining grassland (Strebbing et al., 1954). Initially, however, in areas of high rainfall at either low or high altitude, it is likely that agriculture will be pursued as a regular form of local husbandry. Morgan (1963) has suggested that many of the highland areas of Kenya and Tanzania (now extensively farmed) were not utilized in the past due to the lack of suitable high altitude crops.

It seems likely, therefore, that in the last 500 years or so man has been responsible for the gradual erosion of the forest boundaries and its conversion to wooded savanna, which is subsequently maintained largely by grass burning. The reduction of the high tree standing-crop biomass and its replacement by vegetation of lower stature and shorter generation time also favors the establishment of higher herbivore standing-crop biomasses.

It would be foolhardy to suggest that man has been solely responsible for the formation of the savanna, but it would not seem unreasonable to suggest that he has been the primary agency in the erosion of the forest and dense woodland margins and, therefore, has been a primary force in the extension of the Sahel zone. Man has, so to speak, eroded and extended the boundaries of vegetation zones created by local soil and climatic conditions. Additionally, it would be wrong to suggest that fire induced grasslands are a solely human effect, for natural fires must have existed in the past as suggested by the presence of "pyrophytic" (fire-resistant) trees and shrubs (Cloudsley-Thompson, 1977).

Pastoralism in Arid Habitats

On the African savanna livestock are more numerous than men (Figure 2). Human communities that rely predominantly on their livestock for a living are termed pastoralists. These include both settled communities on the slightly higher rainfall areas and nomads in more arid condi-

FIGURE 2. A herd of Masai cattle, sheep and goats in Southern Kenya.

tions. Sanford (1976) has concluded that some 40 percent of the world's 30 to 40 million pastoralists live in Africa.

It seems likely that man first domesticated cattle around 12,000 years ago, an event which may initially have been more closely connected with religious beliefs than with the need for food or draft animals (Boughey, 1971). Surprisingly, in spite of the wide variety of wild mammals found in Africa, none of them seem to have been regularly domesticated. Indeed the only domestic animal maintained on the African continent before cattle arrived via Nubia and the Horn of Africa about 3,000 years ago, was the Guinea fowl *(Numida meleagris)*.

The practice of animal husbandry in Africa ranges from the maintenance of small numbers of animals in association with settled agriculture to the almost total dependence on these creatures for their blood and milk (which provides a regular supply of renewable protein even though the amount produced varies greatly with the seasons in arid zones). The supplementation of the diet with meat from domestic animals occurs across this spectrum of husbandry practice, although at the arid end of the scale—for ecological and socioeconomic reasons—animals are seldom deliberately killed.

Sandford (1976) considered the fields of argument which may account for the tendency of the pastoralist to overexploit their rangelands. They were: (1) Structural arguments related to social and economic structures and relations; (2) Natural events related to uncontrollable variations in climate and disease; (3) Human fallibility arguments related to either the bad long-term practices of pastoralists or the policies of their governments; and (4) Population arguments related to human populations and concomitant increases in the numbers of their domestic animals. Although he concluded that none of these elements provided satisfactory evidence by themselves—and in most cases need rigorous testing—there is little doubt that pastoral lands are or will become barren landscapes.

Perhaps one of the most valuable contributions to these arguments came when L. H. Brown (1971) published his important paper "The biology of pastoral man as a factor in conservation." I selected this paper because it attempts to produce a balanced view rather than falling into the dangerous and artificial dichotomy produced by livestock and agricultural devotees on one hand, and the often emotional wildlife conservation lobby on the other. Brown and others have pointed out that while the pastoralist who lives almost entirely on the products of his livestock has a great excess of material to sustain him at the height of the rainy season, the number of animals each family unit requires depends far more on their needs at the height of the worst (the dry) season rather than the best. Brown calculated that the basic minimum number of animals needed to support an average family ranges from 2.5 to 4.5 Standard Stock Units (= 500 kg live weight per person) according to the prevailing ecological conditions. When the rising human population becomes too great, damage to the environment is inevitable.

The need to build shelters for himself and his beasts against the climate, and especially, predators, can also have a profound effect on the vegetation. When a small Turkana family of up to 10 people in northern Kenya moves, they need to cut at least 40 trees with which to construct their huts and for bomas (fences) for their stock (Figure 3). If we assume that they will move an average of once a month (sometimes much more often) they are likely to cut down 500 trees a year, so that in an area where vegetation replacement is slow the effect will be rapid. What we must note, though, is that the damage caused is not due to inefficient thoughtless overstocking, but simply to a given number of people requiring a given number of animals.

Is it possible to estimate an "optimum carrying capacity" for wildlife and pastoralists together? Coe et al. (1976) demonstrated that the standing-crop biomass, energy expenditure and production of large mammalian herbivores in the African savannas show a high degree of correlation with mean annual rainfall and the predicted above ground primary production. This relationship is especially clear for areas receiving less than 700

FIGURE 3. A Turkana manyatta in open lava desert in northwestern Kenya. The sparse tree cover results from the rapid rate of cutting to build "bomas" for domestic animals, and huts for themselves. Goats are seen in the foreground, camels in the background.

mm of rainfall per year. Of special interest in this study was the inclusion of data from six predominantly pastoral areas in the arid northern regions of Kenya (Watson, 1972). All six points were found to fall above and nearly parallel to the regression line calculated for wildlife, suggesting that all these pastoral people were exceeding the "optimum carrying capacity" of the environment as defined by the large herbivore wildlife data (Figure 4). It should be pointed out that the authors (Coe et al., 1976) were fully aware that local factors such as soil, ground water, nutrients and rainfall seasonality may affect a specific area's response, but the good fit obtained from the data (derived from a variety of unrelated sources) suggests that mean rainfall may be used as a preliminary yardstick with which to predict "optimum carrying capacity."

Data presented for Ethiopia (Bourn, 1978) indicate that, in general, the cattle biomasses exceed that of the "optimum carrying capacity" with the divergence being greater at the lower rainfall levels. Thus, although the regression for Ethiopian cattle is not parallel to the "optimum carrying capacity" line of Coe et al. (1976), it is probable that at higher rainfall levels and altitudes cattle biomasses are lowered by increased agriculture production; at lower rainfall levels a great percentage of the total biomass is actually represented by cattle. Large areas of Ethiopia are shown to exceed the "optimum carrying capacity" by over 100 percent. It is interesting to note that the large areas in the west where the biomass is less than predicted include a very large region which is infested with tsetse flies. Phillipson (1977) reached similar conclusions.

The tsetse fly *(Glossina* spp.) is considered by many wildlife conservationists to be an important element in conserving large areas for wildlife (Figure 5). This vector of trypanosomiasis of man and cattle occurs over some 10 million km² of the continent (Bourn, 1978) and has prevented

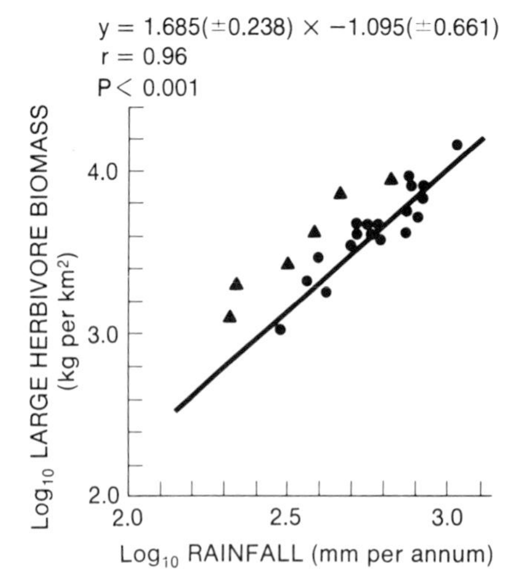

FIGURE 4. The "optimum carrying capacity" of 20 African wildlife areas (●) expressed in terms of standing crop biomass and mean annual precipitation. The points above the line (▲) which are not included in the regression are for six pastoral systems in northern Kenya—note that they all lie above and nearly parallel to the regression for wildlife systems. (Modified from Coe et al., 1976)

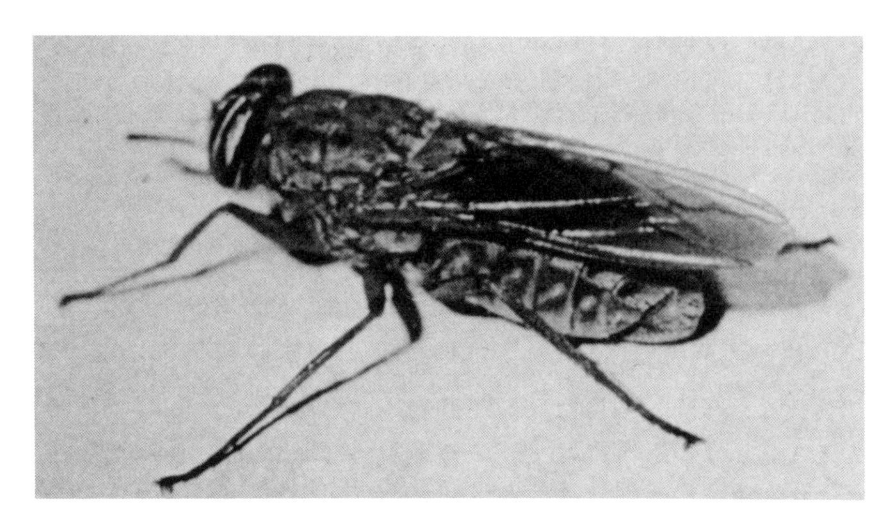

FIGURE 5. The tsetse fly; the vector of trypanosomiasis in man and cattle, but an important agency in preventing overstocking of domestic animals over much of semi-arid Africa.

agricultural development. Indeed, in many cases it is not an exaggeration to say that the establishment of many of Africa's savanna national parks has been a direct result of the presence of this insect, and the land's consequent low potential. Recently it has been stated that one of the aims of the United Nations Food and Agricultural Organization is to eradicate the tsetse fly from the continent (UNFAO, 1974).

While there may be strong socio-economic arguments for such a policy, it is obvious that by clearing low rainfall areas of tsetse and releasing these lands for pastoralism and marginal subsistence agriculture, the land will rapidly be overstocked and degraded by the domestic stock biomass levels sustained in other tsetse free arid regions. Before such a large scale and potentially damaging operation is carried out, it is clear that the ecological consequences as well as the social and political context need to be considered. For example, Ormerod (1976) in a study of trypanosomiasis and cattle in West Africa, has suggested that although long-term fluctuations in climate have been responsible for local droughts, the more recent widespread droughts and the spread of desertification may be to a large extent due to tsetse eradication programs which have allowed cattle standing crop biomasses to build up beyond the optimum carrying capacity. The increase in desertification may be accounted for by population increases (due to reduction of disease), increased herding and agriculture and the provision of water which concentrates the domestic stock and leads to extensive but local overgrazing.

Today we are faced with a rapidly expanding population in Africa; it has been suggested that already two-fifths of Africa south of the Sahara is overcrowded (Myers, 1972). The semiarid country of the Masai tribe in Kenya and Tanzania is probably too thickly populated with as little as 8 people per square mile (L. H. Brown, 1971.) Natural ecosystems seem to function within broad limits, but man, especially in the more extreme climatic zones, appears to be a historic overexploiter who pushes ecosystems beyond the thresholds of resiliancy. To help African peoples and their cultures to survive, we must ask ourselves how the indigenous biota can be utilized without exceeding the "optimal carrying capacity." As Janzen (1973) puts it, "They need a means to integrate what is already known into the process of developing sustained yield tropical agroecosystems."

THE ROLES OF NATIONAL PARKS AND RESERVES

It is common to equate the term wildlife simply with mammals—or even just large mammals, yet as every ecologist is well aware, any natural habitat will continue to function only if all the trophic levels from primary producers to decomposers fulfill their independent and interdependent roles. Herein lies one of the most important values of a wildlife resource. In establishing national parks we are attempting to preserve areas as close as possible to the state of natural stability, and "as outdoor laboratories where research should be relevant beyond the needs of wildlife alone" (Phillipson, 1977). For the parks act as yardsticks against which we can measure human exploitative activities.

As an adjunct to the value of ecological research, the very important educational importance of setting aside national parks and reserves must be mentioned. The young people of the world can be taught the value of the information contained in these undisturbed environments—places where diversity flourishes and relative stability reigns. Above all, there must be training grounds where the young can learn that sense of enquiry without which there will be no researchers, wildlife managers, administrators or park wardens to guard what remains of wildlife resources in the future.

The most difficult task of all is to justify the aesthetic value of wild animals. We might consider this a privilge that only members of the developed world can afford. Certainly the small African farmer with his two acres of maize finds it difficult to appreciate the natural beauty of the African elephant when it comes silently in the night and destroys his work of the last few months and next year's food in a few short minutes. Yet in spite of the many areas where men and animals come into conflict in Africa, we have only to look at the customs of tribes the length of the continent to see a knowledge and affection for wildlife built into their dances and their folklore. The only basic difference is the context and the perspective with which we view the attitudes of a proud man who was yesterday a brave hunter, who killed to provide for his family's survival, but who today is branded a poacher because a businessman in the city has found a ready market for animal products.

The African nations have set aside large areas of their lands as national parks and reserves. Although many of them have only come into being since World War II (and an even larger number in the last decade) the history of many of the reserves is a lot older. The first reserves were set aside in Natal and the Transvaal (South Africa) in 1897–98. In 1925 in the Belgian Congo, the Albert National Park (now called the Virunga National Park) was Africa's first in which complete protection was guaranteed to the flora and fauna. In South Africa the Sabi Game Reserve was upgraded to the Kruger National Park in 1926 (Curry-Lindahl, 1972*a)*. Since these early times the national parks of East and Southern Africa have grown to an area in excess of 200,000 km^2 with a smaller, but none the less vital area being set aside in West African territories.

In the developing nations of Africa where pressures on land for agricultural and settlement purposes is high, we can only view the establishment of national parks in socioeconomic terms. Without outside assistance from world bodies such as the World Heritage Trust System (Myers, 1972), these nations cannot afford the privilege of conserving wildlife for its own sake. Fortunately, tourism has shown that the spectacle of Africa's wildlife has economic value. In 1968 Kenya's tourist income surpassed that of the coffee export (valued at $44 million in foreign exchange). By 1971 this figure had risen to $55 million (Curry-Lindahl, 1972*a*; Myers, 1972). Tourism in Kenya is expected to grow at up to 20 percent per year. At the present time this industry is growing at a rate three times faster than that of the economy as a whole (Myers, 1972).

Not that we must assume that the national parks can take increasing numbers of tourists without very careful management and control. The construction of roads, excessive vehicular traffic and general disturbance to vegetation and animals could prove a blessing in foreign exchange, but a death knell for the very spectacular sights that in large part generate the handsome tourist income.

Whatever we do, and however long we deliberate, the national parks and reserves of Africa must be seen to be relevant to the everyday lives of the rural Africans who live around their boundaries. For most of these Africans, the parks are neocolonial structures that have been built for busloads of tourists and their more affluent fellow countrymen. National parks do not allow visitors on foot. Unless some (usually charitable) body has provided a bus for the carless visitor, it is as remote to a farmer living on the boundary of the park as is the city, two hundred miles away.

When the Nakuru National Park in Kenya was established the local people who had previously walked into the Park found themselves blocked by a "vehicles only" notice (Myers, 1972).

This is the point of contact that we must foster and establish if we are to succeed in helping Africa to help itself. If we spent a little less on research and more on communicating with the people who surround the parks, the chances of the protected animals' survival would be increased out of all proportion to the amount of money spent on the park.

In some cases local administrations that border the national parks have a financial interest in its operation and profits, which adds a whole new dimension to local involvement. Western's (1973) almost lone study of the Masai and the Amboseli Game Reserve in Kenya looks in detail not only at the wildlife but at the people who surround the Park and who have been dispossessed by its presence. The exclusion of the Ik from their former hunting grounds in the Kidepo Reserve in Uganda has almost certainly led to their near or complete extinction (Turnbull, 1973), and illustrates the need to view the local community in the context of wildlife management.

WILDLIFE RESOURCES OUTSIDE NATIONAL PARKS AND RESERVES

Besides scientific, educational and tourism values, the utilization of wildlife as purely a protein resource has extremely important implications to the preservation of wildlife and to local economies. Thus, much of the future of wildlife will depend on what happens outside the boundaries of national parks and reserves. With changing patterns of land ownership from communal to individual title deeds, the fate of wildlife will be in the hands of new individual land owners. In this situation, wildlife are only likely to be preserved if their maintenance, which is often in competion with crops and domestic animal needs, can be shown to be economically advantageous.

Programs of economic utilization of wildlife as a protein resource fall into three categories: (1) the operation of sustained yield culling of free ranging wild species; (2) the selection of suitable wild species and their management in enclosed ranging operations; and (3) the domestication of large African herbivores.

Culling of Free Ranging Wild Species

One must distinguish between the ranching of game, which refers specifically to the utilization of wild herbivores on enclosed land, and the "cropping" of free ranging wild animals. Cropping operations are usually carried out as a management program rather than as a regular or sustained yield exercise (like ranching). There is, however, no special reason why cropping cannot be carried out regularly if management specialists decide it would benefit the wildlife population.

The production of bush meat, as distinct from the more highly organized forms of culling and ranching, clearly requires strict controls over health and veterinary inspection, especially when such a form of natural resource husbandry develops to the point where the material travels from distant rural areas to towns or cities. Additionally, closed seasons would need to be instituted and enforced locally so that a check could be kept on the rate of removal and recruitment. This is possible only if local authorities and governments ensure that similarly priced replacement supplies of protein are available during times when it is ecologically unsound to remove bush meat. Nonetheless, bush meat production, its importance having now been recognized, needs to be developed and enlarged, but not to the point where production becomes expensive and where middle men intervene to where the product is no longer the reasonably priced source of protein that it was intended to be.

Most of the data on bush meat applies to areas in West Africa where the availability of the larger species of wild mammalian herbivores is limited. In the African savanna lands there is little doubt that game meat also comprises an important source of animal protein, even where there is specific legislation against its open sale. Richter (1969) believes that the people of the Kalahari region of Botswana derived up to 16.5 kg of meat per year per person from game animals between 1965 and 1967. While in Rhodesia, Roth (1970) calculated that the value of game (both meat and skins) was in excess of $1 million. Since a number of small cropping schemes have been carried out in East Africa, special licenses have been issued for this purpose and its seems likely that this market will expand.

While local bush meat may satisfy local, rural needs, such commerce is far removed from the large scale culling or ranching programs that have been advocated (Talbot et al., 1965) and carried out (Blower and Brooks, 1963; Parker, 1964) within the lower rainfall savanna zones. These zones have been recorded as producing the highest standing-crop herbivore biomass of any known terrestrial environment. As we pass from low to higher (below 200 mm) rainfall levels there is a tendency for the mean size of mammals to increase (Coe, 1976). This is also accompanied by an increase in generation time, which will in turn affect the rate at which the standing crop biomass will turn over.

Any attempt to organize a system of hunting, culling or ranching should clearly be based upon an area's annual production. In addition, calculations of the potential profit to be made from culling must be based on a sound knowledge of the demography of the species to be utilized (Chapter 10). A useful figure is the ratio of production over biomass or

P/B. Coe et al. (1976) estimated the production to biomass ratios for three size categories of herbivores. These were: large herbivores ($\gtrsim$880 kg) 0.05; intermediate herbivores (100 to 750 kg) 0.20; and small herbivores 0.35. Thus, depending on the proportion of large or small mammals in a given community, the annual production will vary greatly. It should, however, be pointed out that a significant semi-logarithmic relationship was shown to exist between production and rainfall for 24 African wildlife ecosystems (r = 0.882; $P \lesssim 0.001$) (Coe et al., 1976).

One of the first big East African culling operations was carried out in the Queen Elizabeth National Park (now the Ruwenzori National Park) in Uganda by Laws and his colleagues (1968). The area concerned was overpopulated with hippopotamus *(Hippopotamus amphibius)* which were causing considerable damage to local vegetation. The goal of this operation was to reduce the population to a more acceptable level and to provide scientific data on the population dynamics of the hippo population. A nearby market provided a good outlet for hippo meat and a profit was made on the operation. One advantage of this particular cropping exercise was that the amphibious hippo tends to remain in one more or less circumscribed area. This involves quite different problems than harvesting wild terrestrial mammals in the open where they are often widely scattered.

A similar operation was carried out in the Murchison Falls National Park (now Kabalega Falls) on both elephant *(Loxodonta a. africana)* and hippo, which again provided both valuable scientific data on the state of the populations and was financially profitable. Parker (1964) reported on the experimental cropping of elephants in the Galana River Game Management Scheme (which lies at the northeast corner of the Tsavo—East—National Park). This operation encountered difficulties with regard to transporting and marketing the large amount of meat that became available, even when it was dried. Such difficulties could be overcome.

Similar schemes have been carried out in southern Africa. A well organized central slaughterhouse has been operated in the Luangwa valley in Zambia, and local meat has been provided on a fairly large scale (Bainbridge and Hammond-Tooke, 1967). The culling of elephants in the Kruger National Park in the Transvaal has probably been as contentious as the failure to cull them in Kenya. Yet, once the authorities decided that they needed to reduce the numbers of elephants, they did it cleanly and efficiently and even arranged for the meat to be canned.

Large animals yield a great deal of meat from a small number of kills, whereas the nervous, smaller antelopes do not, and might well present much greater difficulties in the free state. Nonetheless, there are large areas of Africa which are infested with tsetse flies and which would almost certainly return more meat and animal products if regularly cropped than could be produced on the land on a sustained yield basis in any other way. The necessary transportation, veterinary inspection facilities, refrigeration units, hunters and handlers will take capital, but the potential is so great that it should, with proper leadership, be a commercial success. At the same time it should improve the very bush that it seeks to exploit. All that the arid regions of Africa are waiting for is a body with the finances and vision to set up the very large operation that would be necessary to test these ideas on a large scale.

We can also include under this heading wild animals that can be raised either for meat or other products, or merely to be released when the level of neonatal predation is lowered (such operations apply particularly well to reptiles, such as crocodiles and turtles). The Nile crocodile *(Crocodylus niloticus)* has been pursued in the past as a source of food, but more recently has been hunted for its valuable hide and because of the damage it can do to fish nets and, it is said, to fish stocks. In areas where they have been virtually eliminated, the local fisheries have deteriorated due to the expansion of predatory fish species which were previously eaten and limited by the crocodile (Curry-Lindahl, 1972). It is well known that the highest mortality for crocodiles is in the egg and early hatchery stages (Cott, 1969; Modha, 1967) and it has therefore been suggested that the eggs might be raised artificially and released (Cott, 1961).

Pooley (1962; 1969; 1971; 1973) has been a pioneer in the breeding of crocodiles in captivity in Africa. Eggs are collected in the wild and incubated in artificial conditions, or the young are wholly produced by captive adults. Since the mortality of natural populations is very high—survival rates are as low as one to five percent according to Pooley (1973) and Graham and Beard (1973)—rearing crocodiles over the first few months would greatly increase the survival of young hatchlings. Captive crocodiles have been shown to increase their length twice as fast as natural hatchlings, thus captive rearing greatly increases their chances of survival (Blake and Loveridge, 1975). Provided high growth rates can be achieved and a local source of cheap fish food is available, there is little doubt that crocodile farming has great potential. The increased hatchling success of eggs and enhanced survival of hatchlings in captivity can be used as a valuable means of replenishing stocks in the wild (Blake, 1974). A similar story can be told about the sea turtles of the Indian ocean which historically have provided a valuable source of food for the people of the coast and islands (Frazier, 1970). But, unless they and their nesting areas are protected their extinction is only a matter of time.

Game Ranching

The South African and Rhodesian efforts in game ranching are the most well organized in Africa; we have a great deal to learn from their

FIGURE 6. Some of the African mammalian herbivores that are potential candidates for game ranching. Each is a source of low-fat meat. Facing page, Top: a herd of blesbok on a private game reserve in Northern Transvaal, South Africa; Middle, left: male springbok in the Northern Cape, South Africa; Middle, right: white bearded wildebeeste; Bottom, left: Coke's hartebeeste or kongoni; Bottom, right: Thomson's gazelle; Above, Top: semi-domesticated eland on the Galana Ranch, Kenya; Bottom: semi-domesticated Beisa oryx on the Galana Ranch, Kenya.

operations (Child, 1970; Joubert, 1968; Skinner, 1966; 1967; 1973*a;* Von La Chevallierie, 1970). The main species that form a regular part of game farming in South Africa (Figure 6) are springbok, eland *(Taurotragus oryx),* blesbok, impala *(Aepyceros melampus)* and greater kudu *(Tragelaphus strepsiceros).* To these, of course, should be added the ostrich *(Struthio camelus),* which is common on enclosed land and is the only vertebrate species that can be said to have recently beeen domesticated in Africa. An analog of a mammalian herbivore, the ostrich's potential for meat production would make an interesting study, though it is already well known for its good "biltong" (strips of dried meat) in Southern Africa.

If we examine the main species utilized on game ranches we see that they have a consistently good dressing percentage and a low fat content (Table II). It is interesting to note that Von La Chevallierie (1971) considered that the increased fat content of castrated eland enhanced carcass quality (2.4 to 12.1 percent). Tests of various meats conducted by Von la Chevallierie showed that springbok was the best on a series of measures of quality. The carcasses of wild ungulates contain about 43 percent of lean meat compared with only 23 percent of domestic animal carcasses (Ledger, 1963).

TABLE II. Meat production data for African antelopes.

Species	Unit weight*	Dressing percentage	Percent fat content
Springbok	26	57.9	<4.0
Eland	340	57.4 castrates	12.1
		53.0 bulls	2.4
Impala	40	58.8	1.8
Blesbok	53	52.9	1.8
Kudu	136	50.6	1.8

*Unit weights for species after Coe, Cumming and Phillipson, 1976.
(Modified from Skinner, 1973*b)*

Although arguments will no doubt continue into the future over the relative merits of domestic and wild herbivores, the fact remains that farmers in South Africa, whose ancestors did so much to exterminate their game, have now learned that game meat can be big business. In the semi-arid areas of the Northern Cape, large estates are now being operated with springbok production as an important part of their diversification. In 1977 some of these estates produced meat whose value was in excess of 50,000 rand–one rand equaled 1.4 U.S. dollars in 1975 (Liversidge, personal communication).

Riney and Kettlitz (1964) described in some detail the situation in a number of estates in the Transvaal and illustrated the reasonable income that could be made from animals shot under permit on private estates; in 1959 this income was valued at 1/4 million rand. The most important observation in this study was the remark: "... there is a tendency for herds of wild animals to increase, while being used economically." Dasmann (1964) and Dasmann and Mossman (1960) have also described the feasibility and common sense of some ranching schemes that have been operating in Rhodesia for many years to the great benefit of local meat supplies and to the financial advantage of those far sighted enough to try ranching of large wild herbivores.

A great deal of space has been devoted to arguments between wildlife ecologists, on the one hand, who extoll the virtue of game, and the range managers, on the other, who advocate the superior production and easier handling of their cattle. It has been suggested that the diversity of game animals in most East African habitats makes better use of vegetation than cattle. Although this is to a degree true, the spectrum of use for the full range of domestic mammals (including cattle, sheep, goats, donkeys and camels) will be wide, even if not quite as wide as that for the richest wild fauna (Pratt and Gwynne, 1977; Longhurst and Heady, 1968). It is important to remember that, while one may discuss the relative monetary values of these animals, to the pastoralist they are far more than chattels to be disposed of for cash. They are the basis of his wealth, power, social customs–indeed the very foundation of his society. Therefore, to discuss alternative ways of using or operating in the savannas without taking the necessary social changes into consideration, might result in the preservation of the antelope and the savanna at the expense of a whole variety of tribes and their way of life.

Many arguments have been made in defense of game meat versus cattle. Physiologists have studied the thermoregulatory abilities of game animals and found that arid country species like the oryx *(Oryx beisa)* can withstand high ambient temperatures by raising and lowering their body temperature, and are water independent, drawing virtually all the water they need from eating dry hygroscopic vegetation in the early morning and retaining body fluids (Taylor, 1969). At the same time, when the eland and the Hereford steer were compared by Taylor and Lyman (1967) eland were found to be better adapted to arid environments; in terms of production, though, the eland appeared to be less efficient than the Hereford. This result has been questioned by Crawford (1974) who quotes the results of Rogerson (1968) as evidence that the low conversion efficiency of the eland (Taylor and Lyman, 1967) may well be related to its efficiency of water utilization, rather than to problems of low conversion efficiency for nitrogen.

An additional argument in favor of game advanced by Crawford and Crawford (1972) is that cattle produce up to three times as much fat as solid nutrients. Their argument in defense of low fat on game meat is related to medical concern over the increased incidence of heart disease in young people, in which animal fats have been implicated. The fact remains, however, that the most favored part of the carcass is the fat. In a butchery large lumps of lean meat will invariably be thrust aside in favor of meat with a good supply of fat. As Crawford points out (1974) the fat is really a source of stored and utilizable energy and perhaps this is exactly what the African needs in his diet.

Such arguments over which solution is best are rather simplistic. Clearly there are environments in which cattle will be a far better means of producing meat, butter, milk and hides, while in lower-elevation dry environments the game (or a game and cattle mix) will fare better. Certainly there are a wide variety of game animals (especially smaller grazers and a variety of browsers) that will not compete with cattle at all. In terms of diet, activity and habitat, Stanley-Price (1974) has shown that the only two herbivore species that show appreciable overlap in their requirements, and can therfore be presumed to compete, are the cattle and the Coke's hartebeeste or kongoni *(Alcelaphus bucelaphus cokei)*. Indeed, this finding was confirmed when, following the removal of Somali cattle from the Nairobi National Park, the kongoni quickly responded by increasing their numbers.

Crawford (1974) has proposed a polyculture in which he envisages a wide variety of wild herbivores being utilized, thus employing a full spectrum of feeding habits from top browsers to low browsers and open ground grazers. Crawford's proposal, however, assumes that this wide spectrum of wild herbivores will be domesticated—a process requiring many years of intensive studies of the animals' physiology and sociobiology.

The biological basis of a polyculture is the ecological differences between species (Gause, 1934; Elton, 1927). Much of the work on wild herbivores in East Africa in the last 20 years has been directed at studying the way in which these species complexes interact—from their spatial and temporal separation (Lamprey, 1963), to the grazing successions (Bell and Gwynne, 1968) and to the more complex structural models (Cobb, 1976) for Tsavo National Park. Thus, although the transition from the enclosed ranch to the open free range is difficult, we have the beginning in East Africa of the technology necessary to make this important step.

Domestication

It is important to distinguish between ranching wild animals on fenced land and the process of domestication. Domestication can only be considered the process of "bringing an animal into subjection to or dependence on man" (Skinner, 1973*a*). There is an area between ranching and domestication where mammals left free to range on comparatively open areas of land will become sufficiently tame to be enclosed, although they could not be herded, nor could they be handled with ease. This applies especially to animals like blesbok *(Damaliscus dorcas phillipsi)* and the springbok (Figure 6) which are common on South African farms today, and which have recently become an important and profitable means of meat production. It is particularly interesting to note that if the blesbok had not been enclosed on farms in this semi-wild state it would by now be extinct, for it has disappeared from most of its former range.

Many authors have claimed that the answer to future protein needs and wildlife conservation lies in the domestication of new species of African herbivores (Crawford, 1940; Kyle, 1972). This is not a new idea; hunter-naturalists at the end of the nineteenth century considered apparently docile species like the eland highly suitable candidates for domestication (Methuen, 1848). Undoubtedly, a number of animals have been tamed as pets in Africa, while others may simply have grown used to the proximity of human beings. However, they are still a long way from being easily herded, handled and milked.

I have already discussed the properties of the meat of the five most likely candidates for game farming in Southern Africa. They show great promise in terms of carcass quality, growth rates and recruitment. There is, however, a great difference between ranching an enclosed herd and the complex taming of an animal to the point of almost complete dependence.

The first real attempt to domesticate eland seems to have been the now famous Askanya Nova herd in Russia which was derived from animals imported between 1892 and 1896. These animals have been used primarily as a source of milk, which is believed to have important medicinal properties (Treus and Kravchenko, 1968). It is difficult to see why the eland was chosen for this purpose since, although it is large, it is also a very nervous animal and choosey about its food in the wild. It also moves large distances each year in pursuit of its normal activities. Whatever the reason, a great deal of work has been done on ranching these beasts and on using them in a semi-domesticated state (Bigalke and Natz, 1954; Huygelen, 1953; Jelliman, 1913; Kerr and Roth, 1970; Kerr, Wilson and Roth, 1970). In the South African and Rhodesian situation the eland seems to be a promising candidate since it is used in a semi-tame state. This means that they can be subject to veterinary examination and treatment especially when they are to come in contact with domestic stock.

One of the most detailed studies of domestication potential was carried out in Kenya at the Galana Game Farm Research Project. A team of workers looked at feeding preferences of eland, oryx, buffalo and cattle. These investigations have recently been concerned with comparative ecology and ecophysiological studies. It seems to be generally agreed that the eland does not do very well in these arid lands but that the oryx does, when compared with either sheep or cattle. The oryx maintained at Galana have been tamed and will walk into a truck on their way to market for auction; whether they can be added to the pastoralists' local stock is another matter (Field, 1975; Lewis, 1977; 1978; Stanley-Price, 1976). These studies show that each climatic zone of Africa has its own peculiarities which need to be taken into account when thinking of ranching or even semi-domestication.

It would seem from what has been said that the three types of programs mentioned above fall into management categories of increasing complexity. The operation of sustained yield culling programs on free ranging wild species would provide much of the knowledge required for the second level—the selection of suitable species for ranching on enclosed land. Out of such long term culling and ranching operations would come the technology and experience necessary to domesticate large African herbivores. Perhaps the most important of all is the need to select and develop technologies that can be integrated with the social customs of the present occupants of Africa's extensive arid lands.

Biologists in West Africa have selected two species for possible domestication. These are the giant rat *(Cricetomys gambianus)* shown in Figure 7 (Ajayi, 1975) and the cane rat mentioned earlier (Asibey, 1974). These species are two of the most common items in bush meat diets. These studies indicate that both these small mammals tame easily and can be bred in captivity. From its ecology and opportunistic scavenging habits, the giant rat (weight about 1 kg), appears to be more likely to succeed than the very popular and larger (4.8 kg) cane rat with its more specialized grass stem diet and smaller number of young (usually only one, versus about three for the giant rat). Examining these small mammals as a protein source would reward further investigation for small-scale local food production and reduce the pressure on some of the medium-sized and now rare duikers.

STRATEGY AND ATTITUDES FOR THE FUTURE

The conservation of wildlife in Africa cannot be considered in isolation from the needs of the developing nations of Africa. Above all, the expanding populations of the continent make it imperative that we take a holistic view, working downwards from problems of human land resource management to those of wildlife.

The human population of Kenya rose from 5.4 million in 1948 to 10.9 million in 1969. Projections for the future suggest that by 1980 50 percent of the population will be under 12 years of age (Republic of Kenya, 1972). Pressures on the land from subsistence agriculture during the last twenty years have led to the extensive utilization of marginal land and the consequent loss of valuable top soil. This, in turn, may have dramatic effects on coastal lagoon fisheries and cause the death of extensive coral reefs, a

FIGURE 7. The giant rat, a suitable candidate for captive breeding in West Africa.

prime tourist resource in coastal areas. With 90 percent of Kenya's population living in rural areas and annual population growth rates as high as 4.3 percent, the increasing pressure on sensitive semi-arid habitats poses a grave threat to both man and wildlife resources.

Little attention was paid in the past to the value of game animals in national parks, but it is becoming increasingly clear that the income from this area of environmental investment far outweighs its value as an agricultural or ranching asset, especially in the semi-arid zones where most of these areas are located. For local people, however, the rush of tourists to their former hunting and grazing grounds fails to satisfy their immediate needs, especially when most of the income goes either to local business or the national treasury. The problem is much less severe in areas where local district councils have been given a stake in the financial rewards derived from tourism and other wildlife resources. Indeed, the fuller and more widespread involvement of the park's human neighbors might well aid in the reduction of poaching by local tribesmen.

With increasing human density the compression of wildlife around national park boundaries rapidly becomes critical. This is especially true during periods of severe drought or intense rain, when short term vegetation changes cause the temporary abandonment by many species of large parts of their normal range. In areas where such phenomena are common and well documented, the establishment of buffer zones around the national park boundaries would not only mitigate the effects of habitat insularization (see Chapter 6), but also provide a region in which either control culling or even sustained yield culling could be carried out. Such multiple land use would serve the aims of management and provide a controllable source of local meat production.

There are few national parks in Africa that can be considered self-sustaining ecological units over the whole range of climate experienced by semi-arid environments. The recognition of the need for multiple use will, in the long-term, protect rather than erode the very boundaries of the areas we seek to give full protection. Perhaps most important of all, such practices would tend to prevent the encroachment of subsistence agriculturalists and pastoralists, and at the same time provide a source of employment and animal products.

Outside the boundaries of national parks, less inviolate areas could, with strict management policies, be utilized for sustained yield culling operations and hunting, and, thus, be considered valuable sources of meat and revenue. Many other habitats, especially those infested with tsetse, could be studied as potential sites for cropping schemes and even semi-domesticated game ranching.

These methods of utilizing wildlife resources, however, will require a large international effort to experiment with pilot operations and to plan the complex support facilities that will be needed. Unless such steps are taken in the near future, not only will the wildlife of Africa be exterminated, but the environments will be degraded to a state where they will produce virtually nothing. The African nations can ill afford to lose their wildlife with their potential nutrients and income.

In the same way, the large areas of land utilized by pastoralists are in urgent need of study to discover the best means of assisting these people and the economies of all countries. Where pastoral peoples utilize grazing and browse on park boundaries, controlled grazing rights could be afforded in those areas that are seasonally under-utilized by wildlife. Such recommendations have been advanced for areas in the Tsavo (West) National Park in Kenya (Cobb, 1976).

In spite of all our efforts to aid the African continent in rationally utilizing her wildlife resources, we have only to look at the recent events in East Africa to realize how rapidly an animal population can be depleted from abundance to the border of extinction. In Uganda, the elephant population of the Ruwenzori and Kabalega Parks was reduced by up to 75 percent and 85 percent, respectively, in less than three years (Eltringham and Malpas, 1976). In Kenya the elephant population is estimated to have been reduced by about 50 percent in little more than five years (Hillman, 1978). In both these instances, the prime cause of this disastrous fall in numbers has been poaching. The value of the ivory alone lost to the Kenya economy during this period is estimated to be close to $8.75 million (Eames, 1978).

While conservation biologists can have little direct influence over such wanton waste of Africa's natural resources, they can, by turning their attention and efforts towards developing rational management policies, convince these nations that wildlife has a part to play in their developing economies. In so doing, they will achieve the one objective that should concern us all: the preservation of the most unique and diverse mammal fauna in the world.

POSTSCRIPT

The wildlife of Africa is a resource that, provided it is managed properly, is capable of renewing itself both to its own benefit and that of the local human population. The vast number of mammal species found on the African continent represents a diversity unequalled in the earth's history. This unique assemblage has not evolved in isolation, for the vegetation which has been modified in response to local soil and climate has also been greatly influenced by the fauna. For example, complex morphological and chemical defenses in plants have evolved to deter grazers and browsers (Whittaker and Feeny, 1971; Janzen, 1975). At the same time the high crowned teeth of the large plains herbivores are a response to the development of savanna plants rich in silica, which would quickly wear down a shorter and softer molarform dentition. Hence we see that there has been a continuous coevolution between the vegetation and the mammal fauna in a system of offense-defense. It is perhaps this simple fact more than any other that tells us that the unique African mammal fauna is a part of the landscape that should not be brushed aside by development plans which are almost wholly homocentric. Man is a part of nature, not apart from it, a fact put a little more succinctly by Leopold (1949) when he talked of the "land ethic" or "ecological conscience" which implies "respect for our fellow members (plants and animals) of the land community and also respect for this biocommunity as such" (Curry-Lindahl, 1972).

SUGGESTED READINGS

Boughey, A. S., 1971, *Man and the Environment,* Macmillan, New York.
Brown, L. H., 1971, The biology of pastoral man as a factor in conservation, *Biol. Conserv.,* 3(2), 93–100.

Cloudsley-Thompson, J. L., 1977, *Man and the Biology of Arid Zones,* Arnold, London.

Coe, M. J., D. H. M. Cumming and J. Phillipson, 1976, Biomass and production of large African herbivores in relation to rainfall and primary production, *Oecologia (Berl.),* 22, 341–354.

Crawford, M., 1974, The case for new domestic animals, *Oryx,* 12, 351–360.

Pratt, D. J. and M. D. Gwynne, 1977, *Rangeland Mangement and Ecology in East Africa,* Hodder and Stoughton, London.

Skinner, J. D., 1973, An appraisal of the status of certain antelope for game ranching in South Africa, *Z. Tierzuchtg. Zuchtsbl.,* 90, 263–277.

CHAPTER 17

THE CONSERVATION OF TROPICAL RAIN FOREST

T. C. Whitmore

The appeal of tropical rain forests is universal. Biologists in particular are attracted to them because of their fabled, but nonetheless real, richness in species, and because they have been the most extensive unexplored ecosystems. Until recently tropical moist forest appeared tractless and limitless. This once seemingly inexhaustible resource of timber, rattans, gums, resins, spices, fruits, seeds, orchids, dyes, drugs, animals, ivory and pelts is disappearing before our eyes at a rate averaging (between 1964 and 1973) 21 hectares a minute. The second half of the twentieth century will stand in history as the brief period during which man reduced the area of the richest and most complex ecosystems in the world by about one third. The onslaught will continue to accelerate, reducing the forests to scattered fragments by A.D. 2000.

What will be the effects of this reduction? Does it matter, and, if so, can it be ameliorated? Conservation is an emotive subject likely to be expressed in dramatic language, but the current and predicted rate of depletion of tropical moist forest is so rapid that it calls for colorful comment. It represents an impact on the world's wild places unprecedented in history. Let us allow the facts and figures to speak for themselves.

THE PROBLEM

Definitions

Our main discussion will center on the tropical rain forests, a group of forest formations developed in the aseasonal or slightly seasonal parts of the tropics. Their global distribution is shown in Figure 1.

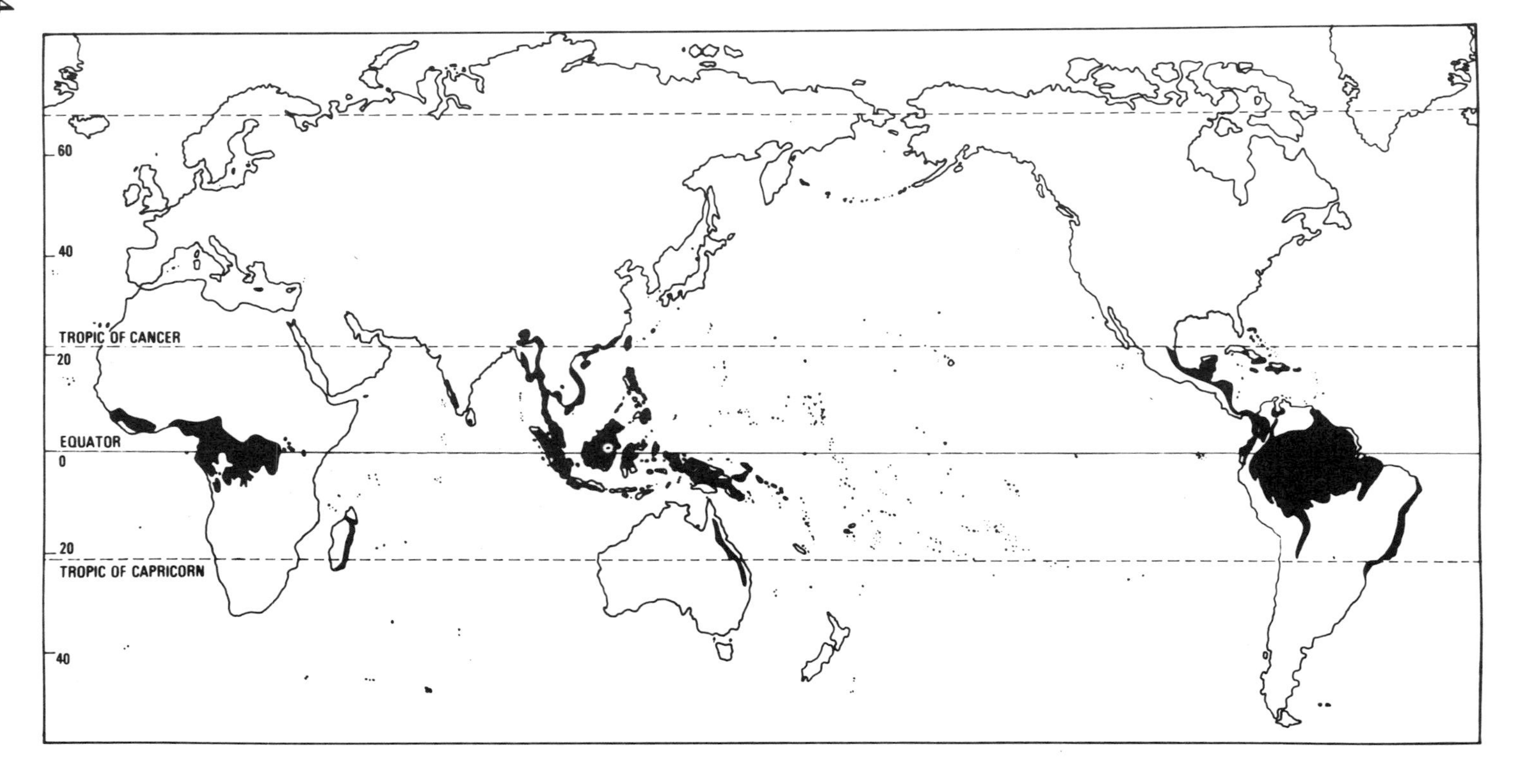
60
40
TROPIC OF CANCER
20
EQUATOR
0
20
TROPIC OF CAPRICORN
40

FIGURE 1. Distribution of tropical rain forests (in black). (From Whitmore, 1975)

In more strongly seasonal tropical climates these are replaced by a further series of forest formations which may collectively be referred to as monsoon forests, an allusion to their extensive occurrence in south Asia. Climates which are drier still carry thorn forest or savanna forest. In the statistics on area and on rates of logging, tropical rain forests and monsoon forests cannot always be disentangled, and so are considered together. The useful term *tropical moist forests* has recently been coined by the United Nations Food and Agriculture Organization (FAO) to describe them. They change polewards into subtropical rain and monsoon forests which are less extensive and not easy to separate. A recent description of the main rain forest formations can be found in Whitmore (1975) and statistics on areas and rates of logging are fully discussed by Pringle (1976) and Sommer (1976).

The Extent of Tropical Moist Forest

It is difficult to assemble consistent figures on the extent of the world's forests. Table I summarizes such an attempt by Persson (1974). Tropical moist forests (1,456 million hectares) comprise just over half the global forest cover and their subtropical fringes cover a further eight percent (224 million hectares). Persson's calculated total area for the world's forest, 2,800 million hectares, is considerably less than the 4,800 million hectares stated by Whittaker (1975), but the relative extents are not dissimilar.

Another analyst, Sommer (1976), lists about the same total area of tropical moist forest as Persson, and provides a reasoned breakdown between tropical rain forest, 1,240 million hectares, and monsoon forest, 360

TABLE I. World forest area by broad formation groups.

	Areas, in millions of hectares	Percentage of world total
Tropical moist forests	1456	52
Subtropical moist forests	224	8
Temperate broadleaved forests Temperate coniferous forests	448	16
Boreal forests	672	24
World total	2800	

(After Persson, 1974)

million hectares. Table II provides the subdivision between continents. Note the preponderance of the neotropics with 506 million hectares out of the 935 million hectares still under forest in 1973.

TABLE II. Summary of various measures of tropical moist forests in millions of hectares.

	Total primeval climax area	As percent of land surface	Actual area in 1973	Area remaining as percent of climax area	Area affected by exploitation in 1964	in 1973	net increase
Africa	362	36	175	48	2.4	3.2	33%
America	803	51	506	63	0.6–1.7	0.8–2.5	46%
Asia-Pacific	435	37	254	58	0.7–1.4	1.3–3.0	120%
(S.E. Asia)	(302)	(67)	(187)	(62)	(0.4–1.1	1.0–2.7)	(144%)
Total	1600	43	935	58	3.7–5.5	5.3–8.7	78%

(From Sommer, 1976)

The Rate of Exploitation of Tropical Moist Forest

By 1973 the actual area still under unexploited tropical moist forest was already only three fifths or less of the potential area for such forest, and everywhere there had been a marked increase in the rate of exploitation during the previous decade, most notably in Asia (Table II).

During the decade 1964 to 1973 the annual loss of forest was reported for 13 countries spanning the tropics. The losses totaled two million hectares every year or 1.2 percent of the actual forest area. Extrapolated to all countries, the estimate of the minimal annual rate of destruction for the whole tropical moist forest is around 11 million hectares. This is equivalent to 21 hectares or 50 acres per minute. The calculation, set out in full in Sommer (1976), is difficult to fault.

Turning now to tropical hardwood production, we can see (Figure 2) that it increased globally fourfold from 1950 to 1973 (with a slight de-

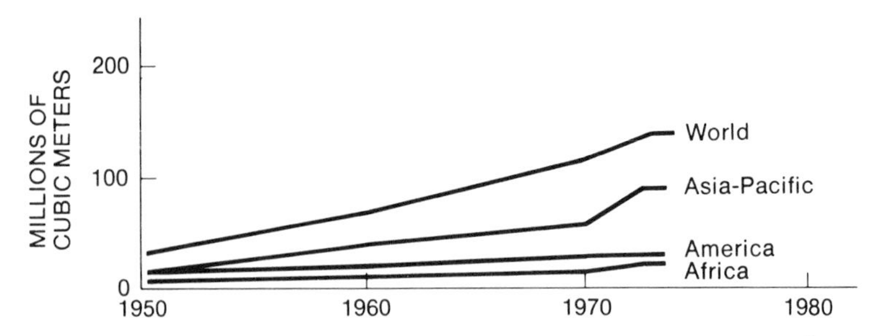

FIGURE 2. Tropical hardwood production 1950–1974, million m^3. (From Pringle, 1976)

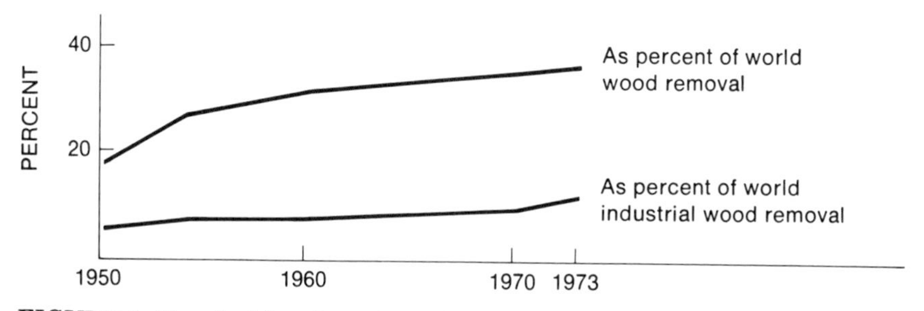

FIGURE 3. Tropical hardwood production 1950–1973. (From Pringle, 1976)

crease from then to 1974—mostly in the Asia-Pacific region). By 1973, tropical hardwoods still only provided a small proportion of world industrial wood (Figure 3), though this was up to 11 percent from five percent in 1953. If we consider fuelwood, which comes mainly from monsoon forest, the tropical moist forest accounted for nearly 40 percent of annual world wood removals by 1973.

Figure 4 is a plot of the predicted tropical hardwood log production from 1973 to 2000, showing a global increase of threefold. This increase is to be accomplished by steadily increasing the rate of extraction in Asia (expected to double in output) and later from America (which is expected to show a sixfold increase over the whole 27 year period). Most hardwood production comes from rain forest, not monsoon forest.

Using these statistics, we can calculate that to meet the demand predicted in Figure 4 for A.D. 2000 at the 10 m^3/hectare yield obtained between 1964 and 1973, consumption will have to be 556 million hectares of forest, or 59 percent of the 935 million hectares remaining in 1973 (Table

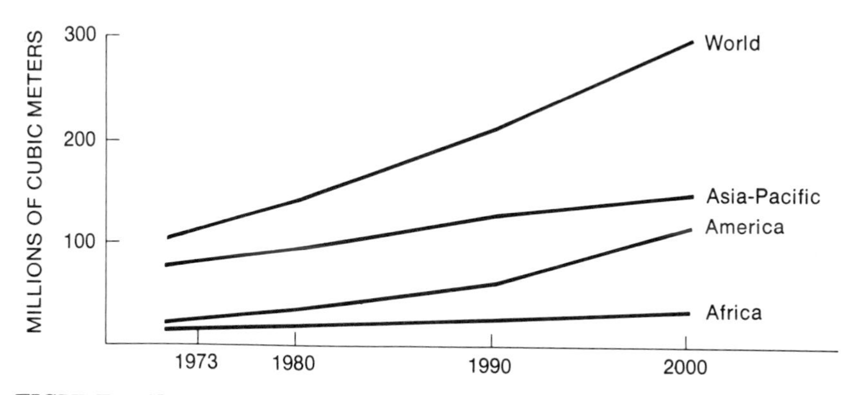

FIGURE 4. Tropical hardwood production, prediction for 1973–2000, million m^3. (From Pringle, 1976)

II). This will leave only 30 percent (480 million hectares) of the total climax area under virgin forest. Utilization will probably become more intensive, and yield per hectare will go up. On the other hand, no allowance has been made for fuelwood production. The prospect is frightening.

The Mode of Utilization

Our discussion thus far has, perforce, considered rain and monsoon forests together. We now continue the argument with a deeper analysis of the tropical rain forest alone.

Tropical rain forest is being "quarried" for timber. The power of modern machinery allows rapacious extractive forestry to be practiced. In economic terms these trees represent idle capital which can be realized in cash to be utilized elsewhere. A forest differs from a mine in that it is potentially inexhaustible, if man so chooses. But nearly everywhere today the rate and intensity of felling demonstrated in Figures 2 and 3 far outstrip the ability of silviculturists to sustain the yield by growing replacement forest.

Tropical rain forest is also being cleared so that the land can be used for agriculture—for example for cattle raising in the southern Amazonian *hylaea,* for oil palm and rubber plantations in Malaysia and Papua New Guinea and for pine tree plantations in Bali. The forest is commonly felled and burned with little or none of the timber being removed first.

Shifting cultivation also continues to make inroads into the rain forest, though two modes must be distinguished. In the long standing traditional agriculture of, for example, much of Southeast Asia, peasant farmers prefer to fell old secondary forest rather than high virgin forest. The system is only out of balance because population increases force enlargement of the total land area under cultivation. Elsewhere, demographic pressures are forcing people into tracts of virgin forest, which become the new frontiers of advancing civilization. An example is the current spread of people from the northern Andes down from the overcrowded altiplano to selvas below. There are similar trends in northern Thailand (Stott, 1978).

The mode of utilization for timber has changed. Everywhere there is (historically) a trend toward increasingly intensive use, well exemplified by the situation in Malaya. Demand there was first for choice cabinet woods and naturally durable timbers. The removal of these scattered trees caused small patches of localized damage to the forest structure. Much of the Amazonian forest is still at this stage of exploitation. In Malaya, the demand shifted after World War II to lighter, paler, nondurable timbers which were marketed as groups of species. Far more trees were removed per unit area and damage to the forest was much greater. This new demand coincided with the development of powerful and reliable extraction vehicles and, in the early 1960's, by the one-man chain saw.

As time goes on "wastage" decreases and increasingly more species are utilized as the accessible forests diminish and the market remains strong.

Thus, ramin *(Gonystylus bancanus),* a common species of the Borneo peat swamp forest, was at one time unmarketable, but later became the most sought after species. Similarly, in the species-poor west Pacific islands an export trade developed in the late 1960's based largely on *Calophyllum, Campnosperma, Pometia* and *Terminalia,* which in West Malaysia are very poorly regarded.

As more and more species are utilized, damage to the forest concomitantly increases. Figure 5 shows the large number of tree species per hectare in various lowland tropical rainforests. Figure 6 shows how that number increases almost exponentially as smaller girth trees are considered. Figure 7 shows the considerable damage which was taking place around 1970 in the Malayan forest, from which just the most valuable of the dipterocarp species were being removed for timber.

At present, extractive forestry is approaching its apogee with plans to remove most of the above-ground biomass for wood chips to be utilized for paper manufacture. In the humid tropics, chip industries are in operation in several places in mangrove forests, but not yet in many dry land rain forests, although the required technology is available (Kyrklund, 1976).

At the same time we may be seeing the dawning of the realization that species-rich tropical rain forests are potential sources of a multitude of products, not just timber and cellulose. For example, the Eighth World Forestry Congress, held in October 1978, was entitled "Forests for Peo-

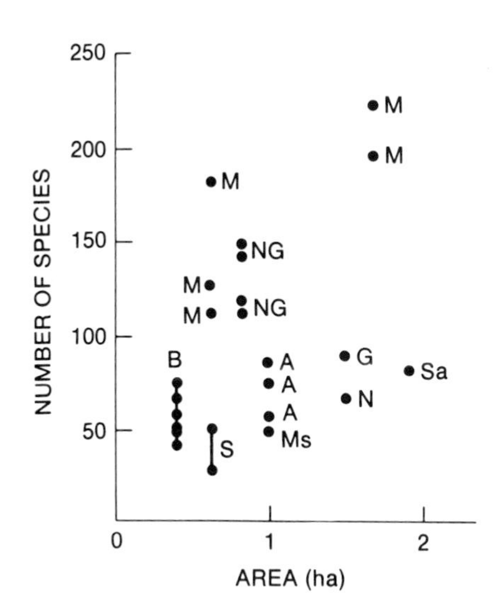

FIGURE 5. Species richness on small plots in lowland tropical rain forest, trees ⩾0.3 m girth. A: Amazon; B: Brunei; G: Guyana; M: Malaya; Ms: Mauritius; N: Nigeria; NG: New Guinea; Sa: Sabah; S: Solomons. Lines connect sample plots which lie near each other. (From Whitmore, 1975)

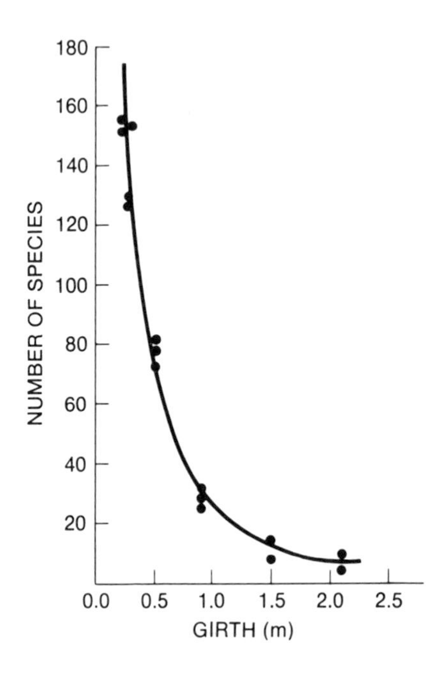

FIGURE 6. Relationship between species number per 0.4 hectares and different minimum sample girths. (From Whitmore, 1975)

ple." Foresters are rediscovering that human societies in forested tropical countries can gain great social benefit from sustained utilization of forests for food, rattans, dyes, drugs, gums and resins.

The wheel has almost turned full circle. Historically, outsiders in tropical rain forests (spearheaded by adventurers and traders) were interested in spices and a multitude of other products. It was this interest which led to the Colonial era as European nations vied with one another to secure and monopolize these distant sources of wealth. It has only been during the present century that timber alone has gained increasing commercial dominance, and its extraction has become ever more concentrated in the hands of big, well-capitalized companies, with the benefits also narrowly focused. The market has been mainly in the industrialized West. Now, however, the interests of local societies are about to be reasserted, assuming the current profligate destruction of tropical forests will slow or halt while enough remains for this to be possible.

Consequences of Rain Forest Exploitation

Two groups of consequences may be distinguished: first, the capacity of a forest to regenerate and second, the broader consequences which may follow the extensive replacement of tropical rain forest by other kinds of land use.

To some extent logging simulates the natural disturbance processes which are occurring all the time in any climax forest (Chapter 5)—gaps

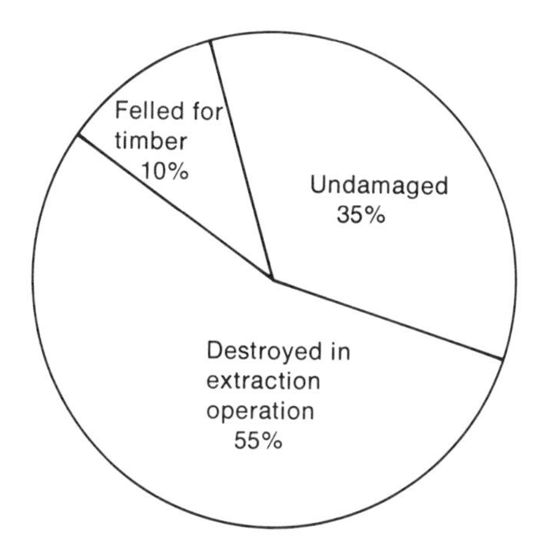

FIGURE 7. (Left) The effect of logging around 1970 on a Malayan lowland dipterocarp rain forest. (From Whitmore, 1975) (Below) Land development, Jerangau State Land Forest, Trengganu, Malaya. (Photo by Ding Hou)

are formed by man instead of by nature. As logging becomes more intensive and bigger gaps are created, the regenerating forest increasingly consists of pioneer species which have well-dispersed seeds or seeds which lie dormant in the soil and, thus, are able to colonize and grow in open conditions. The result is the development of large, structurally uniform, even-aged stands of a few species. This is similar to the effects of natural widescale destructive processes such as cyclones or, in some forests, fire. But modern, heavy extractive machinery and logging methods also cause substantial damage to the soil surface. Besides erosion, extensive compaction delays regrowth of trees and may limit regeneration to a few species. Furthermore, intensive logging over large areas reduces in number or wholly eliminates seed trees. The whole class of species whose seeds have limited dispersability and no dormancy may disappear or be reduced in diversity. Many climax forest trees are in this class.

It can be seen that intensive logging reduces diversity in several different ways. In so far as rain forest animals have highly specialized requirements (Chapter 2), their diversity will also decrease. There is an overall loss of potential for future economic exploitation. There is also a loss of those most complex ecosystems which scientists are only beginning to investigate and which, because they are the most intricate systems ever to have existed on earth, have enormous value to the science of ecology.

The most extreme form of disturbance is the removal of the above-ground biomass for wood chips. The consequences of this include the loss of species diversity and the removal of a high proportion of the total nutrient capital (known to be sequestered mostly in the plants, not the soil, in tropical rain forest). On a local scale, rain forest removal has substantial environmental consequences—mainly by disrupting the hydrological cycle. Rainfall runs off much more quickly; erosion and downstream siltation follow. Richardson (1976) gives this topic a full discussion, though no thorough scientific study of the impact of this new technology has yet been made.

Concern has been expressed that the widespread removal of tropical rain forest could have an effect on the Earth's climate. The likely consequences are discussed by Poore (1976); only a summary is given here. Rain forest has a low albedo and its replacement by another vegetation cover or bare earth would increase reflectivity and lead to an alteration in the heat balance. This would have surprising regional climatic consequences. The extreme case would be global cooling and a decrease in precipitation between 45° and 80°N and 40° and 60°S latitudes (Potter et al., 1975).

Destroying tropical forest would have complex effects on several different parts of the carbon cycle, the details of which are difficult to ascertain. Burning trees would increase atmospheric carbon dioxide, but the oceans act as a gigantic buffer for CO_2. It has been calculated that the burning of all the world's tropical rain forest over the next fifty years would release CO_2 at double the present rate for fossil fuels, assuming new trees do not grow to refix the CO_2. The resulting increase in air temperatures could be important, for example, on the northern margins of the northern hemisphere's wheat belt and could have important agricultural consequences, but the effect is likely to be confused with secular climatic changes from other causes, still not understood. (The suggestion, sometimes made, that atmospheric oxygen levels would be lowered by removal of tropical rain forest rests on a mistaken view of the nature of climax ecosystems.)

The Problem Summarized

Man's current degree of interference with vegetation throughout the world, perhaps at its most acute in the case of tropical rain forest, is comparable to the dramatic upheavals of the geological past, such as the great mountain building period of the Tertiary. But the unprecedented speed and scale of the present disturbance puts a novel strain on the ability of ecosystems to recover. It is tempting to assume, like Mr. Micawber, that something will turn up, but market forces do not work with biological systems. There is a law of the conservation of matter but no law of the conservation of species. A species, like a great art treasure, once destroyed is gone forever.

THE POLITICAL RESPONSE

Arguments about biological conservation fall broadly into three classes, each with a different kind of appeal, as has been well and fully argued by Heslop-Harrison (1973; 1974). First, conservation can be seen as a matter of conscience. Does man's power give him the right to destroy other species? Before we dismiss this ethical appeal as a middle class Western luxury, let it be remembered that reverence for life lies at the heart of the great philosophical and religious systems in the East, and motivates the actions of millions living on the margins of starvation (Heslop-Harrison, 1975; Marshall, 1978). Second, conservation is concerned with the quality of life. By destroying the natural environment in which he evolved, does man provoke uncontainable psychological and sociological tensions? The third class of argument is bluntly materialistic and impinges on economics. It asserts simply that man's dependence on other organisms and especially upon plants is such that unless attack on them is moderated, man's own continued existence is threatened. Appeal to the last argument is most likely to carry weight with hard-pressed decision makers.*

Countries which have, until recently, had huge tracts of untouched tropical rain forest, will soon have a mosaic of land uses like that found in developed countries. The issue is not whether this will happen, but how. It must be accepted as axiomatic that the inevitability of human population growth will increase demands for material goods, including forest products. But land developments should be subject to environmental impact statements in which the costs and benefits are thoroughly exposed. That is to say that rational land use policy is essential. Intensive agriculture should be concentrated on the most suitable land, for example the seasonally flooded Varzea forests of the Amazon and the basalt-derived soils of the tropical Far East.

Timber production from plantations will become increasingly important, although plantations will not replace production from natural and semi-natural forest. Plantations take the pressure off remaining areas of natural forest. The aim should always be to establish them on degraded sites, not to use them to replace remaining rain forest. The high yields per unit area in plantations make feasible expenditures on pruning, thinning, fertilizing, tree selection and breeding.

TROPICAL RAIN FOREST CONSERVATION

The international scientific community has legitimate cause to discuss and develop the biological framework for the conservation of rain forest, drawing on collective expertise in such fields as population biology and island biogeographic theory. Science does not recognize national frontiers. The interplay of minds is global, and valuable contributions to the debate are being made by scientists within the tropical rain forest belt (Gomez-Pompa et al., 1972; Jong et al., 1973). We will, therefore, conclude by focusing exclusively on biological questions and enquire how to conserve in perpetuity the full ecological genetic diversity of forests to make the most of future options. The only realistic answer must be to conserve adequate habitat as I have argued elsewhere (Whitmore, 1975). "Adequate" means the protection of both sufficiently large and sufficiently representative samples of virgin rain forest. The size of national parks is especially critical for those animals which occur at very low density, like the hornbill, the tapir and the tiger; or which have large territories, like the gaur and the elephant (Chapters 3, 6 and 7).

Numerous plant species also occur at very low density. For example, the density of wild fruit trees and their relatives in Malaya was only 3.7

*See Chapters 18 and 19 for contrasting approaches to the politics of conservation (eds.).

to 29 individuals over 0.3 m girth per species on a 10 hectare sample (Table III). Little is known about the genetics and population structure of these trees, but even if we employ Franklin's (Chapter 8) absolute minimum of 500 individuals as the number needed to maintain present genetic diversity into perpetuity, it would appear that a reserve of less than a few square kilometers would gradually become depauperate. For a given species (of plant at least) the maintenance of genetic diversity and its continuing evolution is maximized in a large reserve spanning the full range of environments. It is less likely in a similar area made up of discrete parcels. The reason is that pollination and dispersal of most climax rain forest species probably occur only over short distances, and therefore gene flow between distinct ecotypes is precluded where patches of forest are isolated.

TABLE III. Abundance of fruit trees in a Malayan lowland rain forest.

Species*		a†	b†	c†
Artocarpus	(8 spp.)	58	7.3	6,900
Baccaurea	(8 spp.)	75	9.4	5,300
Durio	(3 spp.)	44	14.7	3,400
Garcinia	(10 spp.)	37	3.7	13,500
Lansium	(2 spp.)	58	29	1,700
Nephelium + *Xerospermum*	(6 + 3 spp.)	44	4.9	10,000
Mangifera	(8 spp.)	39	4.9	10,200
Parkia	(2 spp.)	9	4.5	11,100
Sandoricum sp.		8	8	6,300

* Not all species of these genera have potential for human consumption.

† (a) individuals over 0.3 m girth of each genus on a 10 ha sample, hence (b) average number individuals of each species and by extrapolation (c) area (ha) containing 5000 individuals of each species.

(Pasoh forest data of Ashton from Jong et al. 1973)

Tropical rain forest contains many species with undeveloped economic potential and others whose genetic diversity has been only partly tapped. For the majority there is no practicable alternative to conservation *in situ,* as Thompson (1975) has argued. It should be remembered that there are important differences between the conservation of wild forest plants and the conservation of primitive cultivars of agricultural crops. In the first case conservation of adequate intact forest is all we need. In the second case the candidates for conservation are selected races of relatively few species and which exist due to man.

Ex situ conservation is fraught with the danger of accidental loss. The breeding system and population structure is disrupted and continuing evolution and adjustment to secular environmental change ceases. Storage of seed is expensive and is, at present, technically impossible for the (probably) large fraction of climax forest trees with recalcitrant seeds—unstorable by drying and chilling. Genetic damage can occur. *Ex situ* conservation in botanic gardens raises many similar objections. These can, at best, be living museums, maintaining alive a tiny fraction of a species' diversity—as specimens rather than populations.

Foresters are just starting to explore and utilize the diversity of a few important tropical and subtropical timber tree species. Some of the trees now being used for plantation growth are rain forest *Eucalyptus deglupta* (from Philippines to New Britain), *Araucaria cunninghamii* (from New Guinea, Queensland), *Agathis* spp. (from Sumatra to Fiji), *Cordia alliodora* (from Central and northern South America), a monsoon forest teak *(Tectona grandis)* and *Gmelina arborea* (both from south and continental Southeast Asia) and various species of *Pinus* (mainly in the New World). It has been discovered that the marginal populations adapted to extreme environments, often have silviculturally valuable traits for the sites available for forestry. But foresters are conservative and have yet to investigate many of the species which probably meet their silvicultural criteria. Their education by forest botanists is in hand (Vietmeyer, in press).

So far we have discussed inviolate conservation areas, big enough to allow natural interactions and processes to continue. Thought on the size and compass needed for these areas is still in its early stages. In addition, some facets of conservation are compatible with the utilization of forests, though no research has yet been undertaken on the management of tropical rain forest for multiple purposes, which indeed is a rather new concept in many temperate countries also.

A forest may be only lightly logged to retain a greater part of the population of the species being exploited and of those which would be destroyed by intensive logging. At the same time, light logging maintains part of the canopy so that the habitat remains suitable for the many arboreal animal species. But how much damage can plant and arboreal animal populations tolerate? As a first step research needs to be undertaken simply to monitor the effects of logging operations on populations.

Some animal species are favored by logging. For example the elephant, the gaur, and rhinoceros in Southeast Asia, will prosper on the lush, low growth found naturally on unstable land along rivers and in areas regrowing after logging. Fig trees, especially the banyan and strangling species which are among the largest trees in the eastern rain forests, are a major source of food for many mammals and birds. Most species fruit several times a year, so that at any one time at least one species is bearing. They are always left standing by loggers as survivors in the residual forest. Many casual observations suggest that opening up a forest stimulates more frequent fruiting of the remaining trees, including figs. Is this really so? If so, what effect on animals does the increased fig fruit biomass have? In some parts of the tropical rain forest primitive human societies continue to dwell as nomadic hunter gatherers to this day. These people live in extremely close integration with nature and know their tribal territory intimately (Dunn, 1975). Their knowledge is being increasingly sought by biologists and anthropologists. Their rights are commonly ignored when forests are carved up as logging concessions and they are forced to move elsewhere during logging. But when can they move back? In Southeast Asia pigs provide much of the protein for these people, and regenerating forests provide copious food for the pig. What management regime can capitalize on the timber of virgin forest for the benefit of national development and, at the same time, provide benefits for the nation's jungle dwellers? One possible solution is the concept of an inviolate virgin core of forest surrounded by an area managed partly for timber production and partly for wildlife conservation. This is the design proposed for rain forest Biosphere Reserves in UNESCO's Man and Biosphere program.

Apart from the unexplored biological problems, multiple use presents managerial difficulties. No single objective is maximized. There is a break with tradition if foresters are called upon in this new role, which gives them a conceptual problem involving, among other things, difficulties in applying conventional cost-benefit analysis (Steele, 1974). In sparsely staffed tropical countries, very real problems arise in policing a forest once it has been made accessible by roads, both during logging and afterwards.

Despite difficulties, multiple management clearly has possibilities as the area available for *in situ* conservation continues to diminish. Logged-over forest is an abundant commodity in many tropical countries.

Conservation of rain forest trees should focus on the numerous, diffusely distributed climax forest species. What are the dynamics of long-term succession in forests containing such species? Is a climax equilibrium state ever attained following modern logging, or do big catastrophes inexorably and naturally set the forest back to the species-poor pioneer composition? A few long-term monitoring plots in tropical rain forest are providing some of the answers to these questions. I end with a plea for more such studies of rain forest dynamics, by long-term observation of virgin forests and by careful observation of recovery after different, controlled degrees of perturbation.

SUGGESTED READINGS

Heslop Harrison, J., 1974, Genetic resource conservation: the end and the means, *Proc. Roy. Soc. Arts,* Feb. 1974, 157–69. The various arguments in justification of the conservation of plant species powerfully argued and the steps taken by 1973 reviewed.

Mason, B.J., 1977, Man's influence on weather and climate, *Proc. Roy. Soc. Arts,* Feb. 1977, 150–65. A dispassionate review in which fact is distinguished from several much publicized fictions.

Potter, G.L., H.W. Ellsaesser, M.C. MacCracken and F.M. Luther, 1975, Possible climatic impact of tropical deforestation, *Nature,* 258, 697–8. Unexpected changes revealed by manipulation of a numerical model of world climate.

Sommer, A., 1976, Attempt at an assessment of the world's tropical forests, *Unasylva,* 28, 5–25. The best yet published and difficult to fault in its assessment of destruction rates 1964–73.

Whitmore, T.C., 1975, *Tropical Rain Forests of the Far East,* Clarendon Press, Oxford. A synthesis of present knowledge (700 references) of the world's most studied rainforest region. Emphasizes new insights into the dynamics and the nature of variation in forest; i.e., the biological background to rational conservation.

CHAPTER 18

MANAGEMENT OF NATURE RESERVES

Robert Michael Pyle

A nature reserve can be thought of as an area of land which has effectively been removed from the development stream for the purpose of perpetuating natural conditions. Management, in this context, is the set of means by which that purpose is realized. Since nature reserves are the active elements of diversity conservation, their existence and management requires a rationale. For a relatively small number of cases this is an economic rationale. For the rest, conservation can be said to take place essentially outside the marketplace.

THE RATIONALE FOR NONMARKET CONSERVATION

Perhaps the oldest and most recurrent question in biological conservation is how can nonmarket allocation of natural resources be justified? A current manifestation of this question is the intense public debate over the welfare of endangered species versus public works projects, and wilderness protection versus timber supply. Ehrenfeld (1976) broadly considered the conservation of "non-resources," as he called natural features without conventional economic value. Ehrenfeld showed that these nonresources (such as endangered species) often have economic value, either real or potential. It is possible, therefore, to defend them with arguments reserved for resources, although Ehrenfeld felt that doing so is unreliable. Other biologists, such as Allen (1974), have suggested that using economic grounds for preservation is a necessity in our hardheaded economic climate. Other writers, Stone (1974), Nash (1973) and Douglas (1965), believe that nonresources present fully adequate and defensible reasons for conservation on their own merits, outside of the marketplace.

Indeed, these kinds of intrinsic value arguments have won many proponents and carry substantial weight in some political situations.

Elton (1958) perceived that it is helpful to employ a mixture of market and nonmarket rationales. Conservation, he stated, is justified "because it is a right relation between man and living things, because it gives opportunities for richer experience, and because it tends to promote ecological stability—ecological resistance to invaders and to explosions in native populations." Ehrenfeld does not believe that resource arguments are sufficiently defensible, especially for small or obscure species or natural features (he used the endangered Houston Toad as an example). He believes completely nonresource arguments, however, to be to tenuous as well, because they are not likely to appeal to economically motivated decision-makers. Hence, he supports Elton's mixed—"honest"—approach. In this way one need not rely exclusively to tenuous assertions of supposed economic worth, nor on thoroughly esthetic or moral principles which may be impotent.

There is another way of looking at market versus non-market values of things such as nature reserves—as strictly political management options, regardless of the underlying rationale. Authorities relegate each portion of land to a certain category according to its proposed use, in effect responding to relative political pressure. For example, much of the State of Kansas is now devoted to wheat and cattle farming; zoning and tax structures permit and encourage these uses. When political pressure becomes great enough for the establishment of a "Tall Grass Prairie National Park," the decision will be made by Congress to devote a portion of the Kansas landscape, largely or wholly, to that purpose. Many arguments, both economic and esthetic, will be traded, but the ultimate decision will be based upon relative pressure by constituents, interest groups and others.

More and more land allocation measures are being made in this way—for the most part heedless of strict, traditional resource economics (in the final analysis). Pressure bases, whether composed of dirt bikers, bird watchers, or wool growers, tend to be served when they grow large and loud enough. All of this suggests that decision-makers pay only token attention to the actual rationales profferred by interest groups—what they really hear are volume, votes and voices. This may not be the soundest system for deciding land allocation, but if ecological interests are adequately politicized and well represented in public forums, this method may offer at least as much to diversity conservation as a system whereby all decisions are made strictly on the basis of market considerations.

WHY MANAGE NATURAL AREAS?

If the above argument is to be believed, then setting aside a nature reserve is, in itself, the first management step. Beyond that, why should it be necessary to interfere with wholly natural processes on reserves? It may indeed prove unnecessary for areas of sufficient size and unassailability. However, even wilderness areas of many thousands of square kilometers will require active management to assure the preservation of their wildlife (Schoenfeld and Hendee, 1978; this volume, Chapters 1 through 10). This is especially critical considering the virtual extinction of predators by hunting. National parks are subject to other perturbations which call for manipulative management for restoration and maintenance, as indicated in the U.S. National Park Service Centennial Symposium, *Research in the Parks* (USDI, 1976). Most of the discussion in this book traces management problems in national parks or reserves to human impacts on ecosystems. Small reserves are even less elastic in this respect than large reserves. At the level of hectares (as opposed to square kilometers), management becomes even more necessary to retain or restore the conditions for which the reserve was set aside. Even in the short run, most of our smaller reserves, and many of the larger ones, are not self-contained and are subject to impinging factors from within and without, so that interference becomes necessary to achieve the protection objectives.

OBJECTIVES OF RESERVE MANAGEMENT

The goals of the reserve manager vary with the agency or organization doing the management. The U.S. Fish and Wildlife Service, proprietor of many thousands of hectares of wetlands and other habitats in the system of national wildlife refuges, has long been identified with the production of waterfowl. The agency's scope has broadened lately to incorporate concern for nongame species. Still, the most visible forms of management on the refuges are attempts at maximizing waterfowl habitat and numbers for the benefit of hunters.

Objectives are sometimes paradoxical. The U.S. National Park Service is instructed both to keep the parks "unimpaired for future generations" and to furnish "public pleasuring grounds." Many a management conundrum arises from this set of conflicting mandates, which are all but inimical when applied to sensitive areas. Managing nature reserves has commonly been frustrated in the past by the public's pleasure and its accomodation.

Most biologists agree that the central role of nature reserves is the protection and perpetuation of natural diversity. Indeed, this is the basic tenet of The Nature Conservancy, a nonprofit U.S. body which is the proprietor of the largest private set of nature reserves; it is also the assumption of the International Union for Conservation of Nature and

Natural Resources (IUCN). While this objective leaves room for creative variance and realistic interpretation, "preservation of diversity" sets a clear course for reserve management.

CRITERIA FOR MANAGEABILITY

The likelihood of maintaining the character of a reserve, given the existing threats and available tools, depends largely on its defensibility. This quality (defensibility) should be determined ahead of time as much as possible, so that the scarce resources for management are not applied to pieces of land that cannot be protected over time. The characteristics of defensibility are commonsensical, for the most part. The fewer and less intensive the hostile pressures from outside the reserve, the larger the area set aside, the lower the number of visitors and the friendlier the social climate in which the reserve exists, the greater its defensibility and manageability. Questions such as minimum population size and minimum reserve area are clearly applicable here; these are discussed abundantly in other chapters of this book. On occasion there may be no alternative to setting aside a less-than-manageable reserve, as in the case of the last stand of an endangered species. The assembly of the Redwoods National Park from remaining shards of the coastal redwood forest is such an example.

PRESELECTION OF NATURE RESERVES

Just as manageability can be enhanced by giving careful consideration to potential reserves, so can the significance of reserves gain from advance thought. There are those biologists who think we have no options. They think that all we have left to work with are shattered landscapes and that we should set aside whatever bits we can. They maintain that the choices have been made for us. This is often not true. Utterly undisturbed land may have nearly vanished, but a vast area of the earth retains enough of its natural condition to at least be considered for reservation.

The distribution of nature reserves is uneven, and this is another reason for preselection. For example, the distribution of nature reserves in Washington State bears little relationship to the biogeographical qualities of the region (Pyle, 1976). Using the butterfly fauna as an index of the biogeographical distinctness of regions in Washington, it was shown that there are sixteen butterfly provinces analogous to the biogeographical provinces of Dasmann (1973). Of the sixteen provinces, four were found in the northern mountainous part of the state. Heretofore, these mountainous regions have commonly been considered by managers and agencies to be quite similar in their nature. Because large areas had been set aside in the North Cascades National Park, the Pasayten Wilderness Area and the Glacier Peak Wilderness Area, the agencies and most conservationists thought that this sector of Washington bore adequate biological protection.

The data demonstrated that this is not true. Of the four distinct biogeographical (butterfly) provinces of Washington's northern mountains, the Nooksack Province had 36 percent of its area reserved for purposes of biological conservation; the Yakima-Methow Province, 13.77 percent; the Okanogan Province, 1.16 percent, and the Pend Oreille Province had only 0.19 percent of its area reserved. Similar disproportionalities were found in other parts of the state. Clearly, a more objective selection process would be useful. This has definite implications for management, in that a more equitable distribution of reserves would mean that scarce management funds would be used to sustain a broader array of diversity in the biogeographical context.

A number of systems have been promoted to assure a representative array of ecological reserves. These were summarized by Pyle (1976). Gressitt (1976) discussed such a system of natural area selection for the Pacific island region. The goal of all these approaches is to make sure that the reserves we set aside do indeed represent the widest possible array of organisms and communities, and under the highest quality conditions.

Perhaps the most advanced preselection system in the United States is the Natural Heritage Program of The Nature Conservancy (Sanders, 1978). Operating in concert with about half the states, these programs seek to classify "elements of natural diversity" and to identify the best remaining examples of each. The result is a "shopping list" for the Conservancy's acquisition program and a list of sites of concern for the use of other interested resource agencies and organizations. According to Jenkins (1978), "the purpose [of the Heritage Programs] is to organize what amounts to a dynamic atlas and data base of information on the existence, characteristics, numbers, condition, status, location, and distribution of occurrences of the elements of natural diversity. From such a data base we can finally select our land conservation objectives with rigor and conviction."

SPECIFIC PROBLEMS IN RESERVE MANAGEMENT

Any factor which reduces the natural diversity of a reserve is a problem. In fact, any threat to diversity is a potential problem. The major kinds of reserve stewardship challenges can be grouped as follows: (1) the maintenance of successional stages (Chapters 2 and 5); (2) the removal or mitigation of the effects of alien plants and animals; (3) the repair and prevention of vandalism, poaching, overuse, vehicle trespass and other human impacts; (4) meeting the effects of climatic and hydrologic changes; (5) intercepting or diminishing the effects of incompatible external influences such as air, water or noise pollution; (6) facing legal challenges such as reserved mineral extraction or water rights; (7) dealing with indigenous disease and infestation which would not be a problem on a large reserve, but which could radically alter a small one; (8) coping with the effects of past alterations—particularly soil changes—and the absence of key species such as predators; and (9) resisting further extinctions.

Management techniques which may be employed include fire, mowing, grazing, cautious use of herbicides, regulation of visitor activities, environmental education, public relations, manipulative public interpretation, litigation, restoration of communities, reintroduction and the culling or transfer of individuals. Lengthy discussions could accompany the mention of each of these problems and techniques; each has been met with or employed in many actual nature reserve situations. For smaller reserves, a helpful treatment of available methods may be found in the *Stewardship* manual of The Nature Conservancy (1978). Many of the means presented apply to volunteer as well as professional stewardship efforts. The emphasis is on practicum. Methods of nature interpretation useful in protecting natural areas open to visitors are discussed by Sharpe (1976). The impacts of visitors should be controllable according to how they are distributed over the natural area, and how well the visitors behave. Interpretation can be used to manipulate these factors while enhancing the visitors' experiences.

THE ROLE OF RESEARCH IN MANAGEMENT

Much of the foregoing deals with practical aspects of management. To what extent is reserve management actually conducted or postulated in scientific terms; how important is research to management of natural diversity?

Investigations such as those concerning grizzly bears in Yellowstone National Park (Craighead et al., 1974) and mountain lions in the Idaho Primitive Area (Seidensticker et al., 1973) indicate that rigorous research methods are indeed being applied to difficult reserve conflicts. The African game reserve experience has generated extensive literature dealing with resource use and preservation. It may be too early to tell if research is having a significant impact on reserve problem-solving, but Usher (1973) has attempted to review theory and practice in this area. His examples tend to corroborate the relative lack of sophistication in this field compared with some other areas of applied science, while documenting some of the more successful efforts to date.

Jenkins (personal communication) has pointed out that most reserve management situations are ruled by practical and economic considerations much more than by scientific design. Of 26 reserves currently managed by The Nature Conservancy in six northwest states, serious scientific inquiry is being applied to management in a meaningful way on only five reserves (so far). Nonetheless, the best hope for quality stewardship of diversity refuges still lies in scientific understanding. Ehrenfeld (1970) expressed it this way: "Before a natural community can be managed, its principle elements and their principal interactions must be known. At the least, a managed community should be characterized with respect to species composition, population sizes, and history of fluctuations of these two variables, whenever possible.... community management, to be successful, must strive to base itself on the maximum amount of ecological data, to be responsive to change, not be unduly influenced by rigid ecological theories and models, and must utilize the best historical information available."

It is idealistic to expect that Ehrenfeld's dictum can be met forthwith on many reserves; natural communities are and will continue to be managed in the absence of such information. Clearly, however, it is the proper direction to pursue. Management out of ignorance can conceivably damage resources more than utter negligence. Until we obtain sufficient knowledge to manage all reserves in confidence, a mixture must prevail between common sense, intuitive decision and scientifically enlightened procedure.

Some developing nations are beginning to devote substantial research funding toward the ultimate protection of their natural heritage. Papua New Guinea stands out in its commitment toward these ends. For instance, studies were initiated in 1977 (Pyle and Hughes, 1978) to provide a sound basis for managing the remarkable insect fauna of Papua New Guinea for purposes of conservation and sustained use. Ten sites were suggested for wildlife management areas, and management parameters were proposed. This will lead to increasingly sophisticated experiments devoted to managing the most restricted species and diverse faunas in protected habitats.

Only in Great Britain, however, is the scientific management of nature reserves a broad, if not pervasive, reality. Many of Usher's (1973) examples are drawn from the British experience in fine-tuned management of often small, highly compartmentalized reserves. It is notable that in England a great deal of local reserve management is done by volunteers organized in County Naturalists' Trusts. The volunteers carry out the selection and acquisition of compact reserves and administer their stewardship. Professional scientists in the government service preside over a similar sequence on a national scale. The Nature Conservancy Council is responsible for managing the National Nature Reserves (NNR) of Britain, based largely on data obtained by investigators of the

Institute of Terrestrial Ecology (ITE)—both functions were formerly situated within one body, the Nature Conservancy. Steele and Welch (1973) and Duffey (1971) gave detailed summations of the management of two of the National Nature Reserves, Monks Wood NNR and Woodwalton Fen NNR, respectively. The knowledge generated by extensive interdisciplinary research on these well known sanctuaries has been directly applied to the maintenance of their diversity in numerous specific instances.

The best review of the British approach is Duffey and Watt's (1971) *The Scientific Management of Animal and Plant Communities for Conservation.* In a paper from that symposium Morris (1971) demonstrated the use of interventionist and faunistic experiments to formulate management programs on nature reserves. Morris investigated the abundance and diversity of certain groups of invertebrates, notably leaf hoppers and weevils, under different regimes of management of chalk grassland communities. His findings, and those of others, suggest that the choice among grazing, mowing or burning of grasslands; the time of year in which reduction of biomass is carried out; and particularly, whether it is done continuously, annually or rotationally, can significantly influence the diversity and abundance of organisms on the grassland so managed. The results are complex and vary for different taxa sampled (for example, orchids vs. leafhoppers). Nonetheless, the data unequivocally point to rotational grazing by sheep as the probable best method of maintaining chalk grassland for overall biological richness and interest. Subsequently, the broad field of *Grassland Ecology and Wildlife Management,* as elucidated by British scientists of ITE, has been reviewed by Duffey et al., (1974).

According to Morris (1971), "There is no inherently right or wrong way to manage a nature reserve . . . the aptness of any method of management must be related to the objects of management for any particular site. . . . Only when objects of management have been formulated can results of scientific management research be applied. . . ."

SUMMARY

Nature reserve management is meaningless without clear direction, and meaningless too if it fails to do something different from what is done on nonreserved lands. The essential goal of diversity conservation may not be sufficient in itself, since different groups of organisms respond differently to a given management regime. Ultimately, it is necessary to know as much as possible about a reserve to know how it will work under various management prescriptions. The fine-tuning of a reserve—scientific management, in fact—becomes possible only inasmuch as resources and knowledge can be made available for the process.

It is harder to raise money for management than for acquisition of land, and we have already seen that information is rarely available in the form we would wish to see. The result is that scientific management seldom happens today. Even the British regard nature reserves as "those places where the details of an ultimately complex technology can be put into operation" (Morris, 1971), and in so saying, admit the preliminary nature of the practice.

In the meantime, nature reserve management in most parts of the world is largely an *ad hoc* business, much like fighting brush fires as they occur. The hope is that the principles and practices described elsewhere in this book will give us breathing space, and that by the time management has become a genuinely scientific endeavor, there will still be high quality nature reserves left to manage.

SUGGESTED READINGS

Duffey, E. and A.S. Watt (eds.), 1971, *The Scientific Management of Animal and Plant Communities for Conservation,* Blackwell Scientific Publications Ltd., Oxford. Numerous workers consider the conservation of selected populations through interventionist management.

Duffey, E., N.G. Morris, J. Sheail, L.K. Ward, D.A. Wells and T.C.E. Wells, 1974, *Grassland Ecology and Wildlife Management,* Chapman and Hall, London. Deals specifically with rare organisms and diversity of native or old grasslands. Updates some of the topics discussed in the above volume.

Schoenfeld, C.A. and J.C. Hendee, 1978, *Wildlife Management in Wilderness,* The Boxwood Press, Pacific Grove, California. Deals largely with game species and "unmanaged" lands classified as wilderness.

Usher, M.B., 1975, *Biological Management and Conservation: Ecological Theory, Applications and Planning,* Chapman and Hall, London. Merges contemporary theory with field problems to arrive at application goals and management planning concepts.

CHAPTER 19

THE STRATEGY OF CONSERVATION, 1980–2000

Paul R. Ehrlich

The condition of the world in the year 2000 will be in no small part determined by the success (or failure) of the conservation movement on two levels. The first is the tactical level, at which conservationists deploy their forces in an attempt to preserve both endangered species and endangered ecosystems. Much of this book has focused on tactics. They are largely the tactics of a small and beleagured army fighting rearguard actions on battlegrounds of the enemy's choosing. In the course of its long withdrawal, the conservationist army has become quite adept at harrying the advancing columns of the enemy and carrying out last-ditch defenses. It has honed its politico-legal weapons and made progress in designing its last redoubts–minimum-sized biological preserves and captive breeding programs.

But no army has ever been victorious solely on the basis of a continued high success rate in rearguard actions. For every snail-darter or Furbish lousewort victory–no matter how temporary–our forces suffer crushing, if unheralded, defeats as unknown populations and species are plowed under from Anaheim to the Amazon. The tactical successes of the conservation movement can be appropriately evaluated only against a backdrop of total and continuing strategic disaster. The basic forces leading to extinctions are continuously growing in strength; the enemies of the conservation movement are daily more formidable. For this reason, I would like to focus on the strategic situation, look at the medium-term prospects under a "business as usual" scenario and consider some other scenarios that might represent futures more compatible with the goals of the environmental movement.

SITUATION REPORT—END OF 1970s

The dominant facts of life on our planet are the rapidly increasing size of the population of one species, *Homo sapiens,* and a growing level of affluence of the average member of that species. This has resulted in that species co-opting a large and rapidly increasing share of the Earth's resource flows and, in the process, escalating its assault on the overall life-support systems of the planet. Humanity now turns some five percent of the global photosynthesis to its own use. It has, perhaps, tripled the natural riverborne flow of suspended solids. It mobilizes elements such as sulphur, potassium, iron, aluminum, lead, chromium, arsenic, cadmium, mercury, copper and zinc at rates of the same order of magnitude as those of natural flows. Industrial fixation of nitrogen is now about half that of natural sources of fixation. The burning of fossil fuels and the reduction of the Earth's biomass, primarily through the destruction of forests (especially in the tropics—Woodwell et al., 1978), is causing an increase in the CO_2 content of the atmosphere. This increase may induce rapid climatic changes, which would have catastrophic consequences for some agricultural and natural ecosystems.

Three key assertions can be made about this growing human impact on the biosphere. First, unless these trends can be reversed, the most ingenious tactics on the part of the conservation movement will, at best, slightly delay an unhappy end to the biotic Armageddon now underway. Second, without dramatic changes in the socio-political and, especially, the economic systems that dominate society today, these trends will not be reversed. And third, a nontrivial consequence of the failure to reverse these trends will be the disappearance of civilization as we know it.

POPULATION, RESOURCES, AND ENVIRONMENT IN THE UNITED STATES

Let us now examine these points more closely. The topic of human population growth has been confusingly presented to the public by the media in recent years, so much so that there exists the widespread impression (at least in the United States) that the population explosion is over. Unfortunately, this is not correct. Although the net reproductive rate in the United States has dropped to slightly below one, meaning that each generation is not quite replacing itself, the momentum of growth built into the age composition of the United States' population insures that it will continue to grow for some time. (Details on demographic statistics, projections and how age composition affects future growth rates can be found in Ehrlich et al., 1977.) The exact amount of future growth depends, of course, on what happens to age-specific birth and death rates. If they remain about where they are now, the United States' population will increase from slightly over 220 million in 1980 to about 250 million in the year 2000.

Even this modest projected increase of some 13 percent—based on the optimistic assumption that Americans will not return to larger family size preferences—bodes ill for conservation in North America. The United States has already consumed its easily available resources, cultivated its most suitable land and exploited most of the waste-dilution capacity of its running water. In attempting to maintain a flow of imported petroleum to the inefficient energy system serving only 220 million Americans, the U.S. government has weakened the dollar and generally threatened the financial stability of the Western world. The addition of 30 million Americans, roughly the equivalent of the combined populations of Sweden, Norway, Finland, Denmark and Austria, will make solving the energy, economic and ecological problems facing us at the turn of the century vastly more difficult.

The effects of the U.S. and the world population growth interacting with a "cowboy" economic system (Boulding, 1966) can be seen clearly in Colorado, one of the few remaining states to be thought of as containing relatively unexploited ecosystems. Like other Western states, Colorado is the scene of intense mineral exploration as giant conglomerates are gearing up to supply the uranium, molybdenum and other minerals "needed" by the economy. The need to help the balance-of-payments deficit created by the high cost of imported oil is, for example, one excuse given by Amax Corporation for opening a new molybdenum mine in the state.

Amax is probably the most environmentally conscious mining company in the United States; their Henderson mine, in the Front Range west of Denver, is touted as a model of environmental planning. Nonetheless, extensive areas around the mine are covered by the tailings deposited behind huge dams; more land is occupied by roads, railroads and huge mills; and large amounts of energy must be imported into the area. Of course, every installation like the Henderson mine inevitably leads to the extinction of some populations of organisms and reductions in size of many others.

Amax, operating under the archaic Mining Law of 1872, now wishes to develop their new mine at Mt. Emmons, just outside of the picturesque old mining town of Crested Butte in the Gunnison Basin. This beautiful mountain area is now devoted primarily to "renewable" activities—ranching, recreation and biological research. The 50-year-old Rocky Mountain Biological Laboratory is near Crested Butte, and the town itself is a National Historical Site. Although it is promised (and expected) to be more "environmentally sound" than even their Henderson operation, the new

Amax mine will, nonetheless, involve immense tailings ponds at altitudes over 8,000 feet. At that altitude reclamation is at best problematical and dustlike tailings may be a perpetual source of airborne pollution. The ponds, which will contain hundreds of millions of tons of tailings retained by dams as high as 400 feet, will cover many square miles. Ball mills capable of pulverizing an entire mountain and miles of conveyor belts, access roads and the like will have to be built.

The initiation of this vast industrial project portends a rapid change of the present local rural society to a more or less urban society. Temporary housing will have to be provided for some 13,000 people during the construction phase and for some 8,000 to 10,000 additional people on a permanent basis, doubling the population of Gunnison County. In the view of many, this Amax operation will, by changing the nature of the local population, trigger a wave of mining and other exploitative activities in the area. Most of the mountains nearby are already covered with mining claims, and the mountains still bear the scars of mining operations of nearly a century ago—operations that were orders of magnitude smaller than those contemplated. The potential for a new mineral rush can be seen in the rate at which mining claims are being filed. From January 1 through August 21, 1978, 1,477 new claims had been filed in Gunnison County, 608 in the first three weeks of August.

The great rise both in local population and in mining and ore transport will result in greatly increased travel on local roads, including the network of dirt roads that laces the county. This traffic will undoubtedly have a negative impact on the elk, bear, mountain sheep and other large mammals of the area. Marmot colonies in one local valley, subject of a classic long-term study by Prof. K. B. Armitage of the Rocky Mountain Biological Laboratory and the University of Kansas, have already been badly disturbed by growing recreational use of the local valleys. A mining boom will end this work. Spreading dust from the dirt roads will also have deleterious effects on local insect populations.*

Why should all this worry conservationists? After all, research can be done anywhere, and no prominent endangered species are directly threatened by the proposed Amax operation (although the endangered alpine chickweed, *Stellaria irrigua* Bunge, might be forced to local extinction by related events). The company might be able to meet or better all legal requirements for environmental protection in the area. The answer is that conservationists should worry because there are very few areas left where one can do long-term field research in hopes of understanding how natural systems work and how *Homo sapiens* fits into them. We should also worry because the proposed Amax venture is not an isolated event; it is just one of numerous examples of exploitative operations that are rap-

*Note the similar observations by Coe (Chapter 16) of the rapid disappearance of all large herbivores in an area of 5000 square kilometers surrounding a new mining town (eds.).

idly chewing up the last wilderness and semiwilderness areas of the American West.

The chewing up of the West has been accelerating in recent years. Loggers have struggled to decimate the giant redwoods of Northern California and the forests of the Texas "Big Thicket" before the government could save them for national parks. The Navaho lands were torn up to make way for the Navaho power plant. Dams are everywhere following the Glen Canyon atrocity (the damming of the Grand Canyon was at least temporarily averted).

There is no sign of a letup. If, as seems likely, the extraction of petroleum from oil shales becomes economically viable, large areas of the Colorado basin will become the scene of intensive mining operations. The amount of resultant environmental destruction will be a function of the exact technology employed—especially whether the shale will be processed *in situ,* or whether it will first be mined and then the petroleum extracted, in which case huge amounts of tailings will be generated. Energy and water shortages are bound to continue to create the demand for dams (in addition to being used to generate hydroelectricity, water is required by most technologies for mobilizing energy.) In much of the West, areas with coal deposits seem doomed to be strip-mined, with the original biota going down the drain. Reclamation (to even a vague semblance of the original ecosystems) would seem problematical.

In short, with a business-as-usual scenario, it is clear that many of the remaining native ecosystems of the American West will be destroyed by the end of the century. Some areas in the highest mountains or in national parks will be preserved, and many other areas (pastures, farms, tree farms, "reclaimed" tailings ponds and strip mines, and so forth) will look "natural" to the untrained eye. The loss of species and populations, though, will be immense, although probably largely unnoticed except by professional biologists.

The effects of the loss may be considerably more noticeable as "public service" functions of ecosystems are degraded. The functions include helping to regulate climate, maintaining the quality of the atmosphere and of freshwater supplies, running nutrient cycles, disposing of wastes, producing and maintaining soil, pollinating crops, controlling the vast majority of potential pests of crops and diseases of human beings, providing various aesthetic values and maintaining a "library" of genetic information from which future crops, domestic animals, antibiotics, spices, drugs and tools for medical research can be drawn. Human technology could not begin to replace these public services on the scale required, even if we knew how. In most cases we do not.

It must be remembered that it is not just the decimation of large animals or known endangered species that threaten the systems that support our lives. More important, it is the ploughing under, paving over, strip mining, submerging under tailings, mowing, herbiciding, pesticiding and so forth of a myriad of unsung populations of unspectacular organisms that is the major threat.

The degree to which free environmental services will be compromised in the next two decades is difficult to predict. It seems unlikely that by the year 2000 the United States will be threatened with total environmental breakdown. But we will face some very serious problems. For instance, air quality will probably be badly degraded on a continent-wide basis, with smog spreading over the Rockies, the Great Basin and the desert Southwest, as well as intensifying in the already polluted East and West coasts and Midwest. The quantity and quality of available fresh water will certainly be greatly reduced; water rationing could become a fact of life for most American citizens. Both food and energy might become catastrophically expensive. And then, having ruined the remaining lightly exploited areas of the United States for temporary gain, the growthmanic society will be looking around for a way to do an encore.

From a conservation point of view, the destruction of the Western United States may not be quite finished in the year 2000. But for much of the biota, especially that occurring below the subalpine level, the job may be well on the way to completion. Population sizes will be reduced, ranges fragmented and habitat quality degraded through air, water and soil pollution so that more populations become "marginal" and have an increased probability of extinction. Some organisms, such as coyotes, deer, and alfalfa butterflies, will probably adapt well. Others, including many little known plants and invertebrates, will be forced to extinction. The sad situation now prevailing in much of California, the Great Plains and the East Coast will simply spread across the whole continent.

It should be noted that there is growing evidence that temperate zone insect populations may be subject rather frequently to extinction, especially as a result of short- and medium-term perturbation in the climate. Our research group at Stanford has observed a series of extinctions of populations of *Euphydryas* butterflies, some as a direct result of the dramatic California drought of 1975 to 1977, some apparently due to a gradual drying trend and increased climatic variability on the western fringe of the Great Basin and some without obvious cause (Ehrlich et al., 1975; Ehrlich et al., in preparation).

It is clear from these observations that the size of reserves required to maintain the diversity of herbivorous insects (and by analogy many other inconspicuous temperate zone organisms) may be greatly underestimated unless the "normal" patterns of population extinction and reestablishment are well understood and considered in the planning. It is also clear that the trends described above for the Western United States will be much more serious if the *Euphydryas* patterns are more general (as we believe them to be), than if most temperate populations are essentially permanent on an historical time scale.

POPULATION, RESOURCES, AND ENVIRONMENT OF LESS DEVELOPED COUNTRIES

Depressing as the prognosis may be for the United States by the year 2000, it is cheery when compared with that for the rest of the world. The population of the United States is relatively sparse, and exploitation of soil and other resources began comparatively recently. We have a strong environmental movement and (comparatively) strong environmental laws. If it is lucky, the United States will need only to accomodate about 13 percent growth in population in the next 20 years, and it has the financial and military clout to "export" a considerable part of the environmental impact of its activities. Most of the rest of the world is not so fortunate.

Population growth in the United States and other overdeveloped countries (ODCs) must be considered in developing conservation strategies, but it is not a major component of the problem as it is in less developed countries (LDCs). Characteristically, the human populations of LDCs still have net reproductive rates far above unity—each generation is much larger than the last, and their populations are therefore growing very rapidly. Many LDCs are still increasing their populations at 2.5 percent annually or faster. If this rate persisted, it would mean a doubling of these LCDs' populations every two decades or so.

At present, there are signs that population growth is slowing somewhat in many LDCs as well as in ODCs, which, of course, is a sign of hope. Since vital rates are labile, however, it is very difficult to predict the future course of population growth. But, because of the built-in momentum of growth, even optimistic projections hold little cheer for conservationists. At best, there will be a world population of nearly six billion people by the year 2000—assuming a catastrophic rise in the death rate does not occur. As if this nearly 50 percent increase by the turn of the century were not bad enough, it seems almost certain that, again barring catastrophe, civilization will have to find a way to support 8 to 12 billion people 100 years from now. These projections of 100 to 200 percent increases are based on the assumption that humanity has entered a phase of at least moderately rapid transition to a stationary population—that is, zero population growth (ZPG). They assume that the recent lowering of birth rates represents a permanent decrease in average family size throughout most of the world, and is not merely a temporary phenomenon.

It is important to remember that most of this enormous "minimum" population growth will have to be absorbed in the LDCs, many of which are tropical. Between 1975 and the year 2000, if projections hold, the ODCs and DCs (developed, but not overdeveloped nations—such as Portugal) will have increased by perhaps 10 to 20 percent, and by about 20 to 35 percent before 2075. In contrast, the LDCs as a group will have almost doubled in the last 25 years of the twentieth century; in a hundred years they will have three or four times as many people as they have today.

This potential for population growth in the LDCs has been widely recognized by biologists as having profound implications for those crucial reservoirs of organic diversity—the tropical forests—roughly a third of which have already been destroyed (Chapter 17). Raven (1977) estimated that, at best, only a half million of perhaps three million species of tropical organisms have ever been named, and that it is likely that there are a million waiting to be named in Latin America alone. Let us accept his estimate (which may well be conservative). It means that there are in Latin America two-thirds as many unknown species of organisms as there are species already described and named worldwide—unknown species whose potential or actual benefits to humanity cannot even be guessed. If, as expected, most of the remaining tropical forests of Latin America are destroyed by the end of the century, it is not preposterous to assume that three-quarters of that million unknown species will either be extinct by then, or be condemned to rapid disappearance as the diversity decays away in fragments of remaining habitat (Chapter 6). The loss of unique populations will, of course, be orders of magnitude higher (depending on unknown degrees of geographic variability in genetics and the operational definition of unique).

The changes that the decimation of tropical forests could bring are awesome to contemplate. Their destruction might contribute to climatic shifts that could produce famines and their sequelae, costing hundreds of millions or even a billion lives. As John Holdren put it (1978), "a CO_2/climate-induced famine killing a billion people before 2020 certainly cannot be ruled out." As always, the changes are difficult to predict because of uncertainties in both climatic models and data base. For example, both the precise rate at which tropical forests are now being cut and the potential impact of their destruction on the CO_2 content of the atmosphere are now in dispute. What seems indisputable is that the genetic "library" upon which *Homo sapiens* can draw for future crops (or germplasm for the improvement of present crops), antibiotics, drugs, experimental organisms and so on, will be vandalized, with colossal and irreplaceable losses. In short, the costs of the destruction of tropical forests in loss or modification of ecosystem services may well be more than individual areas, nations or even the entire globe can bear without a catastrophic loss of human life.

IS CONSERVATION A LOST CAUSE?

The picture of the world in the year 2000 under a business-as-usual scenario is hardly an attractive one. One envisages an accelerating disappearance of organic diversity as ODCs try to maintain their profligate ways and as the LDCs struggle to support their rapidly increasing populations. There almost certainly will be deleterious and irreversible changes beyond the extinction of species and the decay of genetic variability. Soils will be eroded away, valuable mineral resources irretrievably dispersed, liquid fossil fuels squandered as energy sources (rather than preserved for use as feedstocks and lubricants), wild rivers dammed and so on. The level of amenity for *Homo sapiens* as a whole will be greatly lowered, as will, in all probability, the average life expectancy.

And all of this will happen even if the conservation movement grows orders of magnitude stronger, as long as it focuses on tactics rather than strategy. Never forget what the attacks on the snail darter and the Furbish lousewort portend for the future. Each population, species, reserve and even zoo will, under the business-as-usual scenario, sooner or later stand in the way of some industrialist's or politician's version of "progress." There will be a Tellico Dam or its equivalent for each nonhuman organism and, indeed, for any human groups whose land and traditions are considered unimportant by those who control the bulldozers.

Is conservation, then, a lost cause? Is the destruction of the majority of the Earth's biota already entrained by the momentum built into population growth and the inertia in our socio-economic systems? Should we give up? I think not for two reasons. The first is somewhat less than cosmic. The tactics of conservationists obviously can delay and have delayed the wiping out of plant and animal populations. Delay will permit us and our descendants to enjoy more biotic diversity than otherwise would be possible. It will also provide more time for a miracle—someday even politicians, economists and engineers might wake up to the facts of life on Earth. There is always the cheering example of the unexpected rapid change of attitudes about family size in the ODCs to demonstrate that social revolutions need not take decades.

The second reason is somewhat more important. Rather than giving up, we can try to convert the conservation movement to a different, much more strategic outlook. It will not be an easy or pleasant task, because it will mean trying to make the movement even more intransigent in the face of already massive criticism that it "prefers dicky-birds to people" (or to "jobs" or "needed energy" or "progress" or what have you).

Sadly, too many conservationists are willing to compromise. You all know the syndrome–if we let you slaughter the snail darter today (or form a committee to determine its fate), will you wait a few years before you exterminate the Furbish lousewort? Conservationists who think it is "reasonable" to exterminate any species could learn a lot about principles from "unreasonable" conservationists like David Brower. They might also remember the words of anthropologist Claude Levi-Strauss, who commented that any species of bug is "an irreplaceable marvel, equal to the works of art which we religiously preserve in museums."

I would like to suggest a new strategy for conservation based on what might be described as the five "Iron Laws of Conservation":

1. In conservation there is only successful defense or retreat, never an advance–a species or an ecosystem once destroyed cannot be restored.

2. Continued human population growth and conservation are fundamentally imcompatible.

3. A growthmanic economic system and conservation are also fundamentally incompatible.

4. The notion that only the short-term goals and immediate happiness of *Homo sapiens* should be considered in making moral decisions about the use of Earth is lethal, not only to nonhuman organisms but to humanity.

5. Arguments about the aesthetic value of nonhuman life forms or their intrinsic interest, or appeals for compassion for what may be our only living companions in the universe, mostly fall on deaf ears. Conservation must be promoted as an issue of human well-being and, in the long run, survival.

The first law should be self-evident. It applies to natural ecosystems as well as individual species. Even if it were possible to reconstitute an ecosystem with precisely the original species composition, it would doubtless evolve differently from the original because the genetic constitution of the member populations would be different from the original.

The second law is obvious to ecologists and most conservationists. But the elementary notion that even the most conservation-minded human population will eventually destroy its habitat if it continues to grow, is still lost on many of the world's decision-makers (or, perhaps more accurately, they erroneously believe that many doublings of the present population will be required before significant impact occurs).

The third law is less widely understood than the second and, thus, is more in need of emphasis in the conservationist camp. Much of the destruction of the world's biota that we now see is attributable as much to the behavior of human populations as to their brute growth rate (for the technical details of assigning responsibility, see Chapter 12 of Ehrlich et al., 1977). For example, the work of political scientist Kathleen Foote Durham (personal communication) shows that a substantial portion of the destruction of the Peruvian part of the Amazon Basin is due to deliberate government and industry policies of exploitation, and is not directly connected with the need to feed and house Peru's burgeoning population. Similarly, Brazil's campaign to destroy Amazonia (an area of over five million square kilometers) uses the need to resettle people from the extremely poor areas of northeastern Brazil as a cover, but is in actuality a program based on mineral exploitation, large-scale cattle ranches and other grandiose schemes that hold little promise of helping the majority of Brazilians.

The links of population growth to the destruction of tropical forests are more direct than those to the pillage of the American West, for example. But, even if LDC populations doubled or tripled, a great deal of the remaining forests could still be preserved if the decision were made to do so. Along with the forests, the ways of life of their remaining indigenous peoples, which are now everywhere threatened (for example, Davis, 1977), could also be saved.

That economic growth rather than population growth *per se* is a potent anticonservation force is perhaps best seen outside of tropical forests. Amax's Mt. Emmons operation in Colorado has little or no connection with the real needs of a slowly growing United States population, beyond the temporary provision of a few thousand jobs (which could easily be supplied in other ways). Indeed, Amax spends a great deal of effort trying to expand the uses of molybdenum and to create a demand for it in order to provide an excuse for Amax's continued rape of the West.

Amax is a fine example of one of the most pernicious human institutions—a conglomerate which is, in essence, a huge clot of capital that must perpetually plunder the Earth in search of a growth rate of 10 percent or so annually. Such conglomerate behavior must be curbed and harnessed to serve genuine human needs (of which there are an abundance) before the biosphere is damaged beyond repair. In particular, the unnecessary and wasteful consumption of resources that typifies ODC economies needs to be controlled. But, such a change would require the will to change the course of society; to alter some of its most basic institutions. That transformation is not yet upon us.

The fourth and fifth laws, I think, hold the key (if there is one) to bringing about that transformation. Large numbers of human beings still are under the impression that their affairs are essentially disconnected from the fate of the rest of the biota. They think that a few jobs or an additional percentage point of profit are perfectly legitimate reasons for wiping a population or species off the face of the Earth. They are deluded enough to think that astro-turf and plastic redwoods can replace plant life and that animal life can be reduced to a few domestic and exploited species with no untoward consequences for humanity. They have no notion of the role micro-organisms (other than "germs") play in their lives. Many engineers doubtless think, as Abbey (1975) has suggested, that Earth should be planed perfectly smooth so that roads can be painted on wherever society wants them.

Even if they see some value in nonexploited species, many people are undisturbed by the prospect of forcing some of them to extinction. After all, they learned in high school or college that extinction is the final fate of all species and that new ones evolve to replace them. For example, Roger Starr (1978), a member of the editorial board of The New York Times, in an ignorant tirade against the snail darter ("four-inch fish") and the Furbish lousewort ("a few specimens of a wild snapdragon"), asserted that the Endangered Species Act "seems to be based on the erroneous notion that man alone endangers other species." He then informs us: "in fact, however, natural history is simply the record of the extinction of species as they fail to adapt themselves to the changing characteristics of the environment. Man himself is the evolutionary product of the endangerment of species. This process will not stop."

He is probably right—it won't stop until the reduction of organic diversity undermines civilization and the human capacity to destroy. Starr could hardly, as an official of a prestigious newspaper, be expected to have any grasp of rate problems. Otherwise, he might have realized that between now and the end of the century the extinction rate of higher vertebrates will probably be 40 to 400 times higher than over most of the history of life (Ehrlich et al., 1977). Rates for lower vertebrates, invertebrates and plants are more difficult to estimate, but are certainly much higher now than a few centuries ago and accelerating quickly in pace with habitat destruction. Recent extinction rates are far too rapid for the normal speciation processes that add to the pool of diversity to make up the loss. *Homo sapiens* may well bring the process of evolution to a virtual halt in the next century, especially in tropical forests (Chapter 9).

It is clear that the Starrs of our world will not hesitate to exterminate any organisms that stand in the way of their concept of progress as long as they believe there will be no unfortunate consequences for their own species. The task of the environmental movement is thus both clear and monumental: somehow to convince a substantial portion of human beings that their own fate is deeply intertwined with the snail darters and Furbish louseworts of our planet.

The fifth law requires little elaboration. After all, simply because someone thinks something unique is pretty or interesting should not prevent someone else from destroying it for a profit. What right does the snail darter have to exist, when by exterminating it a small galaxy of hamburger stands can be illuminated for perhaps a century?

WHAT SHOULD CONSERVATIONISTS DO?

Obviously, the rearguard action of the conservation movement will (and must) continue. But much more effort must go into putting the time thus bought to good use. The principal strategy should be to attack on other fronts—to attempt to change the fundamental assumptions of society about its relationship to Earth's biota and consequently to change its behavior.

It goes almost without saying that the conservation movement must join even more wholeheartedly in the population control movement. The truth of the old saying, "whatever your cause, it's a lost cause without population control," applies nowhere more dramatically than to the cause of conservation. Slowing and then stopping population growth should be followed by a long period when birthrates are kept slightly below death rates, resulting in a gradual decline of the human population to a level well below the long-term carrying capacity. This aspect of the strategy has a time scale of several centuries.

It is equally important to convince society that "useless" organisms like the snail darter and the Furbish lousewort are a part of a giant, complex, poorly understood apparatus that supplies free services crucial to the support of human life. People must come to understand that *Homo sapiens* has already launched a high-level assault on that apparatus, an assault that we continue to our mounting peril. Years ago Aldo Leopold said: "The first rule of intelligent tinkering is to save all the parts." Today we are throwing the parts away ever more rapidly and "tinkering" with sledgehammers. We know there is some redundancy in the apparatus, and that it will continue to function even after sustaining considerable damage. We also know that if it is not protected, sooner or later it will break down with catastrophic consequences. No one knows exactly when that breakdown will occur. Conventional economic wisdom advises us, in the absence of a certain date for collapse, to persist in behavior that involves dealing our life-support apparatus ever stronger blows. It is as if people are prying the rivets, one by one, from the wings of an airplane in which we all are riding. They refuse to stop unless we can prove that the removal of any given rivet will cause the wing to fail. One does not have to be an expert aeronautical engineer to devine the inevitable end to their activities. The snail darter and Furbish lousewort can be thought of as symbolic rivets in the structure of ecosystems.

To relieve pressure on ecosystems will require dramatic changes in attitude, even in countries like the United States (which is relatively environmentally conscious). Not only must population growth be halted and then reduced, but, especially in the United States, the rate of assault by the average individual on the environment must be cut way back.

The best available measure of that assault is per capita energy consumption, since the vast majority of environmentally deleterious activities are energy intensive. Fortunately, we already know how to slow the assault. It can be done by tapping one energy "source" that the United States has in great abundance—the potential for energy conservation. It has clearly been shown possible, even with projected population growth, to reduce the mid-1970's United States energy consumption of about 70 quads (1 quad = 10^{15} Btu) to something less than 60 quads by A.D. 2000 without serious disruption to society and with a doubling of real income (CONAES, 1978).

Such a reduction in energy use would allow our nation to eschew the two most undesirable centralized energy options. We could do without nuclear power, which represents a massive threat both to human life and the democratic system. Nor would we have to replace nuclear power with massive dependence on coal (which threatens both our lives and the health of ecosystems). We could turn instead to those ecologically relatively benign technologies that depend on renewable resources, especially those decentralized ones that conform to Holdren's Rule (Holdren, 1978) of depositing their external costs largely on those who also reap their benefits.

In accordance with all this, conservationists in the ODCs must join the nascent "sustainable society" movement. As population growth slows and eventually stops, there should be a simultaneous transition to a "steady-state economy," as championed by economist Herman Daly (1977). Such an economy would be characterized by no growth in material throughput or energy consumption. Materials would be recycled as much as possible, with mining and refining carried on only to replace losses. Quality of material and human capital would be maximized.

Economic growth, especially in the ODCs, is a major threat to Earth's life-support systems. There is no question that economic growth (like population growth) must one day come to an end. Our choice is whether to wait until an environmental collapse ends growth or to try to end it peacefully and beneficially—first in ODCs and then in LDCs—through social change. Obviously, all sane people would prefer the latter. But as long as the common wisdom argues that economic growth is still desirable, that all ore bodies must inevitably be mined and that energy supplies must always be expanded, then we can kiss the snail darter, the Furbish lousewort, the Sumatran rhino, the gorilla and the whales goodbye.

The problems are knottier in the LDCs where so much more population growth is inevitable and where more economic growth is patently desirable. This leads us to the stickiest wicket of all: the need for the redistribution of wealth within and between nations. No notion of political economy is so discredited by experience as the "trickle-down" theory of development–the idea that by making the rich even richer enough crumbs will trickle down to the poor to lift them out of poverty. Since making the rich even richer further threatens the global environment and thus the carrying capacity of Earth for *Homo sapiens,* it is clear that a real attempt to carve up the resource pie differently is imperative.

Many of the most critical ecosystems, such as the tropical forests, are primarily in the LDCs. It is as if the poor people of the world own the wings of our airplane and one of the things they feel they must do is pry the rivets from the wing one by one and sell them. It would clearly be worth the while of the first class passengers to share their wealth with the owners of the wings so that the poor would not have to sell the rivets to survive. For example, it would be in our self-interest to finance huge reserves of tropical forests, both paying administrative costs and compensating the LDCs for the revenues lost by foregoing the exploitation of the reserves.

It is beyond the scope of this chapter to describe in detail how the economic relations between ODCs and LDCs might be rearranged, but some ideas can be found in Daly (1978), Pirages and Ehrlich (1974) and Ehrlich et al. (1977). My own view is that the first step is for the United States to set an example, both by providing properly for all of its own people and by stopping its disgraceful and economically ruinous pressure on the world's dwindling oil reserves, other nonrenewable resources and renewable resources like tropical forests (which in many cases are being exploited in ways that tend to convert them into nonrenewable resources).

Conservationists can help by initiating an aggressive conservation strategy. To give a specific example, in addition to attempting to minimize damage to the biota in the vicinity of environmentally unsound hard-rock mining operations, they can also vigorously attack the basic piece of legislation encouraging such operations–the Mining Law of 1872. This law, designed to encourage mineral exploration, was appropriate for a century ago when the Western United States seemed still an unlimited wilderness. It allows mining to preempt all other uses of federal lands and (for a nominal fee) gives to anyone who finds valuable minerals on public lands the title to the land required to develop those minerals. There are no environmental safeguards. A successful campaign to modify the law so that development of mineral resources on public land would be prohibited except under the most stringent environmental constraints would do more to protect the biota of the West than a hundred "successful" tactical actions aimed at limiting damage from individual projects.

Similarly, if conservationists could engineer the passage of laws forbidding the import of tropical wood products and beef, this might do more to protect tropical forests than the temporary protection of many thousands of hectares of reserves. As long as there is the economic incentive to clearcut tropical forests for lumber (or woodchips) or to clear them to make pastureland for fast-food chain hamburgers-to-be, the ultimate fate of the forests is sealed.

A final caveat for conservationists: If an aggressive conservation strategy is adopted to go along with the aggressive tactics already in use, you will be confronted by the Starrs and Amaxes of the world—politicians, economists, engineers, developers and so on—asking you to be "reasonable," "responsible" and to make compromises. You will find yourself opposed by people—often intelligent, attractive, well-meaning people—who only want to keep on behaving in a manner that, after all, has been acceptable for a couple of hundred years. Remember always that these people are the enemy. No matter how well meaning, they are unknowingly out for your blood and that of your children and your children's children. That they are bent on their own destruction and that of their descendants makes them no less a threat to the world.

SUGGESTED READING

Ehrlich, P. R., A. H. Ehrlich and J. P. Holdren, 1977, *Ecoscience: Population, Resources, Environment,* W. H. Freeman, San Francisco. Basic background for this chapter can be found in this book.

ACKNOWLEDGMENTS

Gilbert: Food Web Organization and the Conservation of Neotropical Diversity

The support of the Organization for Tropical Studies, the N.S.F. and the Costa Rican Government made this synthesis possible. The author also acknowledges the key role of Daniel Janzen in evolutionary tropical ecology. Special thanks to S. Armbruster and G. Webster for making available an unpublished manuscript, to L. Moore and M. Oldfield for their careful and extensive editorial criticism of the manuscript and to the editors for their help. Finally, the author sincerely thanks Christine Gilbert and Cecilia Marlatt for typing various drafts of the manuscript.

Eisenberg: The Density and Biomass of Tropical Mammals

This chapter has resulted from the cooperation of many colleagues and students. The author is especially indebted to Drs. John Seidenstick-er and R. W. Thorington for advice and sharing data. The research efforts in the Neotropics were dependent on the group efforts of G. G. Montgom-ery, R. Rudran, M. Sunquist, N. Smythe, J. Robinson, P. August, K. Green, C. Brady and M. A. O'Connell. Thanks to one and all. Some por-tions of the research described were supported in part by grants from the World Wildlife Fund and the Smithsonian International Evironmental Science Program.

Diamond: Patchy Distributions of Tropical Birds

The author acknowledges the National Geographic Society and the Lievre Fund for support of field work in the Southwest Pacific.

Foster: Heterogeneity and Disturbance in Tropical Vegetation

The author thanks C. Augspurger and C. Philips and his graduate students for discussion and help in preparation of the chapter.

Wilcox: Insular Ecology and Conservation

The assistance of J. M. Diamond, M. E. Gilpin and M. E. Soulé in critically reading various versions of this chapter is gratefully acknowl-edged. K. S. Miller is thanked for her assistance in the compilation of biogeographic data. The author was supported by NIH grant 6M 07242.

Franklin: Evolutionary Change in Small Populations

Many have read and criticized the original manuscript, or have made important suggestions. They include: F. B. Christiansen, D. S. Falconer,

O. H. Frankel, K. Hammond, W. G. Hill, B. D. H. Latter, B. McGuirk, J. M. Rendel, A. Robertson and M. E. Soulé. The author is grateful to all of them, and especially so to K. Hammond for allowing me to include some data from his Ph.D. thesis.

Soulé: Thresholds for Survival: Maintaining Fitness and Evolutionary Potential

John W. Senner provided invaluable counsel, and he, Bruce Wilcox and Christopher Wills kindly read and criticized the manuscript. The author also wishes to thank Otto Frankel for turning his attention to genetic problems in conservation, and John Eisenberg and Ernst Mayr for their comments on the speciation of mammals and birds relevant to the last section of the chapter.

Goodman: Demographic Intervention for Closely Managed Populations

The author is grateful to T. Smith for making available his analysis of the Pribiloff fur seal life history and for his advice on this topic. Portions of this research were supported by NSF grant DEB 76-21681, and a grant from the Academic Senate of the University of California.

Conway: An Overview of Captive Propagation

The author thanks E. Dolensek, G. Schaller, S. Temple, and D. Western for helpful discussion and T. Foose and W. Henry for unpublished observations.

Benirschke et al.: The Technology of Captive Propagation

Support for the authors' research by USPH Grant #1-S08-RR 09002 is gratefully acknowledged.

Kleiman: The Sociobiology of Captive Propagation

The author's research has been supported by the Smithsonian Research Foundation and NIMH 27241.

Campbell: Is Reintroduction a Realistic Goal?

The author thanks T. Cade, R. Erickson and other participants of the workshop on reintroduction at the First International Conference for Conservation Research, 1978, La Jolla, California.

Coe: African Wildlife Resources

The author thanks Professor J. D. Skinner and Dr. J. Phillipson for valuable discussions during the preparation of this chapter. Above all, the author acknowledges a great debt of gratitude to many colleagues and friends on the African continent who, over 22 years, have stimulated and encouraged his fascination for African biology both in the field and in the laboratory.

Whitmore: The Conservation of Tropical Rain Forest
Thanks are due to Dr. A. G. Marshall and Mr. P. A. Stott who kindly commented on a draft of this chapter and to Dr. M. E. D. Poore and I.U.C.N. who have provided the stimulus and means for my studies in rain forest conservation.

BIBLIOGRAPHY

Abbey, E., 1975, *The Monkey Wrench Gang,* J. B. Lippincott, Philadelphia.

Abplanalp, H. A., 1974, Inbreeding as a tool for poultry improvement, in *First World Congress on Genetics Applied to Livestock Production,* Graficas Orbe, Madrid, Spain, pp. 897–908.

Ajayi, S., 1975, *Domestication of the African Giant Rat* (Cricetomys gombianu, *Waterhouse),* Department of Forest Resource Management, University of Ibadan, Nigeria.

Ali, S. and S. D. Ripley, 1968, *Handbook of the Birds of India and Pakistan,* Oxford University Press, Oxford.

Allen, R., 1974, Does diversity grow cabbages?, *New Scientist,* 71, 35.

Altmann, D., 1974, Beziehungen zwischen sozialer rangordnung und jungenaufzucht bei *Canis lupus* L., *Zool. Gart. N. F. Jena,* 44, 235–236.

Amadon, D., 1953, Avian systematics and evolution in the Gulf of Guinea, *Bull. Amer. Mus. Nat. Hist.,* 100, 399–451.

Anderson, J. F., 1976, Egg parasitoids of forest defoliating Lepidoptera, in *Perspectives in Forest Entomology,* J. F. Anderson and H. K. Kaya (eds.), Academic Press, New York, pp. 233–249.

Anonymous, 1977, World Directory of National Parks and Other Protected Areas. International Union of Conservation of Nature and Natural Resources (IUCN), Morges, Switzerland.

Arman, P. and D. Hopcraft, 1975, Studies on East African herbivores. I. Digestibility of dry matter, crude fibre and crude protein, *Brit. J. Nutr.,* 33, 255–304.

Armbruster, W. S. and G. L. Webster, in press, Pollination of two species of *Dalechampia* (Euphorbiaceae) in Mexico by euglossine bees, *Biotropica.*

Asibey, E. O., 1969, *Grass Cutter* (Thryonomys swinderianus) *as a Source of Bushmeat in Ghana,* Department of Game and Wildlife, Accra, Ghana.

Asibey, E. O., 1974, Wildlife as a source of protein in Africa south of the Sahara, *Biol. Conserv.,* 6, 32–39.

Asibey, E. O., 1977, Expected effects of land-use patterns on future supplies of bushmeat in Africa south of the Sahara, *Environ. Conserv.,* 4, 43–49.

Aubréville, A., 1949, *Climats, forêts et Desertifications de l'Afrique Tropicale,* Société d' Éditions Geographiques Maritimes et Coloniales, Paris.

Avery, P. J., 1978, The effects of finite population size on models of linked overdominant loci, *Genet. Res. Camb.,* 31, 239–254.

Bainbridge, W. R. and A. Hammond-Tooke, 1967, *Report on the Luangwa Game Cropping Scheme. The 1966 season: recommendations for the 1967 and 1968 seasons and an economic analysis,* Game and Fisheries Department, Chilanga, Zambia.

Bawa, K. S., 1974, Breeding systems of tree species of a lowland tropical community, *Evolution,* 28, 85–92.

Beard, J. S., 1945, The progress of plant succession on the Soufriere of St. Vincent, *J. Ecol.,* 33, 1–9.

Beddington, J. R. and D. B. Taylor, 1973, Optimal age specific harvesting of a population, *Biometrics,* 29, 801–809.

Bekoff, M., 1977. Mammalian dispersal and the ontogeny of individual behavioral phenotypes, *Amer. Nat.,* 111, 715–732.

Benson, W. W., K. S. Brown and L. E. Gilbert, 1976, Coevolution of plants and herbivores: passion flower butterflies, *Evolution,* 29, 659–680.

Bentley, B. L., 1977, Extrafloral nectaries and protection by pugnacious bodyguards, *Ann. Rev. Ecol. Syst.,* 8, 407–427.

Bertram, B., 1976. Kin selection in lions and evolution, in *Growing Points in Ethology,* P. P. G. Bateson and R. A. Hinde (eds.), Cambridge University Press, Cambridge, pp. 281–301.

Bigalke, R. C. and W. O. Neitz, 1954, Indigenous ungulates as a possible source of new domesticated animals, *J. S. Afr. Vet. Assoc.,* 25, 45.

Billings, W. D., 1974, Environment: concept and reality, in *Vegetation and Environment,* B. R. Strain and W. D. Billings (eds.), W. Junk, The Hague, pp. 8–35.

Blaffer-Hrdy, S., 1976, The care and exploitation of nonhuman primate infants by conspecifics other than the mother, in *Advances in the Study of Behavior,* Vol. 6, J. S. Rosenblatt et al. (eds.), Academic Press, New York, pp. 101–158.

Blaffer-Hrdy, S., 1977, *The Langurs of Abu,* Harvard University Press, Cambridge, Massachusetts.

Blake, D. K., 1974, The rearing of crocodiles for commercial and conservation purposes in Rhodesia, *Rhod. Sci. News,* 8, 315–324.

Blake, D. K. and J. P. Loveridge, 1975, The role of commercial crocodile farming in crocodile conservation, *Biol. Conserv.,* 8, 261–272.

Bloom, A. L., 1971, Glacial-eustatic and isostatic controls of sea level since the last glaciation, in *The Late Cenozoic Ages,* K. K. Turekian (ed.), Yale University Press, New Haven, pp. 355–379.

Blower, J. H. and A. C. Brooks, 1963, Development and utilization of wildlife resources in Uganda, in *Conservation of Nature and Natural Resources in Modern African States,* G. G. Walterson (ed.), *IUCN Pub. N.S.* 1, pp. 50–101.

Bock, C. E. and L. W. Lepthien, 1976, Synchronous eruptions of boreal seed-eating birds, *Amer. Nat.,* 110, 559–571.

Bogart, M. H., A. T. Kumamoto and B. L. Lasley, 1977, A comparison of the reproductive cycle of three species of lemur, *Folia Primatol.,* 28, 134–143.

Bond, J., 1956, *Check-list of birds of the West Indies,* Academy of Natural Science, Philadelphia.

Bond, J. 1971, *Birds of the West Indies,* 2nd ed., Houghton Mifflin Co., Boston.

Bongso A. and P. K. Basrur, 1977, Bovine fetal fluid cells *in vitro:* fate and fetal sex prediction accuracy, *In Vitro,* 13, 769–776.

Bonnel, M. L. and R. K. Selander, 1974, Elephant seals: genetic variation and near extinction, *Science,* 184, 908–909.

Boughey, A. S., 1971, *Man and the Environment,* Macmillan, New York.

Boulding, K. E., 1966, The economics of coming Spaceship Earth, in *Environmental Quality in a Growing Economy,* H. Jarret (ed.), Johns Hopkins University Press, Baltimore, pp. 3–14.

Bouman, J., 1977, The future of Przewalski horses in captivity, *Inter. Zoo. Yearb.,* 17, 62–68.

Bourlière, F. and J. Verschuren, 1960, *Observation sur l'Ecologie des Ongules du Parc National Albert,* Exploration du Park National Albert, Brussels, Mission F. Bourlière.

Bourn, D., 1978, Cattle, rainfall and tsetse in Africa, *J. Arid. Environ.,* 1, 49–61.

Bowman, J. C. and D. S. Falconer, 1960, Inbreeding depression and heterosis of litter size in mice, *Genet. Res. Camb.,* 1, 262–274.

Brambell, M. R., 1977, Reintroduction, *Inter. Zoo. Yearb.,* 17, 112–116.

Brown, A. H. D., 1978, Isozymes, plant population genetic structure and genetic conservation, *Theor. Appl. Genet.,* 52, 145–157.

Brown, J. H., 1971, Mammals on mountaintops: nonequilibrium insular biogeography, *Amer. Nat.,* 105, 467–478.

Brown, J. H., 1978, The theory of insular biogeography and the distribution of boreal birds and mammals, *Great Basin Nat. Mem.,* 2, 209–227.

Brown, J. H. and A. Kodric-Brown, 1977, Turnover rates in insular biogeography: effect of immigration on extinction, *Ecology,* 58, 445–449.

Brown, K. S., 1972, Maximizing daily butterfly counts, *J. Lepid. Soc.,* 26, 183–196.

Brown, L. H., 1971, The biology of pastoral man as a factor in conservation, *Biol. Conserv.,* 3, 93–100.

Browning, J. A., 1975, Relevance of knowledge about natural ecosystems to development of pest management programs for agro-ecosystems, *Proc. Amer. Phytopath. Soc.,* 1, 191–199.

Brunig, E. F., 1973, Species richness and stand diversity in relation to site and succession of forests in Sarawak and Brunei (Borneo), *Amazoniana,* 4, 293–320.

Bulmer, M. G., 1976, The effect of selection on genetic variability: A simulation study, *Genet. Res. Camb.,* 28, 101–117.

Burgess, P. F., 1972, Studies on the regeneration of the hill forests of the Malay Peninsula. The phenology of dipterocarps, *Malay. Forester,* 35, 103–123.

Burnham, C. P., 1975, The forest environment: soils, in *Tropical Rain Forests of the Far East,* T. C. Whitmore (ed.), Oxford University Press, Oxford, pp. 103–120.

Bush, G. L., 1975, Modes of animal speciation, *Ann. Rev. Ecol. Syst.,* 6, 339–364.

Butler, G. P., 1978, *Realizing the Development Potential Created by an Iron Ore Mining Concession in Liberia,* Sabonah Printing Press, Monrovia, Liberia.

Carlquist, S. J., 1974, *Island Biology,* Columbia University Press, New York.

Case, T. J., 1975, Species numbers, density compensation and colonizing ability of lizards on islands in the Gulf of California, *Ecology,* 56, 3–18.

Caswell, H., 1978, A general formula for the sensitivity of population growth rate to changes in life history parameters, *Theor. Pop. Biol.,* 14, 215–230.

Cates, R. G. and G. H. Orians, 1975, Successional status and the palatability of plants to generalized herbivores, *Ecology,* 56, 410–418.

Cavalli-Sforza, L. L. and W. F. Bodmer, 1971, *The Genetics of Human Populations,* W. H. Freeman and Company, San Francisco.

Chaisson, R. E., L. A. Serunian and T. J. M. Schopf, 1976, Allozyme variation between two marshes and possible heterozygote superiority within a marsh in the bivalve *Modiolus demissus, Biol. Bull.,* 151, 404.

Charter, J. R., 1971, Nigeria's wildlife: a forgotten national asset, in *Wildlife Conservation in West Africa,* D. C. Happold (ed.), *IUCN Pub. N. S.,* 22.

Chasen, F. N., 1940, A handlist of Malaysian mammals, *Bull. Raffles. Mus.,* 15, 1–209.

Cheng, T., 1976, *Distribution of the Birds of China,* 2nd ed., Ko-hsueh chu pan she.

Child, G., 1970, Production and utilization and management in Botswana, *Biol. Conserv.,* 3, 19–22.

Chivers, D. J., 1969, On the daily behavior and spacing of howling monkey groups, *Folia Primat.,* 10, 48–102.

Cicmanec, J. C. and A. K. Campbell, 1977, Breeding the owl monkey *(Aotus trivirgatus)* in a laboratory environment, *Lab. Anim. Sci.,* 27, 512–517.

Clark, A. H., 1949, *The Invasion of New Zealand by People, Plants and Animals, The South Island,* Rutgers University Press, New Brunswick.

Clayton, G. A., G. R. Knight, J. A. Morris and A. Robertson, 1957, An experimental check on quantitative genetical theory: III. Correlated responses, *J. Genet.,* 55, 171–180.

Clegg, M. T. and R. W. Allard, 1972, Patterns of genetic differentiation in the slender wild oat species *Avena barbata, Proc. Natl. Acad. Sci. U.S.A.,* 69, 1820–1824.

Cloudsley-Thompson, J. L., 1977, *Man and the Biology of Arid Zones,* Arnold, London.

Clutton-Brock, T. and P. Harvey, 1977, Primate ecology and social organization, *Journal of Zoology,* 183, 1–40.

Cobb, S., 1976, *The Distribution and Abundance of the Large Herbivore Community of Tsavo National Park,* Kenya, Ph. D. Thesis, University of Oxford.

Coe, M. J., 1975, Mammalian ecological studies on Mount Nimba, Liberia, *Mammalia,* 35, 523–587.

Coe, M. J., 1976, The role of modern ecological studies in the reconstruction of paleoenvironments in sub-saharan Africa, in *Taphonomy and Vertebrate Palaeoecology with Special Reference to the Late Cenozoic in Subsaharan Africa,* Burg Wartenstein Symposium No. 35, Wenner-Grinn Foundation, New York.

Coe, M., D. Bourn and I. R. S. Swingland, in press, The biomass, production and carrying capacity of giant tortoises on Aldabra, *Phil. Trans. Roy. Soc. Lond. B.*

Coe, M. J., D. H. M. Cumming and J. Phillipson, 1976, Biomass and production of large African herbivores in relation to rainfall and primary production, *Oecologia,* 22, 341–354.

Coe, M. J. and K. Curry-Lindahl, 1965, Ecology of a mountain: First report on Liberian Nimba, *Oryx,* 8, 177–184.

Coetzee, J. A., 1964, Evidence for considerable depression of the vegetation belts during the Upper Pleistocene on the East African mountains, *Nature,* 204, 546–66.

Coetzee, J. A., 1967, Pollen analytical studies in East and Southern Africa, *Palaeoecolo. Afr.,* 3.

Cohen, J. E., 1977, Food Webs and the dimensionality of trophic niche space, *Proc. Natl. Acad. Sci. U.S.A.,* 74, 4533–4536.

Cole, L. C., 1954, The population consequences of life history phenomena, *Quart. Rev. Biol.,* 29, 103–137.

Comfort, A., 1962, Survival curves of some birds in the London Zoo, *Ibis,* 104, 115–117.

Comstock, R. E., 1977, Quantitative genetics and design of breeding programs, in *Proceedings of the International Conference on Quantitative Genetics,* E. Pollack, O. Kempthorne and T. B. Bailey (eds.), Iowa State University Press, Ames, pp. 705–718.

CONAES, 1978, Demand and Conservation Panel, U.S. energy demand: some low energy futures, *Science,* 200, 142–152.

Connell, J. H., 1978, Diversity in tropical rain forests and coral reefs: high diversity of trees and corals is maintained only in a nonequilibrium state, *Science,* 199, 1302–1310.

Conway, W. G., 1974, Animal management models and long-term captive propagation, *AAZPA Conference Proceedings,* 141–148.

Conway, W. G., 1978, A different kind of captivity, *Animal Kingdom,* 81, 4–9.

Cott, H. B., 1969, Scientific results of an inquiry into the ecology and economic status of the Nile crocodile *(Crocodylus niloticus)* in Uganda and Northern Rhodesia, *Trans. Zool. Soc. Lond.,* 29, 211–356.

Craighead, J. J. and F. C. Craighead, 1971, Grizzly bear—man relationships in Yellowstone National Park, *Bioscience,* 21, 845–857.

Crawford, M., 1974, The case for new domestic animals, *Oryx,* 12, 351–360.

Crawford, M. A. and S. M. Crawford, 1972, *What We Eat Today: The Food Manipulators vs. The People,* Stein and Day, New York.

Crow, J. F. and M. Kimura, 1970, *An Introduction to Population Genetics Theory,* Harper and Row, New York.

Crow, J. F. and N. E. Morton, 1955, Measurement of gene frequency drift in small populations, *Evolution,* 9, 202–214.

Curry-Lindahl, K., 1969, Report to the Government of Liberia on conservation, management and utilization of wildlife resources, *IUCN Pub. Ser. Supp. Pap.,* 24.

Curry-Lindahl, K., 1972*a, Conservation for Survival: An Ecological Strategy,* William Morrow, New York.

Curry-Lindahl, K., 1972*b, Let Them Live,* William Morrow, New York.

Czekala, N. M. and B. L. Lasley, 1977, A technical note on sex determination in monomorphic birds using faecal steroid analysis, *Inter. Zoo Yearb.,* 17, 209–211.

Dale, W. L., 1959, The rainfall of Malaya I., *J. Trop. Geogr.,* 13, 23–37.

Daly, H., 1977, *Steady State Economics: The Economics of Biophysical Equilibrium and Moral Growth,* W. H. Freeman and Co., San Francisco.

Darlington, P. J., 1957, *Zoogeography: The Geographical Distribution of Animals,* J. Wiley and Sons, New York.

Dasmann, R. F., 1964, *African Game Ranching,* Pergamon Press, London.

Dasmann, R. F., 1968, *Environmental Conservation,* John Wiley and Sons, Inc., New York.

Dasmann, R. F., 1973, A system for defining and classifying natural regions for purposes of conservation, *IUCN Occas. Pap.,* 7, 1–47.

Dasmann, R. F. and A. S. Mossman, 1960, The economic value of Rhodesian game, *Wildlife* (Nairobi), 2, 8–13.

Davis, S. H., 1977, *Victims of the Miracle,* Cambridge University Press, New York.

Delgado-Salinas, A. O. and M. Sousa-Sánchez, 1977, Biologia floral de genero *Cassia* en la region de Lox Tuxtlas, Veracruz, *Bol. Soc. Bot.·Mex.,* 37, 5–52.

Denniston, C., 1978, Small population size and genetic diversity: implications for endangered species, in *Endangered Birds: Management Techniques for Preserving Threatened Species,* S. A. Temple (ed.), University of Wisconsin Press, Madison, pp. 281–289.

Diamond, J. M., 1969, Avifaunal equilibrium and species turnover rates on the Channel Islands of California, *Proc. Natl. Acad. Sci. U.S.A.,* 64, 57–63.

Diamond, J. M., 1971, Comparison of faunal equilibrium turnover rates on a tropical and on a temperate island, *Proc. Natl. Acad. Sci. U.S.A.,* 68, 2742–2745.

Diamond, J. M., 1972*a, Avifauna of the Eastern Highlands of New Guinea,* Nuttall Ornithological Club, Cambridge, Massachusetts.

Diamond, J. M., 1972*b,* Biogeographic kinetics: estimation of relaxation times for avifaunas of southwest Pacific islands, *Proc. Natl. Acad. Sci. U.S.A.,* 69, 3199–3203.

Diamond, J. M., 1973, Distributional ecology of New Guinea birds, *Science,* 179, 759–769.

Diamond, J. M., 1975*a,* Assembly of species communities, in *Ecology and Evolution of Communities,* M. L. Cody and J. M. Diamond (eds.), Harvard University Press, Cambridge, Massachusetts, pp. 342–444.

Diamond, J. M., 1975*b,* The island dilemma: lessons of modern biogeographic studies for the design of natural preserves, *Biol. Conserv.,* 7, 129–146.

Diamond, J. M., 1976, Island biogeography and conservation: strategy and limitations, *Science,* 193, 1027–1029.

Diamond, J. M., in press, Why are many tropical bird species distributed patchily with respect to available habitat?, *Proc. 17th Intern. Ornith. Congr.*

Diamond, J. M. and A. G. Marshall, 1977, Distributional ecology of New Hebridean birds: a species kaleidoscope, *J. Animal Ecology,* 46, 703–727.

Diamond, J. M. and R. M. May, 1976*a,* Species turnover rates on islands: dependence on census interval, *Science,* 197, 266–270.

Diamond, J. M. and R. M. May, 1976*b,* Island biogeography and the design of nature reserves, in *Theoretical Ecology,* R. M. May (ed.), Saunders Co., Philadelphia, pp. 163–186.

Diamond, J. M. and E. Mayr, 1976, Species-area relation for the birds of the Solomon Archipelago, *Proc. Natl. Acad. Sci. U.S.A.,* 73, 262–266.

Dickerson, G. E., C. T. Blunn, A. B. Chapman, R. M. Kottman, J. L. Krider, E. J. Warwick and J. A. Whatley Jr., in collaboration with M. L. Baker, J. L. Lush and L. M. Winters, 1954, Evaluation of selection in developing inbred lines of swine, *Univ. Missouri Coll. Agr. Res. Bull.,* 551.

Dodson, C. H., 1966, Ethology of some bees of the tribe Euglossini, *J. Kansas Ent. Soc.,* 39, 607–629.

Dodson, C. H., 1975, Coevolution of orchids and bees, in *Coevolution of Animals and Plants,* L. E. Gilbert and P. H. Raven (eds.), University of Texas Press, Austin, pp. 91–99.

Doty, R. L., 1976, *Mammalian Olfaction, Reproductive Processes and Behavior,* Academic Press, New York.

Douglas, W. O., 1965, *A Wilderness Bill of Rights,* Little, Brown and Co., Boston.

Dressler, R. L., 1968, Pollination by euglossine bees, *Evolution,* 22, 202–210.

Drummond, B. A., 1976, *Comparative Ecology and Mimetic Relationships of Ithomiine Butterflies in Eastern Ecuador,* Ph.D. thesis, University of Florida.

Drury, W. H., 1974, Rare species, *Biol. Conserv.,* 6, 162–169.

Dudley, D., 1974, Contributions of paternal care to the growth and development of the young in *Peromyscus californicus, Behav. Biol.,* 11, 155–166.

Dudley, J. W., 1977, 76 Generations of selection for oil and protein percentage in maize, in *Proceedings of the International Conference on Quantitative Genetics,* E. Pollak, O. Kempthorne and T. B. Bailey Jr. (eds.), Iowa State University Press, Ames, pp. 459–473.

Dufau, M. L., C. Mendelson and K. J. Catt, 1974, A highly sensitive *in vitro* bioassay for luteinizing hormone and chorionic gonadotropin; testosterone production by dispersed Leydig cells, *J. Clin. Endocrinol. Metab.* 39, 610–613.

Duffey, E., 1971, The management of Woodwalton Fen: a multidisciplinary approach, in *The Scientific Management of Animal and Plant Communities for Conservation,* E. Duffey and A. S. Watt (eds.), Blackwell Scientific Publications, Ltd., London, pp. 581–597.

Duffey, E., 1977, The re-establishment of the large copper butterfly *Lycaena dispar batava* Obth. on Woodwalton Fen National Nature Reserve, Cambridgeshire, England, 1969–1973, *Biol. Conserv.,* 12, 143–158.

Duffey, E., N. G. Morris, J. Sheail, L. K. Ward, D. A. Wells and T. C. E. Wells, 1974, *Grassland Ecology and Wildlife Management,* Chapman and Hall, London.

Duffey, E. and A. S. Watt (eds.), 1971, *The Scientific Management of Animal and Plant Communities for Conservation,* Blackwell Scientific Publications Ltd., London.

Dunlap-Pianka, H., C. L. Boggs and L. E. Gilbert, 1977, Ovarian dynamics in heliconiine butterflies: programmed senescence versus eternal youth, *Science,* 197, 487–490.

Dunn, F. L., 1975, Rainforest collectors and trades: A study of resource utilization in modern and ancient Malaya, *R. Asiat. Soc. Malay. Br. Monogr.,* 5.

Eames, J., 1978, Why, hopefully, Kenya banned the curio trade, *Africana,* 6(8), 9–32.

Eanes, W. F., in press, The relationship between morphological variance and enzyme heterozygosity in the monarch butterfly, *Danaus plexippus, Nature.*

Eberhart, S. A., 1977, Quantitative genetics and practical corn breeding, in *Proceedings of the International Conference on Quantitative Genetics,* E. Pollack, O. Kempthorne and T. B. Bailey (eds.), Iowa State University Press, Ames, pp. 491–502.

Effron, M., H. H. Bogart, A. T. Kumamoto and K. Benirschke, 1976, Chromosome studies in the mammalian subfamily *Antilopinae, Genetica,* 46, 419–444.

Ehrenfeld, D. W., 1970, *Biological Conservation,* Holt, Rinehart and Winston, New York.

Ehrenfeld, D. W., 1976, The conservation of non-resources, *Amer. Scientist,* 64, 648–656.

Ehrlich, P. R., I. L. Brown, D. Murphy, C. Sherwood, M. C. Singer and R. R. White, in preparation, The crash of 1976: checkerspot butterflies and the California drought.

Ehrlich, P. R., A. H. Ehrlich and J. P. Holdren, 1977, *Ecoscience: Population, Resources, Environment,* W. H. Freeman and Co., San Francisco.

Ehrlich, P. R. and L. E. Gilbert, 1973, Population structure and dynamics of the tropical butterfly *Heliconius ethilla, Biotropica,* 5, 69–82.

Ehrlich, P. R. and P. A. Raven, 1969, Differentiation of populations, *Science,* 165, 1228–1232.

Ehrlich, P. R., R. R. White, M. C. Singer, S. W. McKenchnie and L. E. Gilbert, 1975, Checkerspot butterflies: a historical perspective, *Science,* 188, 221–228.

Eidt, R. C., 1968, The climatology of South America, in *Biogeography and Ecology in South America,* E. J. Fittkau, et al. (eds.), W. Junk, The Hague, pp. 54–81.

Eisenberg, J. F., 1967, A comparative study in rodent ethology with emphasis on evolution of social behavior, I, *Proc. U.S. Nat. Mus., 122, 1–51.*

Eisenberg, J. F., 1975, Phylogeny, behavior and ecology in the mammalia, in *Phylogeny of the Primates: An Interdisciplinary Approach,* W. P. Luckett and F. S. Szalay (eds.), Plenum Press, New York, pp. 47–68.

Eisenberg, J. F., 1977, The evolution of the reproductive unit in the class Mammalia, in *Reproductive Behavior and Evolution,* J. S. Rosenblatt and B. R. Komisaruk (eds.), Plenum Publishing Co., New York, pp. 39–71.

Eisenberg, J. F., in press, *a,* Biological strategies of living conservative mammals, in *The Physiology of Primitive Mammals,* K. Schmidt-Nielsen and W. Crompton (eds.), Cambridge University Press, England.

Eisenberg, J. F., in press, *b, The Mammalian Radiations, An Analysis of Trends in Adaptation,* University of Chicago Press, Chicago.

Eisenberg, J. F., in press, *c,* Habitat, economy and society: some correlations and hypotheses for the neotropical primates, in *Ecological Influences on Social Organization: Evolution and Adaptation,* I. Bernstein and F. Smith (eds.), Garland Press, New York.

Eisenberg, J. F. and D. Kleiman, 1977, The usefulness of behavior studies in developing captive breeding programs for mammals, *Inter. Zoo Yearb.,* 17, 81–89.

Eisenberg, J. F. and M. Lockhart, 1972, An ecological reconnaissance of Wilpattu National Park, Ceylon, *Smithsonian Contributions to Zoology,* 101, Washington, D.C.

Eisenberg, J. F. and G. M. McKay, 1974, Comparison of ungulate adaptations in New World and Old World tropical forests with special reference to Ceylon and the rainforest of Central America, in *The Behavior of Ungulates and Its Relation to Management,* Vol. 2, V. Geist and R. Walther (eds.), *IUCN Pub.* No. 24, pp. 585–602.

Eisenberg, J. F., N. A. Muckenhirn and R. A. Rudran, 1972, The relation between ecology and social structure in primates, *Science,* 176, 863–874.

Eisenberg, J. F., M. A. O'Connell and P. August, in press, Density and distribution of mammals in two Venezuelan habitats, in *Vertebrate Ecology in the Northern Neotropics,* J. F. Eisenberg (ed.), Smithsonian Institution Press, Washington, D.C.

Eisenberg, J. F. and J. Seidensticker, 1976, Ungulates in southern Asia: a consideration of biomass estimates for selected habitats, *Biol. Conserv.,* 10, 293–308.

Eisenberg, J. F. and R. W. Thorington Jr., 1973, A preliminary analysis of a neotropical mammal fauna, *Biotropica,* 5, 150–161.

Elton, C., 1927, *Animal Ecology,* Sidgwick and Jackson Ltd., London.

Elton, C. S., 1958, *The Ecology of Invasions by Animals and Plants,* Methuen and Co., Ltd., London.

Elton, C. S., 1975, Conservation and the low population density of invertebrates inside neotropical rainforests, *Biol. Conserv.,* 7, 3–15.

Eltringham, S. K. and R. C. Malpass, 1976, Elephant slaughter in Uganda, *Oryx,* 13, 334–335.

Emmel, T. C. and C. F. Leck, 1970, Seasonal changes in organization of tropical rain forest butterfly populations in Panama, *J. Res. Lep.,* 8, 133–152.

Estes, R. D., 1976, The significance of breeding synchrony in the wildebeest, *E. Afr. Wildl. J.,* 14, 135–152.

Falconer, D. S., 1960, *Introduction to Quantitative Genetics,* Oliver and Boyd Ltd., London.

Falconer, D. S., 1977, Some results of the Edinburgh selection experiments with mice, in *Proceedings of the International Conference on Quantitative Genetics,* E. Pollack, O. Kempthorne and T. B. Bailey (eds.), Iowa State University Press, Ames, pp. 101–116.

Feinsinger, P., 1976, Organization of a tropical guild of nectarivorous birds, *Ecol. Monogr.,* 46, 257–291.

Felsenstein, J., 1971, Inbreeding and variance effective numbers in populations with overlapping generations, *Genetics,* 68, 581–597.

Ffrench, R., 1973, A guide to the birds of Trinidad and Tobago, Livingston, Wynnewood, Pennsylvania.

Field, C. R., 1975, Climate and food habits of ungulates on Galana Ranch, *E. Afr. Wildl. J.,* 13, 303–320.

Fisher, R. A., 1958, *The Genetical Theory of Natural Selection,* 2nd ed., Dover Publications, Inc., New York.

Fittkau, E. J. and H. Klinge, 1973, On biomass and trophic structure of the Central Amazonian rainforest ecosystem, *Biotropica,* 5, 2–14.

Fleming, T. H., 1971, Population ecology of three species of neotropical rodents, *Misc. Pubs. Mus. Zool. Univ. Mich.,* No. 143.

Fleming, T. H., 1973, Numbers of mammal species in North and Central American forest communities, *Ecology,* 54, 555–563.

Fleming, T. H., 1975, The role of small mammals in tropical ecosystems, in *Small Mammals: Their Productivity and Population Dynamics,* F. B. Golley, K. Petrusewicz and L. Ryszkowski (eds.), Cambridge University Press, Cambridge.

Flesness, N., 1977, Gene pool conservation and computer analysis, *Inter. Zoo Yearb.,* 17, 77–81.

Foose, T., 1977, Demographic models for management of captive populations, *Inter. Zoo Yearb.,* 17, 70–76.

Foose, T., 1978, Demographic and genetic models and management for the okapi

(Okapia johnstoni) in captivity, *Acta Zoologica et Pathologica Antverpiensia,* 71, 47–52.

Ford, J., 1971, *The Role of the Trypanosomiases in African Ecology,* Clarendon, Oxford.

Forman, R. T. T., A. E. Galli and C. F. Leck, 1976, Forest size and avian diversity in New Jersey woodlots with some land-use implications, *Oecologia,* 26, 1–8.

Foster, R. B., 1973, *Seasonality of Fruit Production and Seedfall in a Lowland Forest Ecosystem in Panama,* Ph.D. thesis, Duke Univ., Durham, North Carolina.

Fowler, C. W. and T. Smith, 1973, Characterizing stable populations: an application to the African elephant population, *J. Wildl. Mngmt.,* 37, 513–523.

Frame, L. H., J. R. Malcolm, G. W. Frame and H. van Lawick, in preparation, Social organization of African wild dogs *(Lycaon pictus)* on the Serengeti plains.

Frankel, O. H., 1974, Genetic conservation; our evolutionary responsibility, *Genetics,* 78, 53–65.

Frankel, O. H. and M. E. Soulé, in press, *Conservation and Evolution,* Cambridge University Press, New York.

Frankham, R., L. P. Jones and J. S. F. Barker, 1968, The effects of population size and selection intensity in selection for a quantitative character in *Drosophila:* I. Short-term response to selection, *Genet. Res. Camb.,* 12, 237–248.

Frankie, G. W., H. G. Baker and P. A. Opler, 1974, Comparative phenological studies of trees in tropical wet and dry forests in the lowlands of Costa Rica, *J. Ecol.,* 62, 881–919.

Frazier, J., 1970, *Report on Sea Turtles in the Seychelles Area,* Animal Ecology Research Group, Department of Zoology, University of Oxford.

Frazier, J., 1976, Heritage of the sea, *Afr. Wildl. Leadership Found. News,* 11(2), 2–6.

Freese, C., 1976, Censusing *Alouatta palliata, Ateles geoffroyi,* and *Cebus capucinus* in the Costa Rican dry forest, in *Neotropical Primates: Field Studies and Conservation,* R. W. Thorington, Jr., and P. G. Heltne (eds.), National Academy of Sciences, Washington, D.C., pp. 4–9.

Fujino, K. and T. Kang, 1968, Transferrin groups of tuna, *Genetics,* 59, 79–91.

Fuller, J. L., 1969, The Genetics of behavior, in *The Behavior of Domestic Animals,* E. S. E. Hafez (ed.), Balliere, Tindall and Cassell Ltd., London, pp. 45–64.

Futuyma, D., 1973, Community structure and stability in constant environments, *Amer. Nat.,* 107, 443–446.

Galli, A. E., C. F. Leck and R. T. T. Forman, 1976, Avian distribution patterns in forest islands of different sizes in central New Jersey, *Auk,* 93, 356–364.

Garten, C. T., 1976, Relationships between aggressive behavior and genic heterozygosity in the oldfield mouse, *Peromyscus polionotus, Evolution,* 30, 59–72.

Garwood, N. C., 1979, Las limitaciones estacionales que afectan la aparicion de plantulas en un bosque humedo tropical en Panama, *Proc. IV Simp. Int. Ecol. Trop.*, pp. 579–591.

Garwood, N. C., D. P. Janos and N. Brokaw, in press, Earthquake-caused landslides: a major disturbance to tropical forests, *Science.*

Gause, G. F., 1934, *The Struggle for Existance,* Dover Publications, Inc., New York. (Originally published by Williams and Wilkins Co.)

Gentry, A. H., 1976*a,* Bignoniaceae of southern Central America: distribution and ecological specificity, *Biotropica,* 8, 117–131.

Gentry, A. H., 1976*b,* Endangered plant species and habitats of Ecuador and Amazonian Peru, in *Extinction is Forever,* G. Prance and T. Elias (eds.), New York Botanical Garden, Bronx, New York, pp. 136–149.

Gifford, D. P., 1976, Site taphonomy in East Lake, Turkana, Kenya, in *Taphonomy and Vertebrate Palaeoecology with Special Reference to the Late Cenozoic of Subsaharan Africa,* Burg Wartenstein Symposium, No. 69, Wenner-Grinn Foundation, New York.

Gilbert, L. E., 1972, Pollen feeding and the reproductive biology of *Heliconius* butterflies, *Proc. Natl. Acad. Sci. U.S.A.,* 69, 1403–1407.

Gilbert, L. E., 1975, Ecological consequences of a coevolved mutualism between butterflies and plants, in *Coevolution of Animals and Plants,* L. E. Gilbert and P. R. Raven (eds.), University of Texas Press, Austin, pp. 210–240.

Gilbert, L. E., 1977, The role of insect-plant coevolution in the organization of ecosystems, in *Comportement des Insectes et Milieu Trophique,* V. Labyrie (ed.), C.N.R.S., Paris, pp. 399–413.

Gilbert, L. E., 1979, Development of theory in the analysis of insect-plant interactions, in *Analysis of Ecological Systems,* D. Horn, R. Mitchell and G. Stairs (eds.), Ohio State University Press, Columbus.

Gilbert, L. E. and M. C. Singer, 1975, Butterfly ecology, *Ann. Rev. Ecol. Syst.,* 6, 365–397.

Gilbert, L. E. and J. T. Smiley, 1978, Determinants of diversity in phytophagous insects: host specialists in tropical environments, in *Diversity of Insect Faunas,* L. A. Mound (ed.), Blackwell Scientific Publications, Ltd., London, pp. 89–104.

Gilpin, M. E. and T. J. Case, 1976, Multiple domains of attraction in competition communities, *Nature,* 261, 40–42.

Gilpin, M. E. and J. M. Diamond, 1976, Calculation of immigration and extinction curves from the species-area-distance-relation, *Proc. Natl. Acad. Sci. U.S.A.,* 73, 4130–4134.

Glanz, W., 1978, Population ecology of the neotropical red-tailed squirrel, *Sciurus granatensis* in relation to food availability, Abstract 184, in *Abstracts of Technical Papers, 58th Annual Meeting of the American Society of Mammalogists,* Athens, Georgia.

Goldfoot, D. A., 1977, Rearing conditions which support or inhibit later sexual potential of laboratory monkeys: hypothesis and diagnostic behaviors, *Lab. Anim. Sci.,* 27, 548–556.

Gomez-Pompa, A., C. Vazquez-Yanes and S. Guevara, 1972, The tropical rain forest: a non renewable resource, *Science,* 117, 762–765.

Gomez-Pompa, A. and C. Vazquez-Yanes, 1976, Estudios sobre sucesion secundaria en los tropicos calidohumedos: el ciclo de vida de las especies secundarias, en *Regeneracion de Selvas,* A. Gomez-Pompa et al. (eds.), Compania Editorial Continental, Mexico, pp. 579–593.

Goodman, D., 1979, *Management Implications of the Mathematical Demography of Long Lived Animals,* Contract Report, U.S. Marine Mammal Commission.

Goodman, D., in press, Life history analysis of large animals, in *Dynamics of Large Mammal Populations,* C. W. Fowler and T. D. Smith (eds.), John Wiley and Sons, New York.

Gorman, G. L., M. E. Soulé, S. Y. Yang and E. Nevo, 1975, Evolutionary genetics of insular Adriatic lizards, *Evolution,* 29, 52–71.

Gottsberger, G., 1978, Seed dispersal by fish in the inundated regions of Humaita, Amazonia, *Biotropica,* 10, 170–183.

Gould, S. J. and N. Eldredge, 1977, Punctuated equilibria: the tempo and mode of evolution reconsidered, *Paleobiology,* 3, 115–151.

Graham, A., 1968, *The Lake Rudolf Crocodile* (Crocodylus niloticus *Laurenti) Population,* M. Sc. thesis, University of Nairobi.

Graham, A. and P. Beard, 1973, *Eyelids of Morning: The Mingled Densities of Crocodiles and Men,* New York Graphic Society, Greenwich, Connecticut.

Gressitt, J. L., 1976, Natural area (ecological areas) in the Pacific, in *Ecology and Conservation in Papua New Guinea,* K. P. Lamb and J. L. Gressitt (eds.), *Wau Ecol. Inst. Pamph.,* 2, pp. 86–89.

Guillery, R. W. and J. H. Kaas, 1971, Genetic abnormality of the visual pathway in a 'white' tiger, *Science,* 180, 1287–1289.

Gustavsson, I., 1969, Cytogenetics, distribution and phenotypic effects of a translocation in Swedish cattle, *Hereditas,* 63, 68–169.

Gwynne, M. D. and R. V. H. Bell, 1968, Selection of grazing components by grazing ungulates in the Serengeti National Park, *Nature,* 220, 390–393.

Haber, W. A., 1978, *Evolutionary Ecology of Tropical Mimetic Butterflies (Lepidoptera: Ithomiinae),* Ph.D. thesis, University of Minnesota.

Haffer, J., 1969, Speciation in Amazonian forest birds, *Science,* 165, 131–137.

Hagmeier, E. M. and C. Dexter Stults, 1964, A numerical analysis of the distributional patterns of North American mammals, *Syst. Zool.,* 13, 125–155.

Haines, B. and R. B. Foster, 1977, Energy flow through litter in a Panamanian forest, *J. Ecol.,* 65, 147–155.

Hale, E. B., 1969, Domestication and the evolution of behavior, in *The Behavior of Domestic Animals,* E. S. E. Hafez (ed.), Bailliere, Tindall and Cassell Ltd., London, pp. 22–44.

Hall, B. P. and R. E. Moreau, 1970, *An Atlas of Speciation in African Passerine Birds,* Trustees of the British Museum (Natural History), London.

Hammen, T. H. Van der and E. Gonzales, 1960, Upper Pleistocene and Holocene climate and vegetation of the "Sabanade Bogota" (Colombia, South America), *Leidse. Geol. Meded.,* 25, 261–315.

Hammond, K., 1973, *Population Size, Selection Response and Variation in Quantitative Inheritance.* Ph.D. thesis, Sydney University.

Hanrahan, J. P., E. J. Eisen and J. E. Legates, 1973, Effects of population size and selection intensity on short-term response to selection for post weaning gain in mice, *Genetics,* 73, 513–530.

Harlan, J. R., 1975, Our vanishing genetic resources, *Science,* 188, 618–621.

Hartshorn, G. S., 1978, Treefalls and tropical forest dynamics, in *Tropical Trees as Living Systems,* P. B. Tomlinson and M. H. Zimmerman (eds.), Cambridge University Press, New York, pp. 617–638.

Hassan, E., 1967, Untersuchungen über die Bedeutund der Krautund Strauchschicht Nahrungsquelle für Imagines Entomophagen Hymenopteren, *Z. Ang. Entomol.,* 60, 238–265.

Hearn, J. P., 1977, The endocrinology of reproduction in the common marmoset, *Callithrix jacchus,* in *The Biology and Conservation of the Callitrichidae,* D. G. Kleiman (ed.), Smithsonian Institution Press, Washington, D.C., pp. 163–171.

Heithaus, E. R., T. H. Fleming and P. A. Opler, 1975, Foraging patterns and resource utilization in 7 spp. of bats in a seasonal tropical forest, *Ecology,* 56, 841–854.

Hendrichs, H. and U. Hendrichs, 1972, *Dikdik und Elefanten: Ökologie und Soziologie zweier Afrikanischer Huftiere,* R. Piper and Co., Munich.

Henry, G. M. R., 1971, *A Guide to the Birds of Ceylon,* 2nd ed., Oxford University Press, London.

Heslop-Harrison, J., 1973, The plant kingdom: an exhaustible resource?, *Trans. Bot. Soc. Edinburgh,* 42, 1–15.

Heslop-Harrison, J., 1974, Genetic resource conservation: the end and the means, *Proc. Roy. Soc. Arts Feb.,* 157–169.

Heslop-Harrison, J., 1975, Man and the endangered plant, *Inter. Yearb.,* London, pp. xii–xvi.

Hill, R. E., 1977, Feeding trials on eland at Ol'morogi, *Afr. Wildl. Leadership Found. News,* 12, 7–10.

Hill, W. G., 1972, Estimation of genetic change: I. General theory and design of control populations, *Anim. Breed. Abst.,* 40, 1–15.

Hill, W. G., 1977, Selection with overlapping generations, in *Proceedings of the International Conference on Quantitative Genetics,* E. Pollack, O. Kempthorne and T. B. Bailey (eds.), Iowa State University Press, Ames, Iowa, pp. 367–378.

Hillman, K., 1978, Stock-taking 5: elephant survey in Kenya, *Africana,* 6, 18–22, 32–33.

Hills, H. G., N. H. Williams and C. H. Dodson, 1972, Floral fragrances and isolating mechanisms in the genus *Catasetum* (Orchidaceae), *Biotropica,* 4, 61–76.

Hirst, S. M., 1975, Ungulate-habitat relationships in a South African woodland savanna ecosystem, *Wildl. Monogr.,* 44, 1–60.

Hoage, R. J., 1977, Parental care in *Leontopithecus rosalia:* sex and age differences in carrying behavior and the role of prior experience, in *The Biology and Conservation of the Callitrichidae,* D. G. Kleiman (ed.), Smithsonian Institution Press, Washington, D.C., pp. 293–305.

Hoage, R. J., 1978, *Biosocial Development in the Golden Lion Tamarin,* Leontopithecus rosalia rosalia (Primates: Callitrichidae), Ph.D. thesis, University of Pittsburgh.

Holdren, J. P., 1978, Observations for the California Energy Futures Conference, Sacramento, May 20, 1978.

Hook, E. B. and W. J. Schull, 1973, Why is the XX fitter? Evidence consistent with an effect of X-heterosis in the human female from sex ratio data in offspring of first cousin marriages, *Nature,* 244, 570–573.

Hooper, M. D., 1971, The size and surroundings of nature reserves, in *The Scientific Management of Animal and Plant Communities for Conservation,* E. Duffy and A. S. Watt (eds.), Blackwell Scientific Publications Ltd., London, pp. 555–561.

Horn, H. A., 1974, The ecology of secondary succession, *Ann. Rev. Ecol. Syst.,* 5, 25–37.

Horwich, R. and D. Manski, 1975, Maternal care and infant transfer in two species of *Colobus* monkeys, *Primates,* 16, 49–74.

Howe, H. F., 1977, Bird activity and seed dispersal of a tropical wet forest tree, *Ecology,* 58, 539–550.

Hsu, T. C. and K. Benirschke, 1967–1977, *An Atlas of Mammalian Chromosomes,* Vols. I–X, Springer-Verlag, New York.

Hubbell, S. P., 1979, Tree dispersion, abundance and diversity in a tropical dry forest, *Science,* 203, 1299–1309.

Hughes, G. R., 1974, *The Sea Turtles of South East Africa. I. Status, Morphology and Distribution,* South African Association for Marine Biological Resources, Investigational Report No. 35.

Huntington, E. I., 1932, A list of the Rhopalocera of Barro Colorado Island Canal zone, Panama, with descriptions of two new species of Theclinae, *Bull. Amer. Mus. Nat. Hist.,* 63, 191–230.

Huygelen, C., 1955, Eland and their possible economic importance (Flem), *Bull. Agric. Congo. Belge.,* 46, 351.

Jackson, J. K., 1957, Changes in the climate and vegetation of the Sudan, *Sudan Notes Rec.,* 38, 47–66.

Jantschke, F., 1973, On the breeding and rearing of bush dogs, *Speothos venaticus,* at Frankfurt Zoo, *Inter. Zoo Yearb.,* 13, 141–143.

Janzen, D. H., 1971, Euglossine bees as long-distance pollinators of tropical plants, *Science,* 171, 203–205.

Janzen, D. H., 1973, Tropical agroecosystems, *Science,* 182, 1212–1219.

Janzen, D. H., 1974, Tropical black water rivers, animals and mast fruiting by the Dipterocarpaceae, *Biotropica,* 6, 69–103.

Janzen, D. H., 1975, *Ecology of Plants in the Tropics,* Arnold, London.

Janzen, D. H., 1977, The interaction of seed predators and seed chemistry, in *Comportement des Insectes et Milieu Trophique,* V. Labyrie (ed.), C.N.R.S., Paris, pp. 415–427.

Janzen, D. H. and T. W. Schoener, 1968, Differences in insect abundance and diversity between wetter and drier sites during a tropical dry season, *Ecology,* 49, 96–110.

Järvinen, O., in press, Dynamics of North European bird communities, in *Proc. 17th Intern. Ornith. Congr.*

Jarvis, C. E., 1967, Census of rare animals in captivity, *Inter. Zoo Yearb.,* 7, 358–377.

Jelliman, A. R., 1913, Game domestication in Rhodesia, *Rhod. Agric. J.,* 10, 719.

Jenkins, R. E., 1978, Heritage classification: the elements of ecological diversity, *The Nature Conservancy News,* 28, 24–25, 30.

Jenni, D. A., 1974, The evolution of polyandry in birds, *Am. Zool.,* 14, 129–144.

Johnson, A. M., 1976, The climate of Peru, Bolivia and Ecuador, in *Climates of Central and South America,* W. Schwerdtfeger (ed.), Elsevier, New York, pp. 147–218.

Jones, H. L. and J. M. Diamond, 1976, Short-time-base studies of turnover in breeding bird populations on the California Channel Islands, *Condor,* 78, 526–549.

Jones, T. C., R. W. Thorington, M. M. Hu, E. Adams and R. W. Cooper, 1973, Karyotypes of squirrel monkeys *(Salimiri sciureus)* from different geographic regions, *Am. J. Phys. Anthropol.,* 38, 269–277.

Jong, K., B. C. Stone and E. Soepadmo, 1973, Malaysian tropical forest: an underexploited genetic reservoir of edible-fruit tree species, in *Proceedings of the Symposium on Biological Resources and National Development,* Malaya Naturalist Society, University of Malaya, Kuala Lumpur, pp. 113–121.

Joubert, D. M., 1968, An appraisal of game production in South Africa, *Trop. Sci.,* 10, 200–211.

Karr, J. A., 1976, On the relative abundance of migrants from the north temperate zone in tropical habitats, *Wilson. Bull.,* 88, 433.

Kerr, M. A., V. J. Wilson and H. H. Roth, 1970, Studies on the agricultural utilization of semi-domesticated eland *(Taurotragus oryx)* in Rhodesia: 2. Horn development and tooth eruption as indicators of age, *Rhod. J. Agric. Res.,* 8, 71–77.

Keyfitz, N., 1968, *Introduction to the Mathematics of Populations,* Addison Wesley, Reading, Massachusetts.

Kimura, M., 1965, A stochastic model concerning the maintenance of genetic variability in quantitative characters, *Proc. Natl. Acad. Sci. U.S.A.*, 54, 731–736.

Kimura, M. and J. F. Crow, 1963, The measurement of effective population number, *Evolution,* 17, 279–288.

Kimura, M. and T. Ohta, 1971, *Theoretical Aspects of Population Genetics,* Princeton University Press, Princeton.

Kinzey, W. G. and A. H. Gentry, 1979, Habitat utilization in two species of *Callicebus,* in *Primate Ecology: Problem-Oriented Field Studies,* R. W. Sussman (ed.), John Wiley and Sons, New York, pp. 89–100.

Kleiber, M., 1961, *The Fire of Life,* John Wiley and Sons, New York.

Kleiman, D. G., 1977, Monogamy in mammals, *Quart. Rev. Biol.,* 52, 39–69.

Kleiman, D. G., 1978, The development of pair preferences in the lion tamarin *(Leontopithecus rosalia):* male competition or female choice, in *The Biology and Behavior of Marmosets,* H. Rothe, H. J. Wolters and J. P. Hearn (eds.), Eigenverlag Roth, Göttingen, pp. 203–207.

Kleiman, D. G., in press, Correlations among life history characteristics of mammalian species exhibiting two extreme forms of monogamy, in *Natural Selection and Social Behavior,* R. D. Alexander and D. W. Tinkle (eds.), Chiron Press, New York.

Kleiman, D. G. and C. A. Brady, 1978, Coyote behavior in the context of recent canid research: problems and perspectives, in *Coyotes: Behavior and Management,* M. Bekoff (ed.), Academic Press, New York, pp. 163–188.

Kleiman, D. G. and J. F. Eisenberg, 1973, Comparisons of canid and felid social systems from an evolutionary perspective, *Anim. Behav.,* 21, 637–659.

Kleiman, D. G. and D. S. Mack, in press, The effects of age, sex and reproductive status on scent marking frequencies in the golden lion tamarin, *Leontopithecus rosalia, Folia Primat.*

Klein, L. L. and D. J. Klein, 1976, Neotropical primates: aspects of habitat usage, population density, and regional distribution in La Macarena, Columbia, in *Neotropical Primates: Field Studies and Conservation,* R. W. Thorington Jr. and P. G. Heltne (eds.), National Academy of Sciences, Washington, D.C., pp. 70–78.

Koehn, R. K., F. J. Turano and J. B. Mitton, 1973, Population genetics of marine pelecypods: II. Genetic differences in microhabitats of *Modiolus demissus, Evolution,* 27, 100–105.

Koehn, R. K., R. Milkman and J. B. Mitton, 1976, Population genetics of marine pelecypods: IV. Selection, migration and genetic differentiation in the blue mussel *Mytilus edulis, Evolution,* 30, 2–32.

Koelmeyer, K. O., 1959, The periodicity of leaf change and flowering in the principal forest communities of Ceylon, *The Ceylon Forester,* 4, Part I, 157–189, Part KK, 308–364.

Kosuda, K., 1972, Synergistic effect of inbreeding on viability in *Drosophila virilis, Genetics,* 72, 461–468.

Kühme, W., 1965, Freilandstudien zur Soziologie des Hyänenhundes (*Lycaon pictus lupinsu* Thomas, 1902), *Z. Tierpsychol.,* 22, 495–541.

Kulman, H. M., 1971, Effects of insect defoliation on growth and mortality of trees, *Ann. Rev. Ent.,* 16, 289–324.

Kulovich, M. V., 1978, Cesarean delivery in a gorilla, *JAVMA,* 173, 1137–1140.

Kyle, R., 1972, *Meat Production in Africa–the Case for New Domestic Species,* Bristol Veterinary School, Bristol University, England.

Kyrkland, B., 1976, Paper from mixed tropical forests, *Unasylva,* 28, 86–92.

Lack, D., 1954, *The Natural Regulation of Animal Numbers,* Clarendon Press, Oxford.

Lack, D., 1976, *Island Biogeography Illustrated by the Birds of Jamaica,* University of California Press, Berkeley.

Lamprey, H. F., 1963, Ecological separation of large mammal species in the Tarangire Game Reserve, Tanganyika, *E. Afr. Wildl. J.,* 1, 63–92.

Lamprey, H. F., 1964, Estimation of the large mammal densities, biomass and energy exchange in the Tarangire Game Reserve and the Masai Steppe in Tanganyika, *E. Afr. Wildl. J.,* 2, 1–46.

Lande, R., 1976, The maintenance of genetic variability by mutation in a polygenic character with linked loci, *Genet. Res. Camb.,* 26, 221–235.

Lasley, B. L., N. M. Czekala, K. C. Nakakura, S. Amara and K. Benirschke, in press, Armadillos for studies of delayed implantation, quadruplets, uterus simplex and fetal adrenal physiology, in *Animal Models for Contraceptive and Fertility Research,* N. J. Alexander (ed.), Harper and Row, New York.

Latter, B. D. H., 1970, Selection in finite populations with multiple alleles: II. Centripetal selection, mutation and isoallelic variation, *Genetics,* 66, 165–186.

Latter, B. D. H. and A. Robertson, 1962, The effects of inbreeding and artificial selection on reproductive fitness, *Genet. Res. Camb.,* 3, 110–138.

Laurie, A. and J. Seidensticker, 1977, Behavioral ecology of the sloth bear *(Melursus ursinus), J. of Zoology,* 182, 187–204.

Laws, R. M., 1968, Interactions between elephant and hippopotamus populations and their environments, *E. Afr. Agric. For. J.,* 33, 140–147.

Laws, R. M., 1970, Elephants as agents of habitat and landscape change, *Oikos,* 21, 1–15.

Ledger, H. P., 1963, Animal husbandry research and wildlife in East Africa, *E. Afr. Wildl. J.,* 1, 1–12.

Leigh, E. G. Jr., 1975, Population fluctuations, community stability and environmental variability, in *Ecology and Evolution of Communities,* M. L. Cody and J. M. Diamond (eds.), Belknap Press, Cambridge, Massachusetts, pp. 51–73.

Leopold, A., 1933, *Game Management,* C. Scribners Sons, New York.

Leopold, A., 1949, *A Sand County Almanac,* Oxford University Press, London.

Leopold, A. S., 1966, Adaptability of animals to habitat change, in *Future Envi-*

ronments of North America, F. F. Darling and J. P. Milton (eds.), Doubleday, New York.

Lerner, I. M., 1954, *Genetic Homeostasis,* Oliver and Boyd Ltd., Edinburgh.

Leslie, P. H., 1945, On the use of matrices in certain population mathematics, *Biometrika,* 33, 213–245.

Leston, D., 1978, A neotropical ant mosaic, *Ann. Ent. Soc. Amer.,* 71, 649–653.

Leuthold, W., 1977, *African Ungulates: A Comparative Review of Their Ethology and Behavioral Ecology,* Springer-Verlag, Berlin.

Levin, D. A., 1976, The chemical defenses of plants to pathogens and herbivores, *Ann. Rev. Ecol. Syst.,* 7, 121–159.

Lewis, J. G., 1977, Game domestication for animal production in Kenya: activity patterns of eland, oryx, buffalo and zebu cattle, *J. Agric. Sci. Camb.,* 89, 551–563.

Lewis, J. G., 1978, Game domestication for animal production in Kenya: shade behavior and factors affecting the herding of eland, oryx, buffalo and zebu cattle, *J. Agric. Sci. Camb.,* 90, 587–595.

Lewontin, R. C., 1965, Selection for colonizing ability, in *The Genetics of Colonizing Species,* H. G. Baker and G. L. Stebbins (eds.), Academic Press, New York, pp. 77–94.

Lewontin, R. C., 1974, *The Genetic Basis of Evolutionary Change,* Columbia University Press, New York.

Lewontin, R. C., 1978, Adaptation, *Scientific American,* 239, 156–169.

Livingstone, D. A., 1952, Age of deglaciation of the Rwenzori range, Uganda, *Nature,* 194, 859–860.

Livingstone, D. A., 1975, Late quaternary climatic change in Africa, *Ann. Rev. Ecol. Syst.,* 249–280.

Longhurst, W. M. and H. F. Heady (eds.), 1968, *Report of a Symposium on East African Range Problems, Villa Servelloni, Lake Como, Italy, June, 1968,* New York, Rockefeller Foundation.

Lotka, A. J., 1939, A contribution to the theory of self-renewing aggregates, with special reference to industrial replacement, *Ann. Math. Stat.,* 10, 1–25.

Lugo, A. E., 1973, The impact of the leaf-cutter and *Atta colombica* on the energy flow of a tropical wet forest, *Ecol.,* 54, 1292–1301.

Lunn, S. F., 1978, Urinary oestrogen excretion in the common marmoset, *Callithrix jacchus,* in *The Biology and Behavior of Marmosets,* H. J. Wolters and J. P. Hearn (eds.), Eigenverlag Roth, Göttingen, pp. 67–73.

Lynch, C. B., 1977, Inbreeding effects upon animals derived from wild populations of *Mus musculus, Evolution,* 31, 526–537.

MacArthur, R. H., 1972, *Geographical Ecology,* Harper and Row, New York.

MacArthur, R. H. and E. O. Wilson, 1963, An equilibrium theory of insular zoogeography, *Evolution,* 17, 373–387.

MacArthur, R. H. and E. O. Wilson, 1967, *The Theory of Island Biogeography,* Princeton University Press, Princeton, New Jersey.

MacClintock, L., R. F. Whitcomb and B. L. Whitcomb, 1977, Island biogeography and "habitat islands" of eastern forest: II. Evidence for the value of corridors and minimization of isolation in preservation of biotic diversity, *Amer. Birds,* 31, 6–16.

Macdonald, G. A. and A. T. Abbott, 1970, *Volcanoes in The Sea: The Geology of Hawaii,* University of Hawaii Press, Honolulu.

Mackie, R. J., 1970, Range ecology and relations of mule deer, elk and cattle in the Missouri river breaks, Montana, *Wildl. Monogr.,* No. 20.

Mainardi, D., 1963*a,* Un esperimento sulla parte attiva svolta dalla femmina nella selexione sessuale in *Mus musculus, Arch. Sci. Bio.,* 47, 227–237.

Mainardi, D., 1963*b,* Speciazone nel topo, fattori etologici determinianti barriere reproduttive tra *Mus musculus domesticus* e *M. m. bactrianus, Instituto Lombardo, Rend. Sc.,* B., 97, 135–142.

Mainardi, D., 1963*c,* Eliminazione della barriere etologica all' isolamento riproduttivo tra *Mus musculus domesticus* e *M. m. bactrianus* mediante sull' apprendimento infantile, *Instituto Lombardo, Rend. Sc.,* B, 97, 291–299.

Majer, J. D., 1976, The maintenance of the ant mosaic in Ghana cocoa farms, *J. Appl. Ecol.,* 13, 123–144.

Manton, V. J. A., 1975, Captive breeding of cheetahs, in *Breeding Endangered Species in Captivity,* R. D. Martin (ed.), Academic Press, London, pp. 337–344.

Marlier, G., 1967, Hydrobiology in the Amazon Region, *Atas Simposio Biota Amazonica* 3 (Limnologia), pp. 1–7.

Marshall, A. G., 1978, Man and nature in Malaysia: attitudes to wildlife and conservation, in *Nature and Man in South East Asia,* P. A. Stott (ed.), School of Oriental and African Studies, London, pp. 23–33.

Martin, P. S., 1966, Africa and the pleistocene overkill, *Nature,* 112, 339–342.

Martin, P. S., 1967, Prehistoric overkill, in *Pleistocene Extinctions: The Search for a Cause,* P. S. Martin and H. E. Wright (eds.), Yale University Press, New Haven, pp. 75–120.

Martin, P. S., 1973, The discovery of America, *Nature,* 119, 969–974.

Martin, R. D., 1968, Reproduction and ontogeny in the tree-shrews *(Tupaia belangeri)* with reference to their general behavior and taxonomic relationships, *Z. Tierpsychol.,* 25, 409–495, 505–532.

Martin, R. D., 1975, Breeding tree-shrews, *Tupaia belangeri,* and mouse lemurs, *Microcebus murinus,* in captivity, *Inter. Zoo Yearb.,* 15, 35–41.

Mason, B. J., 1977, Man's influence on weather and climate, *Proc. Roy. Soc. Arts.,* 125, 150–165.

May, R. M., 1975, Patterns of species abundance and diversity, in *Ecology and Evolution of Communities,* M. L. Cody and J. M. Diamond (eds.), Harvard University Press, Cambridge, Massachusetts, pp. 81–120.

Mayr, E., 1963, *Animal Species and Evolution,* Harvard University Press, Cambridge, Massachusetts.

McKay, G. M., 1973, Behavior and ecology of the Asiatic elephant in southeastern Ceylon, *Smithsonian Contributions to Zoology,* 125, 1–113.

McKay, G. M. and J. F. Eisenberg, 1974, Movement patterns and habitat utilization of ungulates in Ceylon, in *The Behavior of Ungulates and Its Relation to Management,* Vol. 2, V. Geist and F. Walther (eds.), *IUCN Pub.,* No. 24.

McKey, D., The ecology of coevolved seed dispersal systems, in *Coevolution of Animals and Plants,* L. E. Gilbert and P. H. Raven (eds.), University of Texas Press, Austin.

McKey, D., P. G. Waterman, C. N. Mbi, J. S. Gartlan and T. T. Struhsaker, 1978, Phenolic content of vegetation in two African rain forests: ecological implications, *Science,* 202, 61–63.

McNab, B. K., 1963, Bioenergetics and the determination of home range size, *Amer. Natur.,* 97, 130–140.

McNab, B. K., 1974, The energetics of endotherms, *Ohio J. Sci.,* 74, 370–380.

McNab, B. K., 1978, Energetics of arboreal folivores: physiological problems and ecological consequences of feeding on an ubiguitous food supply, in *The Ecology of Arboreal Folivores,* G. G. Montgomery (ed.), Smithsonian Institution Press, Washington, D.C., pp. 153–162.

McPhee, H. C., E. Z. Russel and J. Zeller, 1931, An inbreeding experiment with Poland China swine, *J. Heredity,* 22, 383–403.

Medway, L., 1972, Phenology of a tropical rain forest in Malaya, *Biol. J. Linn. Soc.,* 4, 117–146.

Medway, L. and D. Wells, 1971, Diversity and density of birds and mammals at Kuala Lompat, Pahang, *Malay. Nat. J.,* 24, 238–247.

Meggars, B. J., 1971, *Amazonia,* Aldine, Chicago.

Mertz, D. B., 1971, The mathematical demography of the California Condor population, *Amer. Natur.,* 105, 437–453.

Methuen, H. H., 1848, *Life in the Wilderness: Or Wanderings in South Africa,* Richard Bentley, London.

Mettler, L. E. and T. G. Gregg, 1969, *Population Genetics and Evolution,* Prentice Hall, Inc., Englewood Cliffs, New Jersey.

Meyer de Schaunesee, R. and W. H. Phelps Jr., 1978, *A Guide to the Birds of Venezuela,* Princeton University Press, Princeton, New Jersey.

Miller, R. S. and D. B. Botkin, 1974, Endangered species: models and predictions, *Amer. Sci.,* 62, 172–181.

Mitton, J. B., 1978, Relationship between heterozygosity for enzyme loci and variation of morphological characters in natural populations, *Nature,* 273, 661–662.

Modha, M. L., 1967, *The ecology of the Nile crocodile* Crocodylus niloticus

Laurenti 1768 *on Central Island, Lake Rudolf,* M. Sc. Thesis, University of Nairobi, Kenya.

Montgomery, G. G. and Y. D. Lubin, 1977, Prey influences on movements of neotropical anteaters, in *Proceedings of the 1975 Predator Symposium,* R. L. Phillips and C. Jonkel (eds.), Montana Forest and Conservation Experiment Station, University of Montana, pp. 103–131.

Montgomery, G. G. and Y. D. Lubin, 1978, Impact of anteaters (*Tamandua* and *Cyclopes,* Edentata, Myrmecophagidae) on arboreal ant populations, Abstract 111, in *Abstracts of Technical Papers 58th Annual Meeting of The American Society of Mammalogists,* Athens, Georgia.

Montgomery, G. G. and M. E. Sunquist, 1975, Impact of sloths on neotropical forest energy flow and nutrient cycling, in *Tropical Ecological Systems: Trends in Terrestrial and Aquatic Research,* F. B. Golley and E. Medina (eds.), Ecological Studies 11, Springer-Verlag, New York, pp. 69–98.

Montgomery, G. G. and M. E. Sunquist, 1978, Habitat selection and use by two-toed and three-toed sloths, in *The Ecology of Arboreal Folivores,* G. G. Montgomery (ed.), Smithsonian Institution Press, Washington, D.C., pp. 329–359.

Moore, N. W., 1962, The heaths of Dorset and their conservation, *J. Ecol.,* 50, 369–391.

Moore, N. W. and M. D. Hooper, 1975, On the number of bird species in British woods, *Biol. Conserv.,* 8, 239–250.

Moreau, R. E., 1963, Vicissitudes of the African biomes in the late Pleistocene, *Proc. Zool. Soc. Lond.,* 141, 395–421.

Moreau, R. E., 1966, *The Bird Faunas of Africa and its Islands,* Academic Press, London.

Morgan, W. T. W., 1963, The "White Highlands" of Kenya, *Geogr. J.,* 129, 140–155.

Morgan, W. T. W., 1968, The role of temperate crops in the Kenya Highlands, *Acta. Geogr.,* 20, 273–278.

Morgan, W. T. W., 1974, Sorghum gardens in South Turkana: Cultivation among a nomadic pastoral people, *Geogr. J.,* 140, 80–93.

Morris, M. G., 1971, The management of grassland for the conservation of invertebrate animals, in *The Scientific Management of Animal and Plant Communities for Conservation,* E. Duffey and A. S. Watt (eds.), Blackwell Scientific Publications Ltd., London, pp. 527–552.

Morton, N. E., J. F. Crow and H. J. Muller, 1956, An estimate of the mutational damage in man from data on consanguineous marriages, *Proc. Natl. Acad. Sci. U.S.A.,* 42, 855–863.

Muckenhirn, N. A. and J. F. Eisenberg, 1973, Home ranges and predation in the Ceylon leopard, in *The World's Cats,* Vol. 1, R. L. Eaton (ed.), Ecology and Conservation, World Wildlife Safari, Winston, Oregon and ISCES, Athens, Georgia, pp. 142–175.

Muul, I. and Lim Boo Liat, 1978, Comparative morphology, food habits and ecology of some Malaysian arboreal rodents, in *The Ecology of Arboreal Folivores,*

G. G. Montgomery (ed.), Smithsonian Institution Press, Washington, D.C., pp. 361–368.

Myers, N., 1972, National Parks in Savanna Africa, *Science,* 178, 1255–1263.

Myers, N., 1976, An expanded approach to the problem of disappearing species, *Science,* 193, 198–202.

Nash, R., 1973, *Wilderness and the American Mind,* Yale University Press, New Haven.

Nature Conservancy, The, 1978, Stewardship, TNC, Arlington, Virginia.

Nei, M., T. Maruyama and R. Chakraborty, 1975, The bottleneck effect and genetic variability in populations, *Evolution,* 29, 1–10.

Nevo, E., D. Zohary, A. H. D. Brown and M. Haber, in press, Allozyme-environment relationships in natural populations of wild barley in Israel, *Evolution.*

Odum, E. P., 1971, *Fundamentals of Ecology,* 3rd ed., Saunders, Philadelphia.

O'Grady, J. P., E. C. Davidson, W. D. Thomas, G. N. Esra, L. Gluck and J. Ojasti, 1973, *Estudio biologico del Chiguire o capibara,* Fundo Nacional de Investigaciones Agropecuarias, Caracas.

Oldfield, M. L., 1976, *The utilization and Conservation of Genetic Resources: An Economic Analysis,* M. A. Thesis, Pennsylvania State University.

Olney, P. J. S., 1976*a,* Birds bred in captivity and multiple generation births, 1974, *Inter. Zoo. Yearb.,* 16, 340–375.

Olney, P. J. S., 1976*b,* Mammals bred in captivity and multiple generation births, 1974, *Inter. Zoo Yearb.,* 16, 375–410.

Olney, P. J. S., 1977*a,* Birds bred in captivity and multiple generation births, 1975, *Inter. Zoo Yearb.,* 17, 260–296.

Olney, P. J. S., 1977*b,* Mammals bred in captivity and multiple generation births, 1975, *Inter. Zoo Yearb.,* 17, 296–334.

Opler, P. A., H. G. Baker and G. W. Frankie, 1977, Recovery of tropical lowland forest ecosystems, in *Recovery and Restoration of Damaged Ecosystems,* J. Cairns Jr. et al. (eds.), University Press of Virginia, Charlottesville.

Oppenheimer, J. R., 1968, *Behavior and Ecology of the White-faced Monkey,* Cebus capucinus, *on Barro Colorado Island, C. Z.,* Ph. D. Dissertation, University of Illinois.

Ormerod, W. E., 1956, Ecological effect of control of African trypanosomiasis, *Science,* 191, 815–821.

Paine, R. T., 1966, Food web complexity and species diversity, *Amer. Natur.,* 100, 65–75.

Parker, I. S. C., 1964, The Galana game management scheme, *Bull. Epizoot. Dis. Afr.,* 12, 21–31.

Parker, I. S. C. and A. D. Graham, 1971, The ecological and economic basis for game ranching in Africa, in *The Scientific Management of Animal and Plant Communities for Conservation,* E. Duffy and A. S. Watt, eds., Blackwell Scientific Publications Ltd., London, pp. 393–404.

Passmore, J., 1974, *Man's Responsibility for Nature,* Gerald Duckworth and Co., Ltd., London.

Patterson, B., and R. Pasqual, 1972, The fossil mammal fauna of South America, in *Evolution, Mammals and Southern Continents,* A. Keast et al. (eds.), State University of New York Press, Albany, pp. 247–309.

Perry, J. and P. B. Kibbee, 1974, The capacity of American zoos, *Inter. Zoo Yearb.,* 14, 240–247.

Persson, R., 1974, World forest resources, *Repportes och Uppsatser,* Institutionen für Skogstatzering Pubn., 17, Stockholm.

Petrides, G. A. and W. G. Swank, 1958, Management of the big game resource in Uganda, East Africa, *Trans. N. Amer. Wildl. Conf.,* 23, 461–477.

Phillipson, J., 1977, Wildlife—a clue to balancing the environmental budget in Kenya, *Post,* (Kenya), II(6), 3–8.

Pickett, S. T. A. and J. N. Thompson, 1978, Patch dynamics and the design of nature reserves, *Biol. Conserv.,* 13, 27–37.

Pinder, N. J. and J. P. Barkham, 1978, An assessment of the contribution of captive breeding to the conservation of rare mammals, *Biol Conserv.,* 13, 187–245.

Pirages, D. C. and P. R. Ehrlich, 1974, *Ark II: Social Response to Environmental Imperatives,* W. H. Freeman and Co., San Francisco.

Pirchner, F., 1969, *Population Genetics in Animal Breeding,* W. H. Freeman and Co., San Francisco.

Pliske, T., 1975, Attraction of Lepidoptera to plants containing pyrrolizidine alkaloids, *Environ. Entomol.,* 4, 455–473.

Pliske, T. and T. Eisner, 1969, Sex pheromone of the queen butterfly: biology, *Science,* 164, 1170–1172.

Pooley, A. C., 1962, The Nile Crocodile, *Crocodylus niloticus, Lammergeyer,* 2, 1–55.

Pooley, A. C., 1969, Preliminary studies in the breeding of the Nile crocodile, *Crocodylus niloticus* in Zululand, *Lammergeyer,* 10, 22–44.

Pooley, A. C., 1971, Crocodile rearing and restocking, *IUCN Pub.,* N. S., 32, 104–30.

Pooley, A. C., 1973, Conservation and management of crocodiles in Africa, *J. S. Afri. Wildl. Mgmt. Assn.,* 3, 101–103.

Poore, M. E. D., 1974, A conservative viewpoint, *Proc. Roy. Soc., A,* 339, 395–410.

Poore, M. E. D., 1976, The value of tropical moist forest ecosystems and the environmental consequences of their removal, *Unasylva,* 28, 127–143.

Posselt, J., 1963, The domestication of the eland, *Rhod. J. Agric. Res.,* 1, 81–87.

Potter, G. L., H. W. Ellsaesser, M. C. MacCracken and F. M. Luther, 1975, Possible climatic impact of tropical deforestation, *Nature,* 258, 697–698.

Prance, G. T., 1976, The phytogeographic subdivisions of Amazonia and their influence on the selection of biological reserves, in *Extinction is Forever,* G. Prance and T. S. Elias (eds.), New York Botanical Garden, Bronx, New York, pp. 195–212.

Pratt, D. J. and M. D. Gwynne (eds.), 1977, *Rangeland Management and Ecology in East Africa,* Hodder and Stoughton, London.

Preston, F. W., 1960, Time and space and the variation of species, *Ecology,* 41, 612–627.

Preston, F. W., 1962, The canonical distribution of commonness and rarity: Part I, *Ecology,* 43, 185–215; Part II, 43, 410–432.

Prout, T., 1962, The effects of stabilizing selection on the time of development in *Drosophila melanogaster, Genet. Res.,* 3, 364–382.

Pyle, R. M., 1976, *The Eco-geographic Basis for Lepidoptera Conservation,* Ph.D. thesis, Yale University.

Pyle, R. M. and S. A. Hughes, 1978, *Conservation and Utilization of the Insect Resources of Papua New Guinea,* Consultants' Report to the Wildlife Branch, Government of Papua New Guinea, Waigani.

Rabb, G. B., J. H. Woolpy and B. E. Ginsburg, 1967, Social relationships in a group of captive wolves, *Am. Zool.,* 7, 305–312.

Racey, P. A., 1973, Environmental factors affecting the length of gestation in heterothermic bats, *J. Reprod. Fert. Suppl.,* 19, 175–189.

Rathbun, C. D., in press, Descriptions and analysis of the arch display in the golden lion tamarin, *Leontopithecus rosalia rosalia, Folia Primat.*

Raven, P. H., 1977, The systematics and evolution of higher plants, in *Changing Scenes in Natural Sciences, 1776–1976,* C. E. Goulden (ed.), *Acad. Nat. Sci. Spec. Pub.,* 12, Philadelphia, pp. 59–83.

Rendel, J. M., 1943, Variations in the weights of hatched and unhatched duck's eggs, *Biometrika,* 33, 48–56.

Rendel, J. M., 1960, Selection for canalisation of the scute phenotype in *Drosophila melanogaster, Aust. J. Biol. Sci.,* 13, 36–47.

Republic of Kenya, 1972, *Kenya's National Report to the United Nations on the Human Environment,* Professional Printers and Stationers, Nairobi.

Resnick, R., P. T. Robinson, B. Lasley and K. Benirschke, 1978, Intrauterine fetal demise associated with consumption coagulopathy in a Douc langur monkey *(Pygathrix nemaeus nemaeus), J. Med. Primatol.,* 7, 249–253.

Richards, P. W., 1977, Tropical forests and woodlands: an overview, *Agro-Ecosystems,* 3, 225–238.

Richardson, S. D., 1976. Foresters and the Faustian bargain, *Y. Coedwigs,* 28, 41–48.

Richter, W. von, 1969, Report to the government of Botswana on a survey of the wild animal trade and skin industry, UNDP/FAO No. TA 2637, United Nations Food and Agriculture Organization, Rome.

Rick, C. M., J. F. Fobes and M. Holle, 1977, Genetic variation in *Lycopersicon pimpinellifolium;* evidence of evolutionary change in mating systems, *Plant Syst. Evol.,* 127, 139–170.

Ricklefs, R. E., Chrmn., D. Amadon, W. Conway, R. Miller, I. Nisbet, R. Schreiber, U. Seal, R. Selander and S. Temple, 1978, *Report of the Advisory Panel on the California Condor,* Audubon Conservation Report No. 6, National Audubon Society, New York.

Ridpath, M. G. and R. E. Moreau, 1966, The birds of Tasmania: ecology and evolution, *Ibis,* 108, 348–393.

Rijksen, H. D., 1978, *A Fieldstudy on Sumatran Orangutans* (Pongo pygmaeus abelii *Lesson* 1827): *Ecology, Behaviour and Conservation,* Mededelingen Landbouwhogeschool, H. Veenman and B. V. Zonen, Wageningen.

Riney, T. and W. L. Kettlitz, 1964, Management of large mammals in the Transvaal, *Mammalia,* 28, 189–248.

Robertson, A., 1964, The effect of nonrandom mating within inbred lines on the rate of inbreeding, *Genet. Res., Camb.,* 5, 164–167.

Robinson, P. T., 1978, Veterinary dentistry in the zoo: new insights, *Zoonooz,* 51, 4–10.

Rogerson, A., 1968, Energy utilization by eland and wildebeeste, in *Comparative Nutrition of Wild Animals,* M. A. Crawford (ed.), *Zool. Soc. Lond. Symp.,* 21, 153–161.

Root, R. B., 1973, Organization of a plant-arthropod association in simple and diverse habitats: the fauna of collards *(Brassica olevaceae), Ecol. Monogr.,* 43, 95–124.

Rose, R. M., I. S. Bernstein and T. P. Gordon, 1975, Consequences of social conflict on plasma testosterone levels in rhesus monkeys, *Psychosom. Med.,* 37, 50–61.

Roth, H. H., 1970, Studies on the agricultural utilization of semi-domesticated eland *(Taurotragus oryx)* in Rhodesia, *Rhod. J. Agric. Res.,* 8, 67–70.

Roughgarden, J., in press, Patchiness in the spatial distribution of a population caused by stochastic fluctuation in resources, *Oikos.*

Rowell, T. E., 1973, Social organization of wild talapoin monkeys, *Am. J. Phys. Anthrop.,* 38, 593–598.

Rudran, R., 1973, Adult male replacement in one male troops of purplefaced langurs *(Presbytis senex senex)* and its effect on population structure, *Folia primat.,* 19, 166–192.

Rutter, R. J. and D. H. Pimlott, 1968, *The World of the Wolf,* J. B. Lippincott Co., Philadelphia.

Ryder, O. A., 1977, Genetic monitoring of endangered species, *Zoonooz,* 50, 15.

Ryder, O. A., N. C. Epel and K. Benirschke, 1978, Chromosome banding studies of the Equidae, *Cytogenet. Cell Genet.,* 20, 323–350.

Sanders, R., 1978, The state of natural heritage programs: a partnership to preserve diversity, *The Nature Conservancy News,* 28, 13–19.

Sandford, S., 1976, Pastoralism under pressure, *ODI Rev.,* 2, 45–68.

Schaal, B. A. and D. A. Levin, 1976, The demographic genetics of *Liatris cylindracea* Michx. (Compositae), *Amer. Natur.,* 110, 191–206.

Schaller, G., 1967, *The Deer and the Tiger,* University of Chicago Press, Chicago.

Schaller, G., 1972, *The Serengeti Lion,* University of Chicago Press, Chicago.

Schneider, D., M. Boppre, H. Schneider, W. R. Thompson, C. J. Borisk, R. L. Petty and J. Meinwald, 1975, A pheromone precursor and its uptake in male *Danaus* butterflies, *J. Comp. Physiol.,* 97, 245–256.

Schoener, T. W., 1976, The species-area-relation within archipelagos: models and evidence from island land birds, in *Proc. 16th Intern. Ornith. Congr.,* H. J. Frith and J. H. Calaby (eds.), Australian Academy of Science, Canberra, pp. 629–642.

Schoenfeld, C. A. and J. C. Hendee, 1978, *Wildlife Management in Wilderness,* The Boxwood Press, Pacific Grove, California.

Schull, W. J. and J. V. Neel, 1965, *The Effects of Inbreeding on Japanese Children,* Harper and Row, New York.

Schull, W. J. and J. V. Neel, 1972, The effects of parental consanguinity and inbreeding in Ilirado, Japan: V. Summary and interpretation. *Amer. J. Hum. Genet.,* 24, 425–453.

Schultz, J. P., 1960, *Ecological Studies on Rain Forest in Northern Surinam,* North-Holland, Amsterdam.

Schwartz, A. and K. Thomas, 1975, A checklist of West Indian amphibians and reptiles, *Carnegie Mus. Nat. His. Special Publ.,* No. 1, Pittsburgh.

Sclater, P., 1970, *A Field Guide to Australian Birds,* Rigby, Adelaide.

Sclater, P., 1974, *A Field Guide to Australian Birds: the Passerines,* Harrowood Books, Newtown Square, Pennsylvania.

Seager, S., C. Platz and W. Hodge, 1975, Successful pregnancy using frozen semen in the wolf, *Inter. Zoo Yearb.,* 15, 140–143.

Seal, U. S., D. G. Makey, D. Bridgewater, L. Simmons and L. Murtfeldt, 1977, ISIS: a computerized record system for the management of wild animals in captivity, *Inter. Zoo Yearb.,* 17, 68–70.

Seidensticker, J., 1976*a*, Ungulate populations in Chitawan Valley, Nepal, *Biol. Conserv.,* 10, 183–210.

Seidensticker, J., 1976*b,* On the ecological separation between tigers and leopards, *Biotropica,* 8, 225–234.

Seidensticker, J. C., M. G. Hornocker, W. V. Wiles and J. P. Messick, 1973, Mountain lion social organization in the Idaho Primitive Area, *Wildl. Monogr.,* No. 35.

Serle, W., 1950, A contribution to the ornithology of the British Cameroons, *Ibis,* 92, 342–376, 602–638.

Serle, W., 1954, A second contribution to the ornithology of the British Cameroons, *Ibis,* 96, 47–80.

Serle, W., 1965, A third contribution to the ornithology of the British Cameroons, *Ibis,* 107, 60–94, 230–246.

Serle, W., and G. J. Morel, 1977, *A Field Guide to the Birds of West Africa,* Collins, London.

Sharpe, G. W. (ed.), 1976, *Interpreting the Environment,* John Wiley and Sons, New York.

Shepard, P., 1978, *Thinking Animals,* Viking Press, New York.

Simberloff, D., 1976, Species turnover and equilibrium island biogeography, *Science,* 154, 572–578.

Simberloff, D. S. and L. G. Abele, 1975, Island biogeography theory and conservation practice, *Science,* 191, 285–286.

Singer, M. C. and L. E. Gilbert, 1978, Ecology of butterflies in the urbs and suburbs, in *Perspectives in Urban Entomology,* C. W. Frankie and C. S. Koehler (eds.), Academic Press, New York, pp. 1–11.

Singer, P., 1975, *Animal Liberation,* The New York Review, New York.

Singh, S. M. and E. Zouros, 1978, Genetic variation associated with growth rate in the American oyster *(Crassostrea virginica), Evolution,* 32, 342–353.

Sittman, K., H. Abplanalp and R. A. Fraser, 1966, Inbreeding depression in Japanese quail, *Genetics,* 54, 371–379.

Skinner, J. D., 1966, An appraisal of the eland *(Taurotragus oryx)* for diversifying and improving animal production in Southern Africa, *Afr. Wildl.,* 20, 29–40.

Skinner, J. D., 1967, An appraisal of the eland as a farm animal in South Africa, *Anim. Breed. Abstr.,* 35, 177–186.

Skinner, J. D., 1973*a,* Technological aspects of domestication and harvesting of certain species of game in South Africa, in *Proceedings of the 3rd World Conference on Animal Production,* R. L. Reid (ed.), Sydney University Press, pp. 752–1126.

Skinner, J. D., 1973*b,* An appraisal of the status of certain antelope for game farming in South Africa, *Z. Tierzuchtg. Zuchtsbiol.,* 90, 263–277.

Slatis, H. M., 1960, An analysis of inbreeding in the European bison, *Genetics,* 45, 275–287.

Slatis, H. M., 1975, *Research in Zoos and Aquariums,* National Academy of Science, Washington, D.C.

Smiley, J. T., 1978, *Host Plant Ecology of* Heliconius *Butterflies in Northeastern Costa Rica,* Ph.D. Dissertation, University of Texas, Austin.

Smith, J. L. D. and Kirti Man Tamang, 1977, *Smithsonian Tiger Ecology Project Report,* No. 12, Smithsonian Institution, Washington, D.C.

Smith, R. L., 1976, Ecological genesis of endangered species: the philosophy of preservation, *Ann. Rev. Ecol. Syst.,* 7, 33–35.

Smith, T. and T. Polacheck, in press, Reexamination of the life table for northern fur seals and implications about population regulatory mechanisms, in *Dynamics of Large Mammal Populations,* C. W. Fowler and T. Smith (eds.), John Wiley and Sons, New York.

Smythe, N., 1970*a, Ecology and Behavior of the Agouti* (Dasyprocta punctata) *and Related Species on Barro Colorado Island, Panama,* Ph.D. Thesis, University of Maryland, College Park.

Smythe, N., 1970*b,* Relationships between fruiting seasons and seed dispersal methods in a neotropical forest, *Amer. Natur.,* 104, 25–35.

Smythe, N., 1978, The natural history of the Central American agouti *(Dasyprocta punctata), Smithsonian Contributions to Zoology,* No. 257.

Snow, D. W. and B. K. Snow, 1964, Breeding seasons and annual cycles of Trinidad land-birds, *Zoologica,* 49, 1–35.

Soemarwoto, O., I. Soemarwoto, E. Karyono, E. M. Soekartadiredja and A. Ramlan, 1975, The Javanese home-garden as an integrated agroecosystem, in *Science for Better Environment, International Conference of Scientists on the Human Environment, Kyoto,* Pergamon Press, New York, pp. 193–197.

Sommer, A., 1976, Attempt at an assessment of the world's tropical forests, *Unasylva,* 28, 5–25.

Soulé, M., 1966, Trends in the insular radiation of a lizard, *Amer. Natur.,* 100, 47–64.

Soulé, M. E., 1973, The epistasis cycle: A theory of marginal populations, *Ann. Rev. Ecol. Syst.,* 4, 165–187.

Soulé, M. E., 1976, Allozyme variation: its determinants in space and time, in *Molecular Evolution,* F. J. Ayala (ed.), Sinauer Assoc., Sunderland, Massachusetts, pp. 60–77.

Soulé, M. E., 1979, Heterozygosity and developmental stability: another look, *Evolution,* 33, 396–401.

Soulé, M. E., G. L. Gorman, S. Y. Yang and J. S. Wyles, in preparation, The Puerto Rican *Anolis* fauna: levels of genetic variation.

Soulé, M. E. and A. J. Sloan, 1966, Biogeography and distribution of the reptiles and amphibians on islands in the Gulf of California, Mexico, *Trans. of the San Diego Society of Natural History,* 14, 137–156.

Soulé, M. E., B. A. Wilcox and C. Holtby, 1979, Benign neglect: a model of faunal collapse in the game reserves of East Africa, *Biol. Conserv.,* 15, 259–272.

Spruce, R., 1908, *Notes of a Botanist on the Amazon and Andes,* Macmillan and Co., New York.

Stanley, S. M., 1975, A theory of evolution above the species level, *Proc. Natl. Acad. Sci. U.S.A.,* 72, 646–650.

Stanley-Price, M. R. S., 1974, *The Feeding Ecology of Coke's Hartebeest,* Alcelaphus buselaphus cokei *Gunther, in Kenya,* Ph.D. Thesis, University of Oxford.

Stanley-Price, M. R. S., 1976, Feeding studies of oryx on the Galana Ranch, *Afri. Wildl. Leader. Found. News.,* 11, 7–11.

Stark, N. M., 1971, Nutrient cycling: II. Nutrient distribution in some Amazonian vegetation, *Trop. Ecol.,* 12, 177–207.

Stark, N. M., 1978, Man, tropical forests and the biological life of a soil, *Biotropica,* 10, 1–10.

Starr, R., 1978, *Zealots fail to separate real threats from trivial ones,* Denver Post, July 9.

Stebbing, E. P., 1954, Forests, aridity and desert, in *Biology of Deserts,* J. Cloudsley-Thompson (ed.), Institute of Biology, London, pp. 123–128.

Steele, R. C., 1974, *Forestry and Wildlife Conservation in the Temperate Zone,* Natural Environment Research Council, London.

Steele, R. C. and R. C. Welch (eds.), 1973, *Monks Wood: A Nature Reserve Record,* The Nature Conservancy, Monks Wood Experimental Station, Huntington, England.

Stephenson, A. B., A. J. Wyatt and A. W. Nordskog, 1953, Influence of inbreeding on egg production in the domestic fowl, *Poultry Sci.,* 32, 510–517.

Stiles, F. G., 1975, Ecology, flowering phenology and hummingbird pollination of some Costa Rican *Heliconia* species, *Ecology,* 56, 285–301.

Stone, C. D., 1974, *Should Trees Have Standing?* William Kaufmann, Los Altos, California.

Stott, P. A., 1978, Nous avons mangé la forêt, in *Man and Nature in South East Asia* (Collected papers in Oriental and African studies), P. A. Stott (ed.), School of Oriental and African Studies, London, pp. 7–22.

Strong, D. R. Jr., 1977, Rolled leaf hispine beetles (Chrysomelidae) and their Zingiberales host plants in Middle America, *Biotropica,* 9, 156–169.

Struhsaker, T. T., 1977, Infanticide and social organization in the red tail monkey *(Cercopithecus ascanius schmidti)* in the Kibala Forest, Uganda, *Z. Tierpsychol.,* 45, 75–84.

Sugiyama, Y., 1965, On the social change of Hanuman langurs *(Presbytis entellus)* in their natural conditions, *Primates,* 6, 381–429.

Sullivan, A. L. and M. C. Shaffer, 1975, Biogeography of the megazoo, *Science,* 189, 13–17.

Sved, J. A. and F. J. Ayala, 1970, A population cage test for heterosis in *Drosophila pseudoobscura, Genetics,* 66, 97–113.

Syme, P. D., 1977, Observations on the longevity and fecundity of *Orgilus obscurator* (Hymenoptera: Brachonidae) and the effects of certain foods on longevity, *Can. Ent.,* 109, 995–1000.

Talbot, L. M., W. J. A. Payne, H. P. Ledger, L. D. Verdcourt and M. H. Talbot, 1965, The meat production potential of wild animals in Africa: a review of biological knowledge, *Tech. Commun. Commonw. Bur. Anim. Breed. Genet.,* 16, 1–42.

Tanaka, K., 1977, Genetic effects of maternal inbreeding in man on congenital abnormality, mental defect, infertility and prenatal death, *Jap. J. Human Genet.,* 22, 55–72.

Taylor, C. R., 1969, The eland and the oryx, *Scient. Amer.,* 88–95.

Taylor, C. R. and C. N. Lyman, 1967, A comparative study of the environmental physiology of an East African antelope, the eland, and the Hereford steer, *Physiol. Zool.,* 40, 280–295.

Taylor, H. M., R. S. Gourley, C. E. Lawrence and R. S. Kaplan, 1974, Natural selection of life history attributes: an analytical approach, *Theor. Pop. Biol.,* 5, 104–122.

Temple, S. A., 1977, Plant-animal mutualism: coevolution with dodo leads to near extinction of plant, *Science,* 197, 885–886.

Terborgh, J., 1974*a,* Preservation of natural diversity: the problem of extinction prone species, *Bioscience,* 24, 715–722.

Terborgh, J., 1974*b,* Faunal equilibria and the design of wildlife preserves, in *Tropical Ecological Systems: Trends in Terrestrial and Aquatic Research,* F. Golly and E. Medina (eds.), Springer-Verlag, New York.

Terborgh, J., 1975, Faunal equilibria and the design of wildlife preserves, in *Trends in Tropical Ecology,* Academic Press, New York, pp. 369–380.

Terborgh, J. W., 1976, Island biogeography and conservation: strategy and limitations, *Science,* 193, 1029–1030.

Terborgh, J. W. and J. Faaborg, 1973, Turnover and ecological release in the avifauna of Mona Island, Puerto Rico, *Auk,* 90, 759–779.

Thompson, P. A., 1975, Should botanic gardens save rare plants? *New Sci.,* 68, 636–639.

Thompson, P. A., 1976, The collection, maintenance and environmental importance of the genetic resources of wild plants, *Env. Consv.,* 2, 223–228.

Tinkle, D. W. and R. K. Selander, 1973, Age-dependent allozymic variation in a natural population of lizards, *Bioch. Genetics,* 8, 231–237.

Torroja, E., 1964, Genetic load in irradiated experimental populations of *D. pseudoobscura, Genetics,* 50, 1289–1298.

Tracey, M. W., N. F. Bellet and C. B. Graven, 1975, Excess of allozyme homozygosity and breeding population structure in the mussel *Mytilus californianus, Marine Biology,* 32, 303–311.

Treus, D. B. and D. Kravchenko, 1968, Methods of rearing and economic utilization of the eland in Askanya Nova Zoological Park, *Symp. Zool. Soc. Lond.,* 21, 395–411.

Trommershausen-Smith, A., O. A. Ryder and Y. Suzuku, in press, Blood typing studies of twelve Przewalski's horses, *Inter. Zoo Yearb.*

Turnbull, C., 1973, *The Mountain People,* Jonathan Cape, London.

United States Department of Agriculture, 1977, *A Directory of Research Natural Areas on Federal Lands of the United States of America,* Federal Commission on Ecological Reserves, Forest Service, USDI, Washington, D.C.

United States Department of the Interior, 1976, *Research in the Parks,* Transactions of the National Park Centennial Symposium, *NPS Symp. Series,* 1.

Usher, M. B., 1975, *Biological Management and Conservation: Ecological Theory, Application and Planning,* Chapman and Hall, London.

van Lawick, H., 1973, *Solo,* Bantam Books, New York.

van Lawick, H. and J. van Lawick-Goodall, 1971, *Innocent Killers,* Houghton Mifflin Co., Boston.

Varona, L. S., 1964, Catalogo de los mamiferos viventes y Extinguidos de las Antillas, *Memoria,* 69, Academia de Ciencias de Cuba.

Vietmeyer, N., in press, *Promising Leguminous Trees for the Tropics,* National Academy of Sciences, Washington, D.C.

Volf, J. (ed.), 1970–1975, *Pedigree Book of the Przewalski Horse,* Zoological Garden Prague, Prague.

von la Chevallierie, M., 1970, Meat production from wild ungulates, *Proc. S. Afr. Soc. Anim. Prod.,* 9, 73–88.

von la Chevallierie, M., 1972, Meat quantity in seven wild ungulate species, *S. Afr. J. Anim. Sci.,* 2, 101–104.

Waddington, C. H., 1957, *The Strategy of the Genes,* Allen and Unwin, London.

Wallace, B. and C. Madden, 1965, Studies on inbred strains of *Drosophila melanogaster, Amer. Natur.,* 99, 495–509.

Walsh, J. and R. Gannon, 1967, *Time is Short and the Water Rises,* Thomas Nelson and Sons, Camden, New Jersey.

Walter, H., 1971, *Ecology of Tropical and Subtropical Vegetation,* Oliver and Boyd, Edinburgh.

Ward, L. K., 1977, The conservation of juniper: the associated fauna with special reference to southern England, *J. Appl. Ecol.,* 14, 81–120.

Ward, L. K. and K. H. Lakhani, 1977, The conservation of juniper: the fauna of food-plant island sites in southern England, *J. Appl. Biol.,* 14, 121–135.

Watson, R. M., 1969, A survey of the large mammal population of South Turkana, The South Turkana Expedition, Scientific Papers, 11, *Geog. J.,* 135, 529–546.

Watson, R. M., 1972, *Results of Aerial Livestock Surveys of Kaputei Division, Samburu District and North-Eastern Province,* Statistic division, Ministry of Finance and Planning, Republic of Kenya, Nairobi.

Watt, W. B., 1977, Adaptation of specific loci: I. Natural selection on phosphoglucose isomerase of *Colias* butterflies: biochemical and population aspects, *Genetics,* 87, 177–194.

Watts, I. E. M., 1954, Line squalls of Malaya, *J. Trop. Geogr.,* 3, 1–14.

Webb, L. J., 1958, Cyclones as an ecological factor in tropical lowland rain forest, North Queensland, *Austr. J. Bot.,* 6, 220–228.

Weiner, J. S., 1971, *Man's Natural History,* Weidenfeld and Nicholson, London.

Weldon, W. F. R., 1901, A first study of natural selection in *Clausilia laminata* (montague), *Biometrika,* 1, 109–124.

Wemmer, C., 1977, Can wildlife be saved in zoos?, *New Scientist,* 75, 585–587.

Wemmer, C. and L. R. Collins, in preparation, Social behavior, social structure and their implications for management, in *The Biology and Captive Management of Père David's Deer,* B. Beck and C. Wemmer (eds.).

Western, D., 1973, *The Structure, Dynamics and Changes of the Amboseli Ecosystem,* Ph.D. Thesis, University of Nairobi, Kenya.

Wetterberg, G. B. 1976. Uma analise de prioridades em conservacao da natureza na Amaxonia, *PNUD/FAO/IBDF/BRA*-45, Serie Tecnica, no. 8.

Whitcomb, R. F., J. F. Lynch, P. A. Opler and C. S. Robbins, 1976, Island biogeography and conservation: strategy and limitations, *Science,* 193, 1030–1032.

White, M. J. D., 1978, *Modes of Speciation,* W. H. Freeman and Co., San Francisco.

Whitmore, T. C., 1975, *Tropical Rain Forests of the Far East,* Clarendon Press, Oxford.

Whitmore, T. C., 1977, A first look at *Agathis., Trop. For. Papers,* 11.

Whitmore, T. C., 1978, Gaps in the forest canopy, in *Tropical Trees as Living Systems,* P. B. Tomlinson and M. H. Zimmerman (eds.), Cambridge University Press, New York, pp. 639–655.

Whittaker, R. H., 1975, *Communities and Ecosystems,* 2nd ed., Macmillan, New York.

Whittaker, R. H. and P. P. Feeney, 1971, Allelochemics: chemical interactions between species, *Science,* 171, 757–770.

Whitten, W. K. and F. H. Bronson, 1970, The role of pheromones in mammalian reproduction, in *Advances in Chemoreception,* Vol. 1, J. W. Johnston, D. G. Moulton and A. Turk (eds.), Appleton, New York, pp. 309–325.

Wicklund, C., 1977, Oviposition, feeding and spatial separation of breeding and foraging habitats in a population of *Leptidae sinapis* (Lepidoptera), *Oikos,* 28, 56–68.

Wilcox, B. A., 1978, Supersaturated island faunas: a species-age relationship for lizards on post-Pleistocene land-bridge islands, *Science,* 199, 996–998.

Wilcox, B. A., in press, Species number, stability and equilibrium status of the reptile faunas of the California Islands, in *The California Islands: Proceedings of a Multidisciplinary Symposium,* D. M. Power (ed.), Santa Barbara Natural History Museum, Santa Barbara.

Wilcox, B. A., in preparation, The comparative island biogeography of the vertebrate faunas of the Gulf of California Islands.

Wilken, G. C., 1977, Integrating forest and small-scale farm systems in middle America, *Agro-ecosystems,* 3, 291–302.

Williams, N. H. and R. L. Dressler, 1976, Euglossine pollination of *Spathiphyllum* (Araceae), *Selbyana,* 1, 349–355.

Williamson, G. B., 1978, A comment on equilibrium turnover rates for islands, *Amer. Natur.,* 112, 241–243.

Williamson, R. O. G. and W. J. A. Payne, 1965, *An Introduction to Animal Husbandry in the Tropics,* 2nd ed., Longmans, London.

Willis, E. O., 1974, Populations and local extinctions of birds on Barro Colorado Island, Panama, *Ecol. Monogr.,* 44, 153–169.

Willis, E. O., in press, The composition of avian communities in remanescent woodlots in southern Brazil, *Papels Avulsos Museu Paulisto.*

Wilson, E. O., 1975, *Sociobiology: The New Synthesis,* Harvard University Press, Cambridge, Massachusetts.

Wilson, E. O. and E. O. Willis, 1975, Applied biogeography, in *Ecology and Evolution of Communities,* M. L. Cody and J. M. Diamond (eds.), Harvard University Press, Cambridge, Massachusetts, pp. 522–534.

Wolda, H., in press, Seasonal fluctuations in rainfall, food and abundance of tropical insects, *J. Anim. Ecol.*

Woodwell, G. M., R. H. Whittaker, W. A. Rainers, G. E. Likens, C. C. Delwiche and D. B. Botkin, 1978, The biota and the world carbon budget, *Science,* 199, 141–146.

Wright, H. E., 1974, Landscape development, forest fires and wilderness management, *Science,* 186, 487–495.

Wright, S., 1977, *Evolution and the Genetics of Populations, Vol. 3, Experimental Results and Evolutionary Deductions,* University of Chicago Press, Chicago.

Wright, S., 1977, Modes of evolutionary change of characters, in *Proceedings of The International Conference on Quantitative Genetics,* E. Pollack, O. Kempthorne and T. B. Bailey Jr., (eds.), Iowa State University Press, Ames, Iowa, pp. 679–702.

Wyatt-Smith, J., 1950, Storm forest in Kelantan, *Malay. Forester,* 17, 5–11.

Yang, S. Y., M. Soulé and G. C. Gorman, 1974, *Anolis* lizards of the Eastern Caribbean: a case study in evolution: I. Genetic relationship, phylogeny, and colonization sequence of the *roquet* group, *Syst. Zool.,* 23, 387–399.

Young, E. G. Jr. and J. C. Murray, 1966, Heterosis and inbreeding depression in the diploid and tetraploid cottons, *Crop Sci.,* 6, 436–438.

Ziswiler, V., 1967, *Extinct and Vanishing Animals,* 2nd ed., Springer-Verlag, New York.

Zoological Society of London (ZSL), *Annual Report,* London, p. 14.

INDEX